Introduction to Telecommunications

Introduction to Telecommunications

Edited by
Kameron Smith

Larsen & Keller
www.larsen-keller.com

Introduction to Telecommunications
Edited by Kameron Smith
ISBN: 978-1-63549-273-6 (Hardback)

© 2017 Larsen & Keller

⊟ Larsen & Keller

Published by Larsen and Keller Education,
5 Penn Plaza,
19th Floor,
New York, NY 10001, USA

Cataloging-in-Publication Data

Introduction to telecommunications / edited by Kameron Smith.
    p. cm.
Includes bibliographical references and index.
ISBN 978-1-63549-273-6
1. Telecommunication. I. Smith, Kameron.
TK5101 .I68 2017
621.38--dc23

The publisher's policy is to use permanent paper from mills that operate a sustainable forestry policy. Furthermore, the publisher ensures that the text paper and cover boards used have met acceptable environmental accreditation standards.

Printed and bound in the United States of America.

For more information regarding Larsen and Keller Education and its products, please visit the publisher's website www.larsen-keller.com

Table of Contents

Preface

This book provides comprehensive insights into the field of telecommunications. It describes in detail about the processes and applications of this subject. Telecommunications refers to the study of the transmission of signals, sounds, messages, images and signs, etc. and the various technologies related to them. The topics introduced in the text cover all the essential concepts of this field. Different approaches, evaluations and methodologies on telecommunications have been included in it. Coherent flow of topics, student-friendly language and extensive use of examples make this textbook an invaluable source of knowledge. Students in the fields of media technologies, wireless technologies and computer networking will find this textbook to be beneficial.

Given below is the chapter wise description of the book:

Chapter 1- Telecommunications helps in the spreading of information, with the help of signs, signals, messaging and radio or other similar electromagnetic systems. Exchange of information between communication applicants which incorporates the use of technology is telecommunications. This chapter will provide an integrated understanding of telecommunications.

Chapter 2- Communication occurs when there is a transfer of information between participants. Some of the types of telecommunications are email, fax, telephony, voice over IP and videoconferencing. All these technologies are widely used in today's times and help in communication across the globe. The aspects elucidated in this chapter are of vital importance, and provide a better understanding of telecommunications.

Chapter 3- Telecommunications have brought a transformation in communication in the lives of people. It has an equal impact on advertising. Communication channels, multiplexing, path protection and electrical length are some of the essential concepts of telecommunications. This text is a compilation of the essential concepts of telecommunications.

Chapter 4- Techniques are an important part of any field of study. The major techniques related to telecommunications are communications satellite, arbitrary slice ordering, diversity combining and phantom circuit. Lesser known techniques of telecommunications include through-the-earth mine communications and two-way communications. The aspects elucidated are of vital importance and provide a better understanding of telecommunications.

Chapter 5- Modulation is the procedure of fluctuating one or more properties of a periodic waveform. A modulator helps in the performance of modulation. There are a number of modulations such as amplitude modulation, quadrature modulation, frequency modulation and demodulation. The major components of modulation are discussed in this section.

Chapter 6- The defining leap into the future of telecommunications occurred with the discovery of optical cables. Optical telecommunications use the medium of light transmitted through glass or plastic for communication purposes. The chapter on optical telecommunications offers an insightful focus, keeping in mind the complex subject matter.

Chapter 7- The collection of terminal nodes, which help in the connection of different parties in telecommunications is known as telecommunications network. Some of the aspects of telecommunications discussed within this text are computer network, public switched telephone network, packet switching, radio network and television network.

At the end, I would like to thank all those who dedicated their time and efforts for the successful completion of this book. I also wish to convey my gratitude towards my friends and family who supported me at every step.

Editor

Telecommunications: An Introduction

Telecommunications helps in the spreading of information, with the help of signs, signals, messaging and radio or other similar electromagnetic systems. Exchange of information between communication applicants which incorporates the use of technology is telecommunications. This chapter will provide an integrated understanding of telecommunications.

Telecommunication is the transmission of signs, signals, messages, writings, images and sounds or intelligence of any nature by wire, radio, optical or other electromagnetic systems. Telecommunication occurs when the exchange of information between communication participants includes the use of technology. It is transmitted either electrically over physical media, such as cables, or via electromagnetic radiation. Such transmission paths are often divided into communication channels which afford the advantages of multiplexing. The term is often used in its plural form, telecommunications, because it involves many different technologies.

Earthstation at the satellite communication facility in Raisting, Bavaria, Germany

Early means of communicating over a distance included visual signals, such as beacons, smoke signals, semaphore telegraphs, signal flags, and optical heliographs. Other examples of pre-modern long-distance communication included audio messages such as coded drumbeats, lung-blown horns, and loud whistles. 20th and 21st century technologies for long-distance communication usually involve electrical and electromagnetic technologies, such as telegraph, telephone, and teleprinter, networks, radio, microwave transmission, fiber optics, and communications satellites.

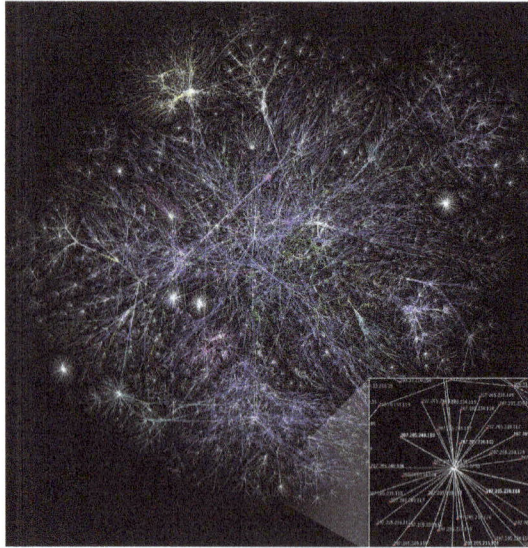
Visualization from the Opte Project of the various routes through a portion of the Internet

A revolution in wireless communication began in the first decade of the 20th century with the pioneering developments in radio communications by Guglielmo Marconi, who won the Nobel Prize in Physics in 1909. Other notable pioneering inventors and developers in the field of electrical and electronic telecommunications include Charles Wheatstone and Samuel Morse (inventors of the telegraph), Alexander Graham Bell (inventor of the telephone), Edwin Armstrong and Lee de Forest (inventors of radio), as well as Vladimir K. Zworykin, John Logie Baird and Philo Farnsworth (some of the inventors of television).

Etymology

The word *telecommunication* was adapted from the French. It is a compound of the Greek prefix *tele-*, meaning «distant», and the Latin *communicare*, meaning "to share", and its written use was recorded in 1904 by the French engineer and novelist Édouard Estaunié. The prefix "tel" means "far, far off, operating over distance"... from Greek tele-, combining form of tele "far off, afar, at or to a distance," related to teleos (genitive telos) "end, goal, completion, result," from PIE root *kwel-"; "tel" also means " "far" in space and time". "Communication" was first used as an En-glish word in the late 14th century. It comes from Old French comunicacion (14c., Modern French communication), from Latin communicationem (nominative communicatio), noun of action from past participle stem of communicare "to share, divide out; communicate, impart, inform; join, unite, participate in," literally "to make common," from communis".

History

Beacons and Pigeons

In the Middle Ages, chains of beacons were commonly used on hilltops as a means of relaying a signal. Beacon chains suffered the drawback that they could only pass a single bit of information, so the meaning of the message such as "the enemy has been sighted" had to be agreed upon in advance. One notable instance of their use was during the Spanish Armada, when a beacon chain relayed a signal from Plymouth to London.

A replica of one of Chappe's semaphore towers

In 1792, Claude Chappe, a French engineer, built the first fixed visual telegraphy system (or semaphore line) between Lille and Paris. However semaphore suffered from the need for skilled operators and expensive towers at intervals of ten to thirty kilometres (six to nineteen miles). As a result of competition from the electrical telegraph, the last commercial line was abandoned in 1880.

Homing pigeons have occasionally been used throughout history by different cultures. Pigeon post is thought to have Persians roots and was used by the Romans to aid their military. Frontinus said that Julius Caesar used pigeons as messengers in his conquest of Gaul. The Greeks also conveyed the names of the victors at the Olympic Games to various cities using homing pigeons. In the early 19th century, the Dutch government used the system in Java and Sumatra. And in 1849, Paul Julius Reuter started a pigeon service to fly stock prices between Aachen and Brussels, a service that operated for a year until the gap in the telegraph link was closed.

Telegraph and Telephone

Sir Charles Wheatstone and Sir William Fothergill Cooke invented the electric telegraph in 1837. Also, the first commercial electrical telegraph is purported to have been constructed by Wheatstone and Cooke and opened on 9 April 1839. Both inventors viewed their device as "an improvement to the [existing] electromagnetic telegraph" not as a new device.

Samuel Morse independently developed a version of the electrical telegraph that he unsuccessfully demonstrated on 2 September 1837. His code was an important advance over Wheatstone's signaling method. The first transatlantic telegraph cable was successfully completed on 27 July 1866, allowing transatlantic telecommunication for the first time.

The conventional telephone was invented independently by Alexander Bell and Elisha Gray in 1876. Antonio Meucci invented the first device that allowed the electrical transmission of voice over a line in 1849. However Meucci's device was of little practical value because it relied upon the electrophonic effect and thus required users to place the receiver in their mouth to "hear" what was being said. The first commercial telephone services were set-up in 1878 and 1879 on both sides of the Atlantic in the cities of New Haven and London.

Radio and Television

In 1832, James Lindsay gave a classroom demonstration of wireless telegraphy to his students. By 1854, he was able to demonstrate a transmission across the Firth of Tay from Dundee, Scotland to Woodhaven, a distance of two miles (3 km), using water as the transmission medium. In December 1901, Guglielmo Marconi established wireless communication between St. John's, Newfoundland (Canada) and Poldhu, Cornwall (England), earning him the 1909 Nobel Prize in physics (which he shared with Karl Braun). However small-scale radio communication had already been demonstrated in 1893 by Nikola Tesla in a presentation to the National Electric Light Association.

On 25 March 1925, John Logie Baird was able to demonstrate the transmission of moving pictures at the London department store Selfridges. Baird's device relied upon the Nipkow disk and thus became known as the mechanical television. It formed the basis of experimental broadcasts done by the British Broadcasting Corporation beginning 30 September 1929. However, for most of the twentieth century televisions depended upon the cathode ray tube invented by Karl Braun. The first version of such a television to show promise was produced by Philo Farnsworth and demonstrated to his family on 7 September 1927.

Computers and the Internet

On 11 September 1940, George Stibitz was able to transmit problems using teletype to his Complex Number Calculator in New York and receive the computed results back at Dartmouth College in New Hampshire. This configuration of a centralized computer or mainframe with remote dumb terminals remained popular throughout the 1950s. However, it was not until the 1960s that researchers started to investigate packet switching — a technology that would allow chunks of data to be sent to different computers without first passing through a centralized mainframe. A four-node network emerged on 5 December 1969; this network would become ARPANET, which by 1981 would consist of 213 nodes.

ARPANET development centered around the Request for Comment process and on 7 April 1969, RFC 1 was published. This process is important because ARPANET eventually merged with other networks to form the Internet and many of the protocols the Internet relies upon today were specified through the Request for Comment process. In September 1981, RFC 791 introduced the Internet Protocol v4 (IPv4) and RFC 793 introduced the Transmission Control Protocol (TCP) — thus creating the TCP/IP protocol that much of the Internet relies upon today.

However, not all important developments were made through the Request for Comment process. Two popular link protocols for local area networks (LANs) also appeared in the 1970s. A patent for the token ring protocol was filed by Olof Soderblom on 29 October 1974 and a paper on the Ethernet protocol was published by Robert Metcalfe and David Boggs in the July 1976 issue of *Communications of the ACM*.

Key Concepts

A number of key concepts reoccur throughout the literature on modern telecommunication theory and systems. Some of these concepts are discussed below.

Basic Elements

Telecommunications is primarily divided up between wired and wireless subtypes. Overall though, a basic telecommunication system consists of three main parts that are always present in some form or another:

- A transmitter that takes information and converts it to a signal.

- A transmission medium, also called the "physical channel" that carries the signal. An example of this is the "free space channel".

- A receiver that takes the signal from the channel and converts it back into usable information for the recipient.

For example, in a radio broadcasting station the station's large power amplifier is the transmitter; and the broadcasting antenna is the interface between the power amplifier and the "free space channel". The free space channel is the transmission medium; and the receiver's antenna is the interface between the free space channel and the receiver. Next, the radio receiver is the destination of the radio signal, and this is where it is converted from electricity to sound for people to listen to.

Sometimes, telecommunication systems are "duplex" (two-way systems) with a single box of electronics working as both the transmitter and a receiver, or a *transceiver*. For example, a cellular telephone is a transceiver. The transmission electronics and the receiver electronics within a transceiver are actually quite independent of each other. This can be readily explained by the fact that radio transmitters contain power amplifiers that operate with electrical powers measured in watts or kilowatts, but radio receivers deal with radio powers that are measured in the microwatts or nanowatts. Hence, transceivers have to be carefully designed and built to isolate their high-power circuitry and their low-power circuitry from each other, as to not cause interference.

Telecommunication over fixed lines is called point-to-point communication because it is between one transmitter and one receiver. Telecommunication through radio broadcasts is called broadcast communication because it is between one powerful transmitter and numerous low-power but sensitive radio receivers.

Telecommunications in which multiple transmitters and multiple receivers have been designed to cooperate and to share the same physical channel are called multiplex systems. The sharing of physical channels using multiplexing often gives very large reductions in costs. Multiplexed systems are laid out in telecommunication networks, and the multiplexed signals are switched at nodes through to the correct destination terminal receiver.

Analog Versus Digital Communications

Communications signals can be sent either by analog signals or digital signals. There are analog communication systems and digital communication systems. For an analog signal, the signal is varied continuously with respect to the information. In a digital signal, the information is encoded as a set of discrete values (for example, a set of ones and zeros). During the propagation and reception, the information contained in analog signals will inevitably be degraded by undesirable physical noise. (The output of a transmitter is noise-free for all practical purposes.) Commonly, the noise in a communication system can be expressed as adding or subtracting from the desirable signal in a completely random way. This form of noise is called additive noise, with the understanding that the noise can be negative or positive at different instants of time. Noise that is not additive noise is a much more difficult situation to describe or analyze, and these other kinds of noise will be omitted here.

On the other hand, unless the additive noise disturbance exceeds a certain threshold, the information contained in digital signals will remain intact. Their resistance to noise represents a key advantage of digital signals over analog signals.

Telecommunication Networks

A telecommunications network is a collection of transmitters, receivers, and communications channels that send messages to one another. Some digital communications networks contain one or more routers that work together to transmit information to the correct user. An analog communications network consists of one or more switches that establish a connection between two or more users. For both types of network, repeaters may be necessary to amplify or recreate the signal when it is being transmitted over long distances. This is to combat attenuation that can render the signal indistinguishable from the noise. Another advantage of digital systems over analog is that their output is easier to store in memory, i.e. two voltage states (high and low) are easier to store than a continuous range of states.

Communication Channels

The term "channel" has two different meanings. In one meaning, a channel is the physical medium that carries a signal between the transmitter and the receiver. Examples of this include the atmosphere for sound communications, glass optical fibers for some kinds of optical communications, coaxial cables for communications by way of the voltages and electric currents in them, and free space for communications using visible light, infrared waves, ultraviolet light, and radio waves. This last channel is called the "free space channel". The sending of radio waves from one place to another has nothing to do with the presence or absence of an atmosphere between the two. Radio waves travel through a perfect vacuum just as easily as they travel through air, fog, clouds, or any other kind of gas.

The other meaning of the term "channel" in telecommunications is seen in the phrase communications channel, which is a subdivision of a transmission medium so that it can be used to send multiple streams of information simultaneously. For example, one radio station can broadcast radio waves into free space at frequencies in the neighborhood of 94.5 MHz (megahertz) while another radio station can simultaneously broadcast radio waves at frequencies in the neighborhood of 96.1 MHz. Each radio station would transmit radio waves over a frequency bandwidth of about 180 kHz (kilohertz), centered at frequencies such as the above, which are called the "carrier frequencies". Each station in this example is separated from its adjacent stations by 200 kHz, and the difference between 200 kHz and 180 kHz (20 kHz) is an engineering allowance for the imperfections in the communication system.

In the example above, the "free space channel" has been divided into communications channels according to frequencies, and each channel is assigned a separate frequency bandwidth in which to broadcast radio waves. This system of dividing the medium into channels according to frequency is called "frequency-division multiplexing". Another term for the same concept is "wavelength-division multiplexing", which is more commonly used in optical communications when multiple transmitters share the same physical medium.

Another way of dividing a communications medium into channels is to allocate each sender a recurring segment of time (a "time slot", for example, 20 milliseconds out of each second), and to allow each sender to send messages only within its own time slot. This method of dividing the medium into communication channels is called "time-division multiplexing" (TDM), and is used in optical fiber communication. Some radio communication systems use TDM within an allocated FDM channel. Hence, these systems use a hybrid of TDM and FDM.

Modulation

The shaping of a signal to convey information is known as modulation. Modulation can be used to represent a digital message as an analog waveform. This is commonly called "keying" – a term derived from the older use of Morse Code in telecommunications – and several keying techniques exist (these include phase-shift keying, frequency-shift keying, and amplitude-shift keying). The "Bluetooth" system, for example, uses phase-shift keying to exchange information between various devices. In addition, there are combinations of phase-shift keying and amplitude-shift keying which is called (in the jargon of the field) "quadrature amplitude modulation" (QAM) that are used in high-capacity digital radio communication systems.

Modulation can also be used to transmit the information of low-frequency analog signals at higher frequencies. This is helpful because low-frequency analog signals cannot be effectively transmitted over free space. Hence the information from a low-frequency analog signal must be impressed into a higher-frequency signal (known as the "carrier wave") before transmission. There are several different modulation schemes available to achieve this [two of the most basic being amplitude modulation (AM) and frequency modulation (FM)]. An example of this process is a disc jockey's voice being impressed into a 96 MHz carrier wave using frequency modulation (the voice would then be received on a radio as the channel "96 FM"). In addition, modulation has the advantage that it may use frequency division multiplexing (FDM).

Society

Telecommunication has a significant social, cultural and economic impact on modern society. In 2008, estimates placed the telecommunication industry's revenue at $4.7 trillion or just under 3 percent of the gross world product (official exchange rate). Several following sections discuss the impact of telecommunication on society.

Economic Impact

Microeconomics

On the microeconomic scale, companies have used telecommunications to help build global business empires. This is self-evident in the case of online retailer Amazon.com but, according to academic Edward Lenert, even the conventional retailer Walmart has benefited from better tele-communication infrastructure compared to its competitors. In cities throughout the world, home owners use their telephones to order and arrange a variety of home services ranging from pizza de-liveries to electricians. Even relatively poor communities have been noted to use telecommunica-tion to their advantage. In Bangladesh's Narshingdi district, isolated villagers use cellular phones to speak directly to wholesalers and arrange a better price for their goods. In Côte d'Ivoire, coffee growers share mobile phones to follow hourly variations in coffee prices and sell at the best price.

Macroeconomics

On the macroeconomic scale, Lars-Hendrik Röller and Leonard Waverman suggested a causal link between good telecommunication infrastructure and economic growth. Few dispute the existence of a correlation although some argue it is wrong to view the relationship as causal.

Because of the economic benefits of good telecommunication infrastructure, there is increasing worry about the inequitable access to telecommunication services amongst various countries of the world—this is known as the digital divide. A 2003 survey by the International Telecommu-nication Union (ITU) revealed that roughly a third of countries have fewer than one mobile sub-scription for every 20 people and one-third of countries have fewer than one land-line telephone subscription for every 20 people. In terms of Internet access, roughly half of all countries have fewer than one out of 20 people with Internet access. From this information, as well as educational data, the ITU was able to compile an index that measures the overall ability of citizens to access and use information and communication technologies. Using this measure, Sweden, Denmark and Iceland received the highest ranking while the African countries Nigeria, Burkina Faso and Mali received the lowest.

Social Impact

Telecommunication has played a significant role in social relationships. Nevertheless, devices like the telephone system were originally advertised with an emphasis on the practical dimensions of the device (such as the ability to conduct business or order home services) as opposed to the so-cial dimensions. It was not until the late 1920s and 1930s that the social dimensions of the device became a prominent theme in telephone advertisements. New promotions started appealing to consumers' emotions, stressing the importance of social conversations and staying connected to family and friends.

Since then the role that telecommunications has played in social relations has become increasingly important. In recent years, the popularity of social networking sites has increased dramatically. These sites allow users to communicate with each other as well as post photographs, events and profiles for others to see. The profiles can list a person's age, interests, sexual preference and re-lationship status. In this way, these sites can play important role in everything from organising social engagements to courtship.

Prior to social networking sites, technologies like short message service (SMS) and the telephone also had a significant impact on social interactions. In 2000, market research group Ipsos MORI reported that 81% of 15- to 24-year-old SMS users in the United Kingdom had used the service to coordinate social arrangements and 42% to flirt.

Other Impacts

In cultural terms, telecommunication has increased the public's ability to access music and film. With television, people can watch films they have not seen before in their own home without having to travel to the video store or cinema. With radio and the Internet, people can listen to music they have not heard before without having to travel to the music store.

Telecommunication has also transformed the way people receive their news. A survey led in 2006 by the non-profit Pew Internet and American Life Project found that when just over 3,000 people living in the United States were asked where they got their news "yesterday", more people said television or radio than newspapers. The results are summarised in the following table (the percentages add up to more than 100% because people were able to specify more than one source).

Local TV	National TV	Radio	Local paper	Internet	National paper
59%	47%	44%	38%	23%	12%

Telecommunication has had an equally significant impact on advertising. TNS Media Intelligence reported that in 2007, 58% of advertising expenditure in the United States was spent on mediums that depend upon telecommunication. The results are summarised in the following table.

	Internet	Radio	Cable TV	Syndicated TV	Spot TV	Network TV	Newspaper	Magazine	Outdoor	Total
Percent	7.6%	7.2%	12.1%	2.8%	11.3%	17.1%	18.9%	20.4%	2.7%	100%
Dollars	$11.31 billion	$10.69 billion	$18.02 billion	$4.17 billion	$16.82 billion	$25.42 billion	$28.22 billion	$30.33 billion	$4.02 billion	$149 billion

Government

Many countries have enacted legislation which conforms to the *International Telecommunication Regulations* established by the International Telecommunication Union (ITU), which is the "leading UN agency for information and communication technology issues." In 1947, at the Atlantic City Conference, the ITU decided to "afford international protection to all frequencies registered in a new international frequency list and used in conformity with the Radio Regulation." According to the ITU's *Radio Regulations* adopted in Atlantic City, all frequencies referenced in the *International Frequency Registration Board*, examined by the board and registered on the *International Frequency List* "shall have the right to international protection from harmful interference."

From a global perspective, there have been political debates and legislation regarding the management of telecommunication and broadcasting. The history of broadcasting discusses some debates in relation to balancing conventional communication such as printing and telecommunication such as radio broadcasting. The onset of World War II brought on the first explosion of international broadcasting propaganda. Countries, their governments, insurgents, terrorists, and militiamen have all used telecommunication and broadcasting techniques to promote propaganda. Patriotic propaganda for political movements and colonization started the mid-1930s. In 1936, the BBC broadcast propaganda to the Arab World to partly counter similar broadcasts from Italy, which also had colonial interests in North Africa.

Modern insurgents, such as those in the latest Iraq war, often use intimidating telephone calls, SMSs and the distribution of sophisticated videos of an attack on coalition troops within hours of the operation. "The Sunni insurgents even have their own television station, Al-Zawraa, which while banned by the Iraqi government, still broadcasts from Erbil, Iraqi Kurdistan, even as coalition pressure has forced it to switch satellite hosts several times."

On 10 November 2014, President Obama recommended the Federal Communications Commission reclassify broadband Internet service as a telecommunications service in order to preserve net neutrality.

Modern Media

Worldwide Equipment Sales

According to data collected by Gartner and Ars Technica sales of main consumer's telecommunication equipment worldwide in millions of units was:

Equipment / year	1975	1980	1985	1990	1994	1996	1998	2000	2002	2004	2006	2008
Computers	0	1	8	20	40	75	100	135	130	175	230	280
Cell phones	N/A	N/A	N/A	N/A	N/A	N/A	180	400	420	660	830	1000

Telephone

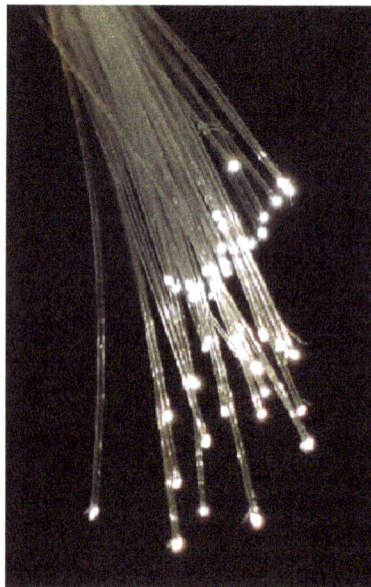

Optical fiber provides cheaper bandwidth for long distance communication.

In a telephone network, the caller is connected to the person they want to talk to by switches at various telephone exchanges. The switches form an electrical connection between the two users and the setting of these switches is determined electronically when the caller dials the number. Once the connection is made, the caller's voice is transformed to an electrical signal using a small microphone in the caller's handset. This electrical signal is then sent through the network to the user at the other end where it is transformed back into sound by a small speaker in that person's handset.

The landline telephones in most residential homes are analog—that is, the speaker's voice directly determines the signal's voltage. Although short-distance calls may be handled from end-to-end as analog signals, increasingly telephone service providers are transparently converting the signals to digital signals for transmission. The advantage of this is that digitized voice data can travel side-by-side with data from the Internet and can be perfectly reproduced in long distance communication (as opposed to analog signals that are inevitably impacted by noise).

Mobile phones have had a significant impact on telephone networks. Mobile phone subscriptions now outnumber fixed-line subscriptions in many markets. Sales of mobile phones in 2005 totalled 816.6 million with that figure being almost equally shared amongst the markets of Asia/Pacific (204 m), Western Europe (164 m), CEMEA (Central Europe, the Middle East and Africa) (153.5 m), North America (148 m) and Latin America (102 m). In terms of new subscriptions over the five years from 1999, Africa has outpaced other markets with 58.2% growth. Increasingly these phones are being serviced by systems where the voice content is transmitted digitally such as GSM or W-CDMA with many markets choosing to depreciate analog systems such as AMPS.

There have also been dramatic changes in telephone communication behind the scenes. Starting with the operation of TAT-8 in 1988, the 1990s saw the widespread adoption of systems based on optical fibers. The benefit of communicating with optic fibers is that they offer a drastic increase in data capacity. TAT-8 itself was able to carry 10 times as many telephone calls as the last copper cable laid at that time and today's optic fibre cables are able to carry 25 times as many telephone calls as TAT-8. This increase in data capacity is due to several factors: First, optic fibres are physically much smaller than competing technologies. Second, they do not suffer from crosstalk which means several hundred of them can be easily bundled together in a single cable. Lastly, improvements in multiplexing have led to an exponential growth in the data capacity of a single fibre.

Assisting communication across many modern optic fibre networks is a protocol known as Asynchronous Transfer Mode (ATM). The ATM protocol allows for the side-by-side data transmission mentioned in the second paragraph. It is suitable for public telephone networks because it establishes a pathway for data through the network and associates a traffic contract with that pathway. The traffic contract is essentially an agreement between the client and the network about how the network is to handle the data; if the network cannot meet the conditions of the traffic contract it does not accept the connection. This is important because telephone calls can negotiate a contract so as to guarantee themselves a constant bit rate, something that will ensure a caller's voice is not delayed in parts or cut off completely. There are competitors to ATM, such as Multiprotocol Label Switching (MPLS), that perform a similar task and are expected to supplant ATM in the future.

Radio and Television

In a broadcast system, the central high-powered broadcast tower transmits a high-frequency electromagnetic wave to numerous low-powered receivers. The high-frequency wave sent by the tower is modulated with a signal containing visual or audio information. The receiver is then tuned so as to pick up the high-frequency wave and a demodulator is used to retrieve the signal containing the visual or audio information. The broadcast signal can be either analog (signal is varied continuously with respect to the information) or digital (information is encoded as a set of discrete values).

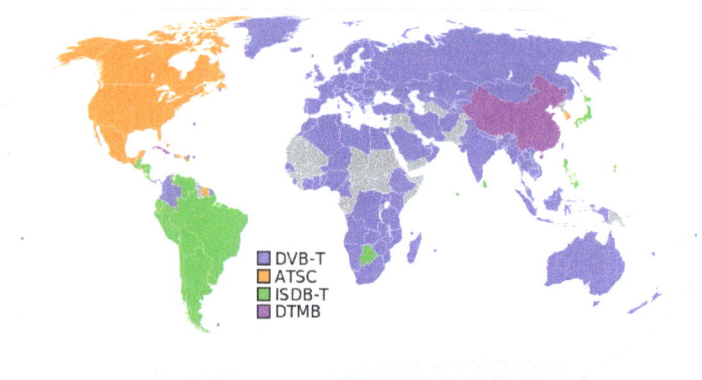

Digital television standards and their adoption worldwide

The broadcast media industry is at a critical turning point in its development, with many countries moving from analog to digital broadcasts. This move is made possible by the production of cheaper, faster and more capable integrated circuits. The chief advantage of digital broadcasts is that they prevent a number of complaints common to traditional analog broadcasts. For television, this includes the elimination of problems such as snowy pictures, ghosting and other distortion. These occur because of the nature of analog transmission, which means that perturbations due to noise will be evident in the final output. Digital transmission overcomes this problem because digital signals are reduced to discrete values upon reception and hence small perturbations do not affect the final output. In a simplified example, if a binary message 1011 was transmitted with signal amplitudes [1.0 0.0 1.0 1.0] and received with signal amplitudes [0.9 0.2 1.1 0.9] it would still decode to the binary message 1011 — a perfect reproduction of what was sent. From this example, a problem with digital transmissions can also be seen in that if the noise is great enough it can significantly alter the decoded message. Using forward error correction a receiver can correct a handful of bit errors in the resulting message but too much noise will lead to incomprehensible output and hence a breakdown of the transmission.

In digital television broadcasting, there are three competing standards that are likely to be adopted worldwide. These are the ATSC, DVB and ISDB standards; the adoption of these standards thus far is presented in the captioned map. All three standards use MPEG-2 for video compression. ATSC uses Dolby Digital AC-3 for audio compression, ISDB uses Advanced Audio Coding (MPEG-2 Part 7) and DVB has no standard for audio compression but typically uses MPEG-1 Part 3 Layer 2. The choice of modulation also varies between the schemes. In digital audio broadcasting, standards are much more unified with practically all countries choosing to adopt the Digital Audio Broadcasting standard (also known as the Eureka 147 standard). The exception is the United States which has chosen to adopt HD Radio. HD Radio, unlike Eureka 147, is based upon a transmission method known as in-band on-channel transmission that allows digital information to "piggyback" on normal AM or FM analog transmissions.

However, despite the pending switch to digital, analog television remains being transmitted in most countries. An exception is the United States that ended analog television transmission (by all but the very low-power TV stations) on 12 June 2009 after twice delaying the switchover deadline,Kenya also ended analog television transmission in December 2014 after multiple delays. For analog television, there are three standards in use for broadcasting color TV. These are known as

PAL (German designed), NTSC (North American designed), and SECAM (French designed). (It is important to understand that these are the ways of sending color TV, and they do not have anything to do with the standards for black & white TV, which also vary from country to country.) For analog radio, the switch to digital radio is made more difficult by the fact that analog receivers are sold at a small fraction of the price of digital receivers. The choice of modulation for analog radio is typically between amplitude (AM) or frequency modulation (FM). To achieve stereo playback, an amplitude modulated subcarrier is used for stereo FM.

Internet

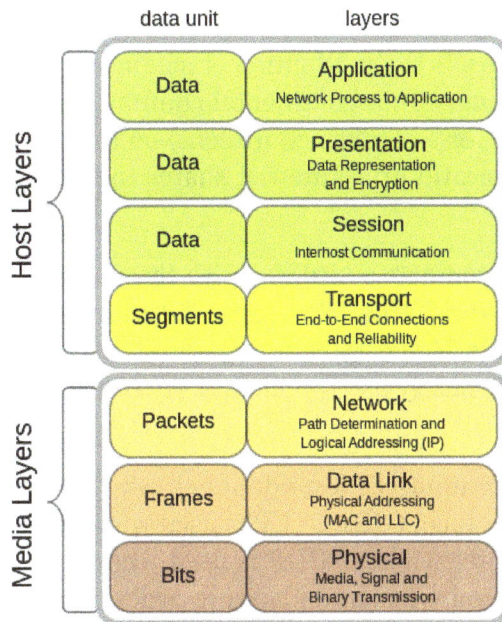

The OSI reference model

The Internet is a worldwide network of computers and computer networks that communicate with each other using the Internet Protocol. Any computer on the Internet has a unique IP address that can be used by other computers to route information to it. Hence, any computer on the Internet can send a message to any other computer using its IP address. These messages carry with them the originating computer's IP address allowing for two-way communication. The Internet is thus an exchange of messages between computers.

It is estimated that the 51% of the information flowing through two-way telecommunications networks in the year 2000 were flowing through the Internet (most of the rest (42%) through the landline telephone). By the year 2007 the Internet clearly dominated and captured 97% of all the information in telecommunication networks (most of the rest (2%) through mobile phones). As of 2008, an estimated 21.9% of the world population has access to the Internet with the highest access rates (measured as a percentage of the population) in North America (73.6%), Oceania/Australia (59.5%) and Europe (48.1%). In terms of broadband access, Iceland (26.7%), South Korea (25.4%) and the Netherlands (25.3%) led the world.

The Internet works in part because of protocols that govern how the computers and routers communicate with each other. The nature of computer network communication lends itself to a layered approach where individual protocols in the protocol stack run more-or-less independently of other protocols. This allows lower-level protocols to be customized for the network situation while not changing the way higher-level protocols operate. A practical example of why this is important is because it allows an Internet browser to run the same code regardless of whether the computer it is running on is connected to the Internet through an Ethernet or Wi-Fi connection. Protocols are often talked about in terms of their place in the OSI reference model (pictured on the right), which emerged in 1983 as the first step in an unsuccessful attempt to build a universally adopted networking protocol suite.

For the Internet, the physical medium and data link protocol can vary several times as packets traverse the globe. This is because the Internet places no constraints on what physical medium or data link protocol is used. This leads to the adoption of media and protocols that best suit the local network situation. In practice, most intercontinental communication will use the Asynchronous Transfer Mode (ATM) protocol (or a modern equivalent) on top of optic fiber. This is because for most intercontinental communication the Internet shares the same infrastructure as the public switched telephone network.

At the network layer, things become standardized with the Internet Protocol (IP) being adopted for logical addressing. For the World Wide Web, these "IP addresses" are derived from the human readable form using the Domain Name System (e.g. 72.14.207.99 is derived from www.google.com). At the moment, the most widely used version of the Internet Protocol is version four but a move to version six is imminent.

At the transport layer, most communication adopts either the Transmission Control Protocol (TCP) or the User Datagram Protocol (UDP). TCP is used when it is essential every message sent is received by the other computer whereas UDP is used when it is merely desirable. With TCP, packets are retransmitted if they are lost and placed in order before they are presented to higher layers. With UDP, packets are not ordered or retransmitted if lost. Both TCP and UDP packets carry port numbers with them to specify what application or process the packet should be handled by. Because certain application-level protocols use certain ports, network administrators can manipulate traffic to suit particular requirements. Examples are to restrict Internet access by blocking the traffic destined for a particular port or to affect the performance of certain applications by assigning priority.

Above the transport layer, there are certain protocols that are sometimes used and loosely fit in the session and presentation layers, most notably the Secure Sockets Layer (SSL) and Transport Layer Security (TLS) protocols. These protocols ensure that data transferred between two parties remains completely confidential. Finally, at the application layer, are many of the protocols Internet users would be familiar with such as HTTP (web browsing), POP3 (e-mail), FTP (file transfer), IRC (Internet chat), BitTorrent (file sharing) and XMPP (instant messaging).

Voice over Internet Protocol (VoIP) allows data packets to be used for synchronous voice communications. The data packets are marked as voice type packets and can be prioritized by the network administrators so that the real-time, synchronous conversation is less subject to contention with other types of data traffic which can be delayed (i.e. file transfer or email) or buffered in advance (i.e. audio and video) without detriment. That prioritization is fine when the network has sufficient capacity for all the VoIP calls taking place at the same time and the network is enabled for prioritization i.e.

a private corporate style network, but the Internet is not generally managed in this way and so there can be a big difference in the quality of VoIP calls over a private network and over the public Internet.

Local Area Networks and Wide Area Networks

Despite the growth of the Internet, the characteristics of local area networks (LANs)--computer networks that do not extend beyond a few kilometers—remain distinct. This is because networks on this scale do not require all the features associated with larger networks and are often more cost-effective and efficient without them. When they are not connected with the Internet, they also have the advantages of privacy and security. However, purposefully lacking a direct connection to the Internet does not provide assured protection from hackers, military forces, or economic powers. These threats exist if there are any methods for connecting remotely to the LAN.

Wide area networks (WANs) are private computer networks that may extend for thousands of kilometers. Once again, some of their advantages include privacy and security. Prime users of private LANs and WANs include armed forces and intelligence agencies that must keep their information secure and secret.

In the mid-1980s, several sets of communication protocols emerged to fill the gaps between the data-link layer and the application layer of the OSI reference model. These included Appletalk, IPX, and NetBIOS with the dominant protocol set during the early 1990s being IPX due to its popularity with MS-DOS users. TCP/IP existed at this point, but it was typically only used by large government and research facilities.

As the Internet grew in popularity and its traffic was required to be routed into private networks, the TCP/IP protocols replaced existing local area network technologies. Additional technologies, such as DHCP, allowed TCP/IP-based computers to self-configure in the network. Such functions also existed in the AppleTalk/ IPX/ NetBIOS protocol sets.

Whereas Asynchronous Transfer Mode (ATM) or Multiprotocol Label Switching (MPLS) are typical data-link protocols for larger networks such as WANs; Ethernet and Token Ring are typical data-link protocols for LANs. These protocols differ from the former protocols in that they are simpler, e.g., they omit features such as quality of service guarantees, and offer collision prevention. Both of these differences allow for more economical systems.

Despite the modest popularity of IBM Token Ring in the 1980s and 1990s, virtually all LANs now use either wired or wireless Ethernet facilities. At the physical layer, most wired Ethernet implementations use copper twisted-pair cables (including the common 10BASE-T networks). However, some early implementations used heavier coaxial cables and some recent implementations (especially high-speed ones) use optical fibers. When optic fibers are used, the distinction must be made between multimode fibers and single-mode fibers. Multimode fibers can be thought of as thicker optical fibers that are cheaper to manufacture devices for, but that suffers from less usable bandwidth and worse attenuation – implying poorer long-distance performance.

Transmission Capacity

The effective capacity to exchange information worldwide through two-way telecommunication networks grew from 281 petabytes of (optimally compressed) information in 1986, to 471 petabytes in 1993, to 2.2 (optimally compressed) exabytes in 2000, and to 65 (optimally compressed) exabytes in 2007. This is the informational equivalent of two newspaper pages per person per day in 1986, and six entire newspapers per person per day by 2007. Given this growth, telecommunications play an increasingly important role in the world economy and the global telecommunications industry was about a $4.7 trillion sector in 2012. The service revenue of the global telecommunications industry was estimated to be $1.5 trillion in 2010, corresponding to 2.4% of the world's gross domestic product (GDP).

References

- Blechman, Andrew (2007). Pigeons-The fascinating saga of the world's most revered and reviled bird. St Lucia, Queensland: University of Queensland Press. ISBN 9780702236419.

- Ambardar, Ashok (1999). Analog and Digital Signal Processing (2nd ed.). Brooks/Cole Publishing Company. pp. 1–2. ISBN 0-534-95409-X.

- Wood, James & Science Museum (Great Britain) "History of international broadcasting". IET 1994, Volume 1, p.2 of 258 ISBN 0-86341-302-1, ISBN 978-0-86341-302-5. Republished by Googlebooks. Accessed 21 July 2009.

- Stallings, William (2004). Data and Computer Communications (7th edition (intl) ed.). Pearson Prentice Hall. pp. 337–366. ISBN 0-13-183311-1.

- Stallings, William (2004). Data and Computer Communications (7th edition (intl) ed.). Pearson Prentice Hall. ISBN 0-13-183311-1.

- Martin, Michael (2000). Understanding the Network (The Networker's Guide to AppleTalk, IPX, and NetBIOS), SAMS Publishing, ISBN 0-7357-0977-7.

- Wyatt, Edward (10 November 2014). "Obama Asks F.C.C. to Adopt Tough Net Neutrality Rules". New York Times. Retrieved 15 November 2014.

- NYT Editorial Board (14 November 2014). "Why the F.C.C. Should Heed President Obama on Internet Regulation". New York Times. Retrieved 15 November 2014.

- Lazar, Irwin (22 February 2011). "The WAN Road Ahead: Ethernet or Bust?". Telecom Industry Updates. Retrieved 22 February 2011.

Types of Telecommunications

Communication occurs when there is a transfer of information between participants. Some of the types of telecommunications are email, fax, telephony, voice over IP and videoconferencing. All these technologies are widely used in today's times and help in communication across the globe. The aspects elucidated in this chapter are of vital importance, and provide a better understanding of telecommunications.

Email

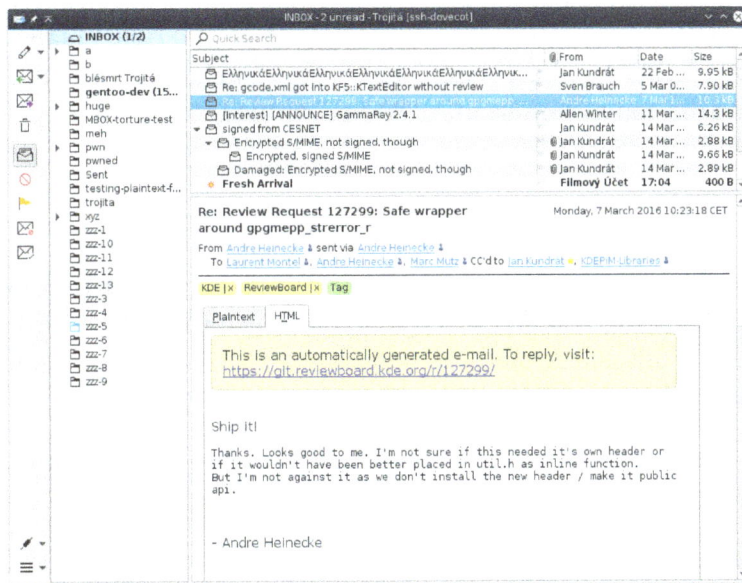

This screenshot shows the "Inbox" page of an email system, where users can see new emails and take actions, such as reading, deleting, saving, or responding to these messages.

Electronic mail is a method of exchanging digital messages between computer users; Email first entered substantial use in the 1960s and by the mid-1970s had taken the form now recognised as email. Email operates across computer networks, which in the 2010s is primarily the Internet. Some early email systems required the author and the recipient to both be online at the same time, in common with instant messaging. Today's email systems are based on a store-and-forward model. Email servers accept, forward, deliver, and store messages. Neither the users nor their computers are required to be online simultaneously; they need to connect only briefly, typically to a mail server, for as long as it takes to send or receive messages.

Originally an ASCII text-only communications medium, Internet email was extended by Multipurpose Internet Mail Extensions (MIME) to carry text in other character sets and multimedia content attachments. International email, with internationalized email addresses using UTF-8, has been standardized, but as of 2016 not widely adopted.

The at sign, a part of every SMTP email address.

The history of modern Internet email services reaches back to the early ARPANET, with standards for encoding email messages published as early as 1973 (RFC 561). An email message sent in the early 1970s looks very similar to a basic email sent today. Email played an important part in creating the Internet, and the conversion from ARPANET to the Internet in the early 1980s produced the core of the current services.

Terminology

Historically, the term *electronic mail* was used generically for any electronic document transmission. For example, several writers in the early 1970s used the term to describe fax document transmission. As a result, it is difficult to find the first citation for the use of the term with the more specific meaning it has today.

Electronic mail has been most commonly called email or e-mail since around 1993, but various variations of the spelling have been used:

- *email* is the most common form used online, and is required by IETF Requests for Comments and working groups and increasingly by style guides. This spelling also appears in most dictionaries.

- *e-mail* has long been the form that appears most frequently in edited, published American English and British English writing as reflected in the Corpus of Contemporary American English data, but is falling out of favor in style guides.

- *mail* was the form used in the original RFC. The service is referred to as *mail*, and a single piece of electronic mail is called a *message*.

- *EMail* is a traditional form that has been used in RFCs for the "Author's Address" and is expressly required "for historical reasons".

- *E-mail* is sometimes used, capitalizing the initial *E* as in similar abbreviations like *E-piano*, *E-guitar*, *A-bomb*, and *H-bomb*.

Origin

The AUTODIN network, first operational in 1962, provided a message service between 1,350 terminals, handling 30 million messages per month, with an average message length of approximately 3,000 characters. Autodin was supported by 18 large computerized switches, and was connected to the United States General Services Administration Advanced Record System, which provided similar services to roughly 2,500 terminals. By 1968, AUTODIN linked more than 300 sites in several countries.

Host-based Mail Systems

With the introduction of MIT's Compatible Time-Sharing System (CTSS) in 1961 multiple users could log in to a central system from remote dial-up terminals, and to store and share files on the central disk. Informal methods of using this to pass messages were developed and expanded:

- 1965 – MIT's CTSS MAIL.

Developers of other early systems developed similar email applications:

- 1962 – 1440/1460 Administrative Terminal System.

- 1968 – ATS/360.

- 1971 – *SNDMSG*, a local inter-user mail program incorporating the experimental file transfer program, *CPYNET*, allowed the first networked electronic mail.

- 1972 – Unix mail program.

- 1972 – APL Mailbox by Larry Breed.

- 1974 – The PLATO IV Notes on-line message board system was generalized to offer 'personal notes' in August 1974.

- 1978 – *Mail* client written by Kurt Shoens for Unix and distributed with the Second Berkeley Software Distribution included support for aliases and distribution lists, forwarding, formatting messages, and accessing different mailboxes. It used the Unix *mail* client to send messages between system users. The concept was extended to communicate remotely over the Berkley Network.

- 1979 – *EMAIL* written by V.A. Shiva Ayyadurai to emulate the interoffice mail system of the University of Medicine and Dentistry of New Jersey.

- 1979 – MH Message Handling System developed at RAND provided several tools for managing electronic mail on Unix.

- 1981 – PROFS by IBM.

- 1982 – ALL-IN-1 by Digital Equipment Corporation.

- 1982 – HP Mail (later HP DeskManager) by Hewlett-Packard.

These original messaging systems had widely different features and ran on systems that were incompatible with each other. Most of them only allowed communication between users logged into the same host or "mainframe", although there might be hundreds or thousands of users within an organization.

LAN Email Systems

In the early 1980s, networked personal computers on LANs became increasingly important. Server-based systems similar to the earlier mainframe systems were developed. Again, these systems initially allowed communication only between users logged into the same server infrastructure. Examples include:

- cc:Mail

- Lantastic

- WordPerfect Office

- Microsoft Mail

- Banyan VINES

- Lotus Notes

Eventually these systems too could link different organizations as long as they ran the same email system and proprietary protocol.

Email Networks

To facilitate electronic mail exchange between remote sites and with other organizations, telecommunication links, such as dialup modems or leased lines, provided means to transport email globally, creating local and global networks. This was challenging for a number of reasons, including the widely different email address formats in use.

- In 1971 the first ARPANET email was sent, and through RFC 561, RFC 680, RFC 724, and finally 1977's RFC 733, became a standardized working system.

- PLATO IV was networked to individual terminals over leased data lines prior to the implementation of personal notes in 1974.

- Unix mail was networked by 1978's uucp, which was also used for USENET newsgroup postings, with similar headers.

- BerkNet, the Berkeley Network, was written by Eric Schmidt in 1978 and included first in the Second Berkeley Software Distribution. It provided support for sending and receiving messages over serial communication links. The Unix mail tool was extended to send messages using BerkNet.

- The delivermail tool, written by Eric Allman in 1979 and 1980 (and shipped in 4BSD), provided support for routing mail over dissimilar networks, including Arpanet, UUCP, and BerkNet. (It also provided support for mail user aliases.)

- The mail client included in 4BSD (1980) was extended to provide interoperability between a variety of mail systems.

- BITNET (1981) provided electronic mail services for educational institutions. It was based on the IBM VNET email system.

- 1983 – MCI Mail Operated by MCI Communications Corporation. This was the first commercial public email service to use the internet. MCI Mail also allowed subscribers to send regular postal mail (overnight) to non-subscribers.

- In 1984, IBM PCs running DOS could link with FidoNet for email and shared bulletin board posting.

Email Address Internationalization

Globally countries started adopting IDN registrations for supporting country specific scripts (non-English) for domain names. In 2010 Egypt, the Russian Federation, Saudi Arabia, and the United Arab Emirates started offering IDN registrations. The government of India also registered .bharat in 8 languages/scripts in 2014.

Attempts at Interoperability

Early interoperability among independent systems included:

- ARPANET, a forerunner of the Internet, defined protocols for dissimilar computers to exchange email.

- uucp implementations for Unix systems, and later for other operating systems, that only had dial-up communications available.

- CSNET, which initially used the UUCP protocols via dial-up to provide networking and mail-relay services for non-ARPANET hosts.

- Action Technologies developed the Message Handling System (MHS) protocol (later bought by Novell, which abandoned it after purchasing the non-MHS WordPerfect Office—renamed Groupwise).

- HP OpenMail was known for its ability to interconnect several other APIs and protocols, including MAPI, cc:Mail, SMTP/MIME, and X.400.

- Soft-Switch released its eponymous email gateway product in 1984, acquired by Lotus Software ten years later.

- The Coloured Book protocols ran on UK academic networks until 1992.

- X.400 in the 1980s and early 1990s was promoted by major vendors, and mandated for government use under GOSIP, but abandoned by all but a few in favor of Internet SMTP by the mid-1990s.

From SNDMSG to MSG

In the early 1970s, Ray Tomlinson updated an existing utility called SNDMSG so that it could copy messages (as files) over the network. Lawrence Roberts, the project manager for the ARPANET development, took the idea of READMAIL, which dumped all "recent" messages onto the user's terminal, and wrote a programme for TENEX in TECO macros called *RD*, which permitted access to individual messages. Barry Wessler then updated RD and called it *NRD*.

Marty Yonke rewrote NRD to include reading, access to SNDMSG for sending, and a help system, and called the utility *WRD*, which was later known as *BANANARD*. John Vittal then updated this version to include three important commands: *Move* (combined save/delete command), *Answer* (determined to whom a reply should be sent) and *Forward* (sent an email to a person who was not already a recipient). The system was called *MSG*. With inclusion of these features, MSG is considered to be the first integrated modern email programme, from which many other applications have descended.

ARPANET Mail

Experimental email transfers between separate computer systems began shortly after the creation of the ARPANET in 1969. Ray Tomlinson is generally credited as having sent the first email across a network, initiating the use of the "@" sign to separate the names of the user and the user's machine in 1971, when he sent a message from one Digital Equipment Corporation DEC-10 computer to another DEC-10. The two machines were placed next to each other. Tomlinson's work was quickly adopted across the ARPANET, which significantly increased the popularity of email. Tomlinson is internationally known as the inventor of modern email.

Initially addresses were of the form, *username@hostname* but were extended to "username@ host.domain" with the development of the Domain Name System (DNS).

As the influence of the ARPANET spread across academic communities, gateways were developed to pass mail to and from other networks such as CSNET, JANET, BITNET, X.400, and FidoNet. This often involved addresses such as:

hubhost!middlehost!edgehost!user@uucpgateway.somedomain.example.com

which routes mail to a user with a "bang path" address at a UUCP host.

Operation

The diagram to the right shows a typical sequence of events that takes place when sender Alice transmits a message using a mail user agent (MUA) addressed to the email address of the recipient.

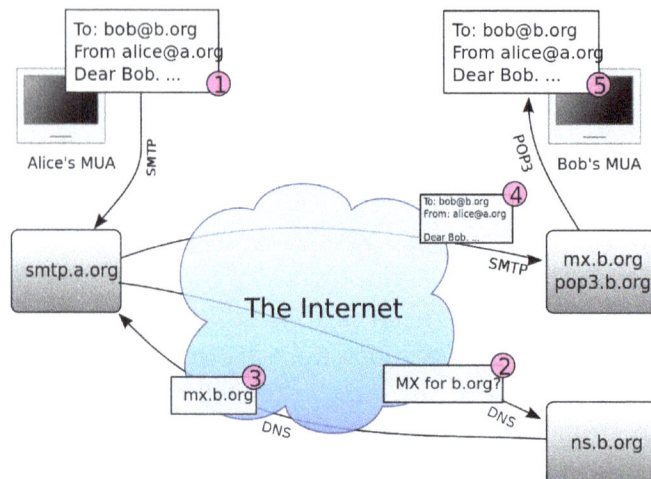

1. The MUA formats the message in email format and uses the submission protocol, a profile of the Simple Mail Transfer Protocol (SMTP), to send the message to the local mail submission agent (MSA), in this case *smtp.a.org*.

2. The MSA determines the destination address provided in the SMTP protocol (not from the message header), in this case *bob@b.org*. The part before the @ sign is the *local part* of the address, often the username of the recipient, and the part after the @ sign is a domain name. The MSA resolves a domain name to determine the fully qualified domain name of the mail server in the Domain Name System (DNS).

3. The DNS server for the domain *b.org* (*ns.b.org*) responds with any MX records listing the mail exchange servers for that domain, in this case *mx.b.org*, a message transfer agent (MTA) server run by the recipient's ISP.

4. smtp.a.org sends the message to mx.b.org using SMTP. This server may need to forward the message to other MTAs before the message reaches the final message delivery agent (MDA).

5. The MDA delivers it to the mailbox of user *bob*.

6. Bob's MUA picks up the message using either the Post Office Protocol (POP3) or the Internet Message Access Protocol (IMAP).

In addition to this example, alternatives and complications exist in the email system:

- Alice or Bob may use a client connected to a corporate email system, such as IBM Lotus Notes or Microsoft Exchange. These systems often have their own internal email format and their clients typically communicate with the email server using a vendor-specific, proprietary protocol. The server sends or receives email via the Internet through the product's Internet mail gateway which also does any necessary reformatting. If Alice and Bob work for the same company, the entire transaction may happen completely within a single corporate email system.

- Alice may not have a MUA on her computer but instead may connect to a webmail service.

- Alice's computer may run its own MTA, so avoiding the transfer at step 1.

- Bob may pick up his email in many ways, for example logging into mx.b.org and reading it directly, or by using a webmail service.

- Domains usually have several mail exchange servers so that they can continue to accept mail even if the primary is not available.

Many MTAs used to accept messages for any recipient on the Internet and do their best to deliver them. Such MTAs are called *open mail relays*. This was very important in the early days of the Internet when network connections were unreliable. However, this mechanism proved to be exploitable by originators of unsolicited bulk email and as a consequence open mail relays have become rare, and many MTAs do not accept messages from open mail relays.

Message Format

The Internet email message format is now defined by RFC 5322, with multimedia content attachments being defined in RFC 2045 through RFC 2049, collectively called *Multipurpose Internet Mail Extensions* or *MIME*. RFC 5322 replaced the earlier RFC 2822 in 2008, and in turn RFC 2822 in 2001 replaced RFC 822 – which had been the standard for Internet email for nearly 20 years. Published in 1982, RFC 822 was based on the earlier RFC 733 for the ARPANET.

Internet email messages consist of two major sections, the message header and the message body. The header is structured into fields such as From, To, CC, Subject, Date, and other information about the email. In the process of transporting email messages between systems, SMTP communicates delivery parameters and information using message header fields. The body contains the message, as unstructured text, sometimes containing a signature block at the end. The header is separated from the body by a blank line.

Message Header

Each message has exactly one header, which is structured into fields. Each field has a name and a value. RFC 5322 specifies the precise syntax.

Informally, each line of text in the header that begins with a printable character begins a separate field. The field name starts in the first character of the line and ends before the separator character ":". The separator is then followed by the field value (the "body" of the field). The value is continued onto subsequent lines if those lines have a space or tab as their first character. Field names and values are restricted to 7-bit ASCII characters. Non-ASCII values may be represented using MIME encoded words.

Header Fields

Email header fields can be multi-line, and each line should be at most 78 characters long and in no event more than 998 characters long. Header fields defined by RFC 5322 can only contain US-ASCII characters; for encoding characters in other sets, a syntax specified in RFC 2047 can be used. Recently the IETF EAI working group has defined some standards track extensions, replacing previous experimental extensions, to allow UTF-8 encoded Unicode characters to be used within the header. In particular, this allows email addresses to use non-ASCII characters. Such addresses are supported by Google and Microsoft products, and promoted by some governments.

The message header must include at least the following fields:

- *From*: The email address, and optionally the name of the author(s). In many email clients not changeable except through changing account settings.

- *Date*: The local time and date when the message was written. Like the *From:* field, many email clients fill this in automatically when sending. The recipient's client may then display the time in the format and time zone local to him/her.

RFC 3864 describes registration procedures for message header fields at the IANA; it provides for permanent and provisional field names, including also fields defined for MIME, netnews, and HTTP, and referencing relevant RFCs. Common header fields for email include:

- *To*: The email address(es), and optionally name(s) of the message's recipient(s). Indicates primary recipients (multiple allowed), for secondary recipients see Cc: and Bcc: below.

- *Subject*: A brief summary of the topic of the message. Certain abbreviations are commonly used in the subject, including "RE:" and "FW:".

- *Cc*: Carbon copy; Many email clients will mark email in one's inbox differently depending on whether they are in the To: or Cc: list. (*Bcc*: Blind carbon copy; addresses are usually only specified during SMTP delivery, and not usually listed in the message header.)

- Content-Type: Information about how the message is to be displayed, usually a MIME type.

- *Precedence*: commonly with values "bulk", "junk", or "list"; used to indicate that automated "vacation" or "out of office" responses should not be returned for this mail, e.g. to prevent vacation notices from being sent to all other subscribers of a mailing list. Sendmail uses this field to affect prioritization of queued email, with "Precedence: special-delivery" messages delivered sooner. With modern high-bandwidth networks, delivery priority is less of an issue than it once was. Microsoft Exchange respects a fine-grained automatic response suppression mechanism, the *X-Auto-Response-Suppress* field.

- *Message-ID*: Also an automatically generated field; used to prevent multiple delivery and for reference in In-Reply-To.

- *In-Reply-To*: Message-ID of the message that this is a reply to. Used to link related messages together. This field only applies for reply messages.

- *References*: Message-ID of the message that this is a reply to, and the message-id of the message the previous reply was a reply to, etc.

- *Reply-To*: Address that should be used to reply to the message.

- *Sender*: Address of the actual sender acting on behalf of the author listed in the From: field (secretary, list manager, etc.).

- *Archived-At*: A direct link to the archived form of an individual email message.

Note that the *To:* field is not necessarily related to the addresses to which the message is delivered. The actual delivery list is supplied separately to the transport protocol, SMTP, which may or may not originally have been extracted from the header content. The "To:" field is similar to the addressing at the top of a conventional letter which is delivered according to the address on the outer envelope. In the same way, the "From:" field does not have to be the real sender of the email message. Some mail servers apply email authentication systems to messages being relayed. Data pertaining to server's activity is also part of the header, as defined below.

SMTP defines the *trace information* of a message, which is also saved in the header using the following two fields:

- *Received*: when an SMTP server accepts a message it inserts this trace record at the top of the header (last to first).

- *Return-Path*: when the delivery SMTP server makes the *final delivery* of a message, it inserts this field at the top of the header.

Other fields that are added on top of the header by the receiving server may be called *trace fields*, in a broader sense.

- *Authentication-Results*: when a server carries out authentication checks, it can save the results in this field for consumption by downstream agents.

- *Received-SPF*: stores results of SPF checks in more detail than Authentication-Results.

- *Auto-Submitted*: is used to mark automatically generated messages.

- *VBR-Info*: claims VBR whitelisting

Message Body

Content Encoding

Email was originally designed for 7-bit ASCII. Most email software is 8-bit clean but must assume it will communicate with 7-bit servers and mail readers. The MIME standard introduced character set specifiers and two content transfer encodings to enable transmission of non-ASCII data: quoted printable for mostly 7-bit content with a few characters outside that range and base64 for arbitrary binary data. The 8BITMIME and BINARY extensions were introduced to allow transmission of mail without the need for these encodings, but many mail transport agents still do not support them fully. In some countries, several encoding schemes coexist; as the result, by default, the message in a non-Latin alphabet language appears in non-readable form (the only exception is coincidence, when the sender and receiver use the same encoding scheme). Therefore, for international character sets, Unicode is growing in popularity.

Plain Text and HTML

Most modern graphic email clients allow the use of either plain text or HTML for the message body at the option of the user. HTML email messages often include an automatically generated plain text copy as well, for compatibility reasons. Advantages of HTML include the ability to include inline links and images, set apart previous messages in block quotes, wrap naturally on any display, use emphasis such as underlines and italics, and change font styles. Disadvantages include the increased size of the email, privacy concerns about web bugs, abuse of HTML email as a vector for phishing attacks and the spread of malicious software.

Some web-based mailing lists recommend that all posts be made in plain-text, with 72 or 80 characters per line for all the above reasons, but also because they have a significant number of readers using text-based email clients such as Mutt. Some Microsoft email clients allow rich formatting using their proprietary Rich Text Format (RTF), but this should be avoided unless the recipient is guaranteed to have a compatible email client.

Servers and Client Applications

Messages are exchanged between hosts using the Simple Mail Transfer Protocol with software programs called mail transfer agents (MTAs); and delivered to a mail store by programs called mail delivery agents (MDAs, also sometimes called local delivery agents, LDAs). Accepting a message obliges an MTA to deliver it, and when a message cannot be delivered, that MTA must send a bounce message back to the sender, indicating the problem.

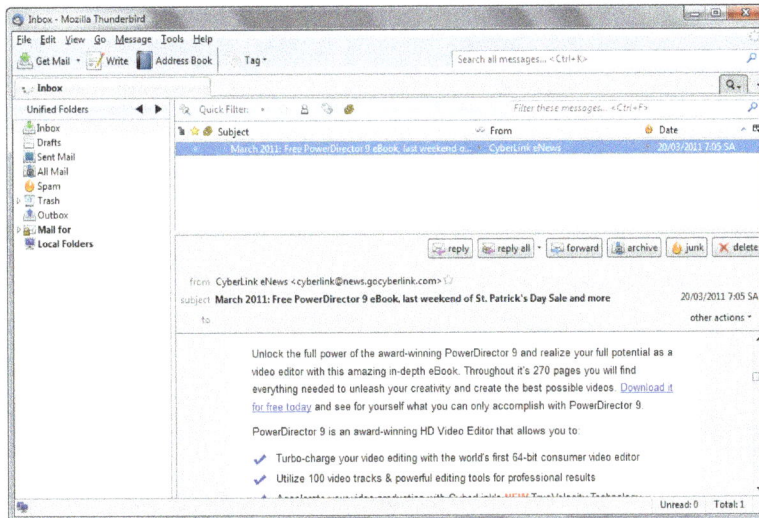

The interface of an email client, Thunderbird.

Users can retrieve their messages from servers using standard protocols such as POP or IMAP, or, as is more likely in a large corporate environment, with a proprietary protocol specific to Novell Groupwise, Lotus Notes or Microsoft Exchange Servers. Programs used by users for retrieving, reading, and managing email are called mail user agents (MUAs).

Mail can be stored on the client, on the server side, or in both places. Standard formats for mailboxes include Maildir and mbox. Several prominent email clients use their own proprietary format and require conversion software to transfer email between them. Server-side storage is often in a proprietary format but since access is through a standard protocol such as IMAP, moving email from one server to another can be done with any MUA supporting the protocol.

Many current email users do not run MTA, MDA or MUA programs themselves, but use a web-based email platform, such as Gmail, Hotmail, or Yahoo! Mail, that performs the same tasks. Such webmail interfaces allow users to access their mail with any standard web browser, from any computer, rather than relying on an email client.

Filename Extensions

Upon reception of email messages, email client applications save messages in operating system files in the file system. Some clients save individual messages as separate files, while others use various database formats, often proprietary, for collective storage. A historical standard of storage is the *mbox* format. The specific format used is often indicated by special filename extensions:

eml

> Used by many email clients including Novell GroupWise, Microsoft Outlook Express, Lotus notes, Windows Mail, Mozilla Thunderbird, and Postbox. The files are plain text in MIME format, containing the email header as well as the message contents and attachments in one or more of several formats.

emlx

Used by Apple Mail.

msg

Used by Microsoft Office Outlook and OfficeLogic Groupware.

mbx

Used by Opera Mail, KMail, and Apple Mail based on the mbox format.

Some applications (like Apple Mail) leave attachments encoded in messages for searching while also saving separate copies of the attachments. Others separate attachments from messages and save them in a specific directory.

URI Scheme Mailto

The URI scheme, as registered with the IANA, defines the mailto: scheme for SMTP email addresses. Though its use is not strictly defined, URLs of this form are intended to be used to open the new message window of the user's mail client when the URL is activated, with the address as defined by the URL in the *To:* field.

Types

Web-based Email

Many email providers have a web-based email client (e.g. AOL Mail, Gmail, Outlook.com and Yahoo! Mail). This allows users to log in to the email account by using any compatible web browser to send and receive their email. Mail is typically not downloaded to the client, so can't be read without a current Internet connection.

POP3 Email Services

The Post Office Protocol 3 (POP3) is a mail access protocol used by a client application to read messages from the mail server. Received messages are often deleted from the server. POP supports simple download-and-delete requirements for access to remote mailboxes (termed maildrop in the POP RFC's).

IMAP Email Servers

The Internet Message Access Protocol (IMAP) provides features to manage a mailbox from multiple devices. Small portable devices like smartphones are increasingly used to check email while travelling, and to make brief replies, larger devices with better keyboard access being used to reply at greater length. IMAP shows the headers of messages, the sender and the subject and the device needs to request to download specific messages. Usually mail is left in folders in the mail server.

MAPI Email Servers

Messaging Application Programming Interface (MAPI) is used by Microsoft Outlook to communicate to Microsoft Exchange Server - and to a range of other e-mail server products such as Axigen

Mail Server, Kerio Connect, Scalix, Zimbra, HP OpenMail, IBM Lotus Notes, Zarafa, and Bynari where vendors have added MAPI support to allow their products to be accessed directly via Outlook.

Uses

Business and Organizational Use

Email has been widely accepted by business, governments and non-governmental organizations in the developed world, and it is one of the key parts of an 'e-revolution' in workplace communication (with the other key plank being widespread adoption of highspeed Internet). A sponsored 2010 study on workplace communication found 83% of U.S. knowledge workers felt email was critical to their success and productivity at work.

It has some key benefits to business and other organizations, including:

Facilitating logistics

> Much of the business world relies on communications between people who are not physically in the same building, area, or even country; setting up and attending an in-person meeting, telephone call, or conference call can be inconvenient, time-consuming, and costly. Email provides a method of exchanging information between two or more people with no set-up costs and that is generally far less expensive than a physical meeting or phone call.

Helping with synchronisation

> With real time communication by meetings or phone calls, participants must work on the same schedule, and each participant must spend the same amount of time in the meeting or call. Email allows asynchrony: each participant may control their schedule independently.

Reducing cost

> Sending an email is much less expensive than sending postal mail, or long distance telephone calls, telex or telegrams.

Increasing speed

> Much faster than most of the alternatives.

Creating a "written" record

> Unlike a telephone or in-person conversation, email by its nature creates a detailed written record of the communication, the identity of the sender(s) and recipient(s) and the date and time the message was sent. In the event of a contract or legal dispute, saved emails can be used to prove that an individual was advised of certain issues, as each email has the date and time recorded on it.

Email Marketing

Email marketing via "opt-in" is often successfully used to send special sales offerings and new product information. Depending on the recipient's culture, email sent without permission—such as an "opt-in"—is likely to be viewed as unwelcome "email spam".

Personal Use

Desktop

Many users access their personal email from friends and family members using a desktop computer in their house or apartment.

Mobile

Email has become widely used on smartphones and Wi-Fi-enabled laptops and tablet computers. Mobile "apps" for email increase accessibility to the medium for users who are out of their home. While in the earliest years of email, users could only access email on desktop computers, in the 2010s, it is possible for users to check their email when they are away from home, whether they are across town or across the world. Alerts can also be sent to the smartphone or other device to notify them immediately of new messages. This has given email the ability to be used for more frequent communication between users and allowed them to check their email and write messages throughout the day. Today, there are an estimated 1.4 billion email users worldwide and 50 billion non-spam emails that are sent daily.

Individuals often check email on smartphones for both personal and work-related messages. It was found that US adults check their email more than they browse the web or check their Facebook accounts, making email the most popular activity for users to do on their smartphones. 78% of the respondents in the study revealed that they check their email on their phone. It was also found that 30% of consumers use only their smartphone to check their email, and 91% were likely to check their email at least once per day on their smartphone. However, the percentage of consumers using email on smartphone ranges and differs dramatically across different countries. For example, in comparison to 75% of those consumers in the US who used it, only 17% in India did.

Issues

Attachment Size Limitation

Email messages may have one or more attachments, which are additional files that are appended to the email. Typical attachments include Microsoft Word documents, pdf documents and scanned images of paper documents. In principle there is no technical restriction on the size or number of attachments, but in practice email clients, servers and Internet service providers implement various limitations on the size of files, or complete email - typically to 25MB or less. Furthermore, due to technical reasons, attachment sizes as seen by these transport systems can differ to what the user sees, which can be confusing to senders when trying to assess whether they can safely send a file by email. Where larger files need to be shared, file hosting services of various sorts are available; and generally suggested. Some large files, such as digital photos, color presentations and video or music files are too large for some email systems.

Information Overload

The ubiquity of email for knowledge workers and "white collar" employees has led to concerns that recipients face an "information overload" in dealing with increasing volumes of email. This can lead to increased stress, decreased satisfaction with work, and some observers even argue it could have a significant negative economic effect, as efforts to read the many emails could reduce productivity.

Spam

Email "spam" is the term used to describe unsolicited bulk email. The low cost of sending such email meant that by 2003 up to 30% of total email traffic was already spam. and was threatening the usefulness of email as a practical tool. The US CAN-SPAM Act of 2003 and similar laws elsewhere had some impact, and a number of effective anti-spam techniques now largely mitigate the impact of spam by filtering or rejecting it for most users, but the volume sent is still very high—and increasingly consists not of advertisements for products, but malicious content or links.

Malware

A range of malicious email types exist. These range from various types of email scams, including "social engineering" scams such as advance-fee scam "Nigerian letters", to phishing, email bombardment and email worms.

Email Spoofing

Email spoofing occurs when the email message header is designed to make the message appear to come from a known or trusted source. Email spam and phishing methods typically use spoofing to mislead the recipient about the true message origin. Email spoofing may be done as a prank, or as part of a criminal effort to defraud an individual or organization. An example of a potentially fraudulent email spoofing is if an individual creates an email which appears to be an invoice from a major company, and then sends it to one or more recipients. In some cases, these fraudulent emails incorporate the logo of the purported organization and even the email address may appear legitimate.

Email Bombing

Email bombing is the intentional sending of large volumes of messages to a target address. The overloading of the target email address can render it unusable and can even cause the mail server to crash.

Privacy Concerns

Today it can be important to distinguish between Internet and internal email systems. Internet email may travel and be stored on networks and computers without the sender's or the recipient's control. During the transit time it is possible that third parties read or even modify the content. Internal mail systems, in which the information never leaves the organizational network, may be more secure, although information technology personnel and others whose function may involve monitoring or managing may be accessing the email of other employees.

Email privacy, without some security precautions, can be compromised because:

- email messages are generally not encrypted.

- email messages have to go through intermediate computers before reaching their destination, meaning it is relatively easy for others to intercept and read messages.

- many Internet Service Providers (ISP) store copies of email messages on their mail servers before they are delivered. The backups of these can remain for up to several months on their server, despite deletion from the mailbox.

- the "Received:"-fields and other information in the email can often identify the sender, preventing anonymous communication.

There are cryptography applications that can serve as a remedy to one or more of the above. For example, Virtual Private Networks or the Tor anonymity network can be used to encrypt traffic from the user machine to a safer network while GPG, PGP, SMEmail, or S/MIME can be used for end-to-end message encryption, and SMTP STARTTLS or SMTP over Transport Layer Security/ Secure Sockets Layer can be used to encrypt communications for a single mail hop between the SMTP client and the SMTP server.

Additionally, many mail user agents do not protect logins and passwords, making them easy to intercept by an attacker. Encrypted authentication schemes such as SASL prevent this. Finally, attached files share many of the same hazards as those found in peer-to-peer filesharing. Attached files may contain trojans or viruses.

Flaming

Flaming occurs when a person sends a message (or many messages) with angry or antagonistic content. The term is derived from the use of the word "incendiary" to describe particularly heated email discussions. The ease and impersonality of email communications mean that the social norms that encourage civility in person or via telephone do not exist and civility may be forgotten.

Email Bankruptcy

Also known as "email fatigue", email bankruptcy is when a user ignores a large number of email messages after falling behind in reading and answering them. The reason for falling behind is often due to information overload and a general sense there is so much information that it is not possible to read it all. As a solution, people occasionally send a "boilerplate" message explaining that their email inbox is full, and that they are in the process of clearing out all the messages. Harvard University law professor Lawrence Lessig is credited with coining this term, but he may only have popularized it.

Tracking of Sent Mail

The original SMTP mail service provides limited mechanisms for tracking a transmitted message, and none for verifying that it has been delivered or read. It requires that each mail server must either deliver it onward or return a failure notice (bounce message), but both software bugs and system failures can cause messages to be lost. To remedy this, the IETF introduced Delivery Status Notifications (delivery receipts) and Message Disposition Notifications (return receipts); however, these are not universally deployed in production.)

Many ISPs now deliberately disable non-delivery reports (NDRs) and delivery receipts due to the activities of spammers:

- Delivery Reports can be used to verify whether an address exists and if so, this indicates to a spammer that it is available to be spammed.

- If the spammer uses a forged sender email address (email spoofing), then the innocent email address that was used can be flooded with NDRs from the many invalid email addresses the spammer may have attempted to mail. These NDRs then constitute spam from the ISP to the innocent user.

In the absence of standard methods, a range of system based around the use of web bugs have been developed. However, these are often seen as underhand or raising privacy concerns, and only work with e-mail clients that support rendering of HTML. Many mail clients now default to not showing "web content". Webmail providers can also disrupt web bugs by pre-caching images.

U.S. Government

The U.S. state and federal governments have been involved in electronic messaging and the development of email in several different ways. Starting in 1977, the U.S. Postal Service (USPS) recognized that electronic messaging and electronic transactions posed a significant threat to First Class mail volumes and revenue. The USPS explored an electronic messaging initiative in 1977 and later disbanded it. Twenty years later, in 1997, when email volume overtook postal mail volume, the USPS was again urged to embrace email, and the USPS declined to provide email as a service. The USPS initiated an experimental email service known as E-COM. E-COM provided a method for the simple exchange of text messages. In 2011, shortly after the USPS reported its state of financial bankruptcy, the USPS Office of Inspector General (OIG) began exploring the possibilities of generating revenue through email servicing. Electronic messages were transmitted to a post office, printed out, and delivered as hard copy. To take advantage of the service, an individual had to transmit at least 200 messages. The delivery time of the messages was the same as First Class mail and cost 26 cents. Both the Postal Regulatory Commission and the Federal Communications Commission opposed E-COM. The FCC concluded that E-COM constituted common carriage under its jurisdiction and the USPS would have to file a tariff. Three years after initiating the service, USPS canceled E-COM and attempted to sell it off.

The early ARPANET dealt with multiple email clients that had various, and at times incompatible, formats. For example, in the Multics, the "@" sign meant "kill line" and anything before the "@" sign was ignored, so Multics users had to use a command-line option to specify the destination system. The Department of Defense DARPA desired to have uniformity and interoperability for email and therefore funded efforts to drive towards unified inter-operable standards. This led to David Crocker, John Vittal, Kenneth Pogran, and Austin Henderson publishing RFC 733, "Standard for the Format of ARPA Network Text Message" (November 21, 1977), a subset of which provided a stable base for common use on the ARPANET, but which was not fully effective, and in 1979, a meeting was held at BBN to resolve incompatibility issues. Jon Postel recounted the meeting in RFC 808, "Summary of Computer Mail Services Meeting Held at BBN on 10 January 1979" (March 1, 1982), which includes an appendix listing the varying email systems at the time. This, in turn, led to the release of David Crocker's RFC 822, "Standard for the

Format of ARPA Internet Text Messages" (August 13, 1982). RFC 822 is a small adaptation of RFC 733's details, notably enhancing the host portion, to use Domain Names, that were being developed at the same time.

The National Science Foundation took over operations of the ARPANET and Internet from the Department of Defense, and initiated NSFNet, a new backbone for the network. A part of the NSFNet AUP forbade commercial traffic. In 1988, Vint Cerf arranged for an interconnection of MCI Mail with NSFNET on an experimental basis. The following year Compuserve email inter-connected with NSFNET. Within a few years the commercial traffic restriction was removed from NSFNETs AUP, and NSFNET was privatised. In the late 1990s, the Federal Trade Commission grew concerned with fraud transpiring in email, and initiated a series of procedures on spam, fraud, and phishing. In 2004, FTC jurisdiction over spam was codified into law in the form of the CAN SPAM Act. Several other U.S. federal agencies have also exercised jurisdiction including the Department of Justice and the Secret Service. NASA has provided email capabilities to astronauts aboard the Space Shuttle and International Space Station since 1991 when a Macintosh Portable was used aboard Space Shuttle mission STS-43 to send the first email via AppleLink. Today astronauts aboard the International Space Station have email capabilities via the wireless networking throughout the station and are connected to the ground at 10 Mbit/s Earth to station and 3 Mbit/s station to Earth, comparable to home DSL connection speeds.

Fax

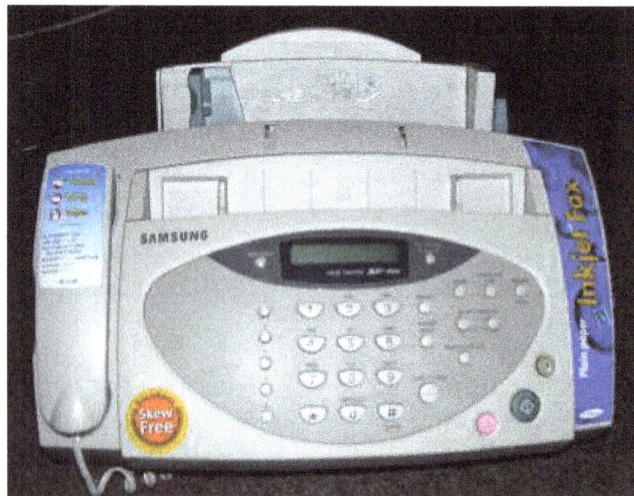

A fax machine from the late 1990s

Fax (short for facsimile), sometimes called telecopying or telefax (the latter short for telefacsimile), is the telephonic transmission of scanned printed material (both text and images), normally to a telephone number connected to a printer or other output device. The original document is scanned with a fax machine (or a telecopier), which processes the contents (text or images) as a single fixed graphic image, converting it into a bitmap, and then transmitting it through the telephone system in the form of audio-frequency tones. The receiving fax machine interprets the tones and recon-structs the image, printing a paper copy. Early systems used direct conversions of image darkness to audio tone in a continuous or analog manner. Since the 1980s, most machines modulate the transmitted audio frequencies using a digital representation of the page which is compressed to quickly transmit areas which are all-white or all-black.

History

Wire Transmission

Scottish inventor Alexander Bain worked on chemical mechanical fax type devices and in 1846 was able to reproduce graphic signs in laboratory experiments. He received British patent 9745 on May 27, 1843 for his "Electric Printing Telegraph." Frederick Bakewell made several improvements on Bain's design and demonstrated a telefax machine. The Pantelegraph was invented by the Italian physicist Giovanni Caselli. He introduced the first commercial telefax service between Paris and Lyon in 1865, some 11 years before the invention of the telephone.

In 1880, English inventor Shelford Bidwell constructed the *scanning phototelegraph* that was the first telefax machine to scan any two-dimensional original, not requiring manual plotting or drawing. Around 1900, German physicist Arthur Korn invented the *Bildtelegraph*, widespread in continental Europe especially, since a widely noticed transmission of a wanted-person photograph from Paris to London in 1908, used until the wider distribution of the radiofax. Its main competitors were the *Bélinographe* by Édouard Belin first, then since the 1930s the *Hellschreiber*, invented in 1929 by German inventor Rudolf Hell, a pioneer in mechanical image scanning and transmission.

The 1888 invention of the telautograph by Elisha Grey marked a further development in fax technology, allowing users to send signatures over long distances, thus allowing the verification of identification or ownership over long distances.

On May 19, 1924, scientists of the AT&T Corporation "by a new process of transmitting pictures by electricity" sent 15 photographs by telephone from Cleveland to New York City, such photos suitable for newspaper reproduction. Previously, photographs had been sent over the radio using this process.

The Western Union "Deskfax" fax machine, announced in 1948, was a compact machine that fit comfortably on a desktop, using special spark printer paper.

Wireless Transmission

As a designer for the Radio Corporation of America (RCA), in 1924, Richard H. Ranger invented the wireless photoradiogram, or transoceanic radio facsimile, the forerunner of today's "fax" machines. A photograph of President Calvin Coolidge sent from New York to London on November 29, 1924 became the first photo picture reproduced by transoceanic radio facsimile. Commercial use of Ranger's product began two years later. Also in 1924, Herbert E. Ives of AT&T Corporation transmitted and reconstructed the first color facsimile, using color separations. Around 1952 or so, Finch Facsimile, a highly developed machine, was described in detail in a book; it was never manufactured in quantity.

By the late 1940s, radiofax receivers were sufficiently miniaturized to be fitted beneath the dashboard of Western Union's "Telecar" telegram delivery vehicles.

In the 1960s, the United States Army transmitted the first photograph via satellite facsimile to Puerto Rico from the Deal Test Site using the Courier satellite.

Radio fax is still in limited use today for transmitting weather charts and information to ships at sea.

Telephone Transmission

In 1964, Xerox Corporation introduced (and patented) what many consider to be the first commercialized version of the modern fax machine, under the name (LDX) or Long Distance Xerography. This model was superseded two years later with a unit that would truly set the standard for fax machines for years to come. Up until this point facsimile machines were very expensive and hard to operate. In 1966, Xerox released the Magnafax Telecopiers, a smaller, 46-pound facsimile machine. This unit was far easier to operate and could be connected to any standard telephone line. This machine was capable of transmitting a letter-sized document in about six minutes. The first sub-minute, digital fax machine was developed by Dacom, which built on digital data compression technology originally developed at Lockheed for satellite communication.

By the late 1970s, many companies around the world (especially Japan), entered the fax market. Very shortly after a new wave of more compact, faster and efficient fax machines would hit the market. Xerox continued to refine the fax machine for years after their ground-breaking first machine. In later years it would be combined with copier equipment to create the hybrid machines we have today that copy, scan and fax. Some of the lesser known capabilities of the Xerox fax technologies included their Ethernet enabled Fax Services on their 8000 workstations in the early 1980s.

Prior to the introduction of the ubiquitous fax machine, one of the first being the Exxon Qwip in the mid-1970s, facsimile machines worked by optical scanning of a document or drawing spinning on a drum. The reflected light, varying in intensity according to the light and dark areas of the document, was focused on a photocell so that the current in a circuit varied with the amount of light. This current was used to control a tone generator (a modulator), the current determining the frequency of the tone produced. This audio tone was then transmitted using an acoustic coupler (a speaker, in this case) attached to the microphone of a common telephone handset. At the receiving end, a handset's speaker was attached to an acoustic coupler (a microphone), and a demodulator converted the varying tone into a variable current that controlled the mechanical movement of a pen or pencil to reproduce the image on a blank sheet of paper on an identical drum rotating at the same rate.

Computer Facsimile Interface

In 1985, Dr. Hank Magnuski, founder of GammaLink, produced the first computer fax board, called GammaFax.

Fax in the 21st Century

Although businesses usually maintain some kind of fax capability, the technology has faced increasing competition from Internet-based alternatives. In some countries, because electronic signatures on contracts are not yet recognized by law, while faxed contracts with copies of signatures are, fax machines enjoy continuing support in business. In Japan, faxes are still used extensively for cultural and graphemic reasons and are available for sending to both domestic and international recipients from over 81% of all convenience stores nationwide. Convenience-store fax machines commonly print the slightly re-sized content of the sent fax in the electronic confirmation-slip, in A4 paper size.

In many corporate environments, freestanding fax machines have been replaced by fax servers and other computerized systems capable of receiving and storing incoming faxes electronically, and then routing them to users on paper or via an email (which may be secured). Such systems have the advantage of reducing costs by eliminating unnecessary printouts and reducing the number of inbound analog phone lines needed by an office.

The once ubiquitous fax machine has also begun to disappear from the small office and home office environments. Remotely hosted fax-server services are widely available from VoIP and e-mail providers allowing users to send and receive faxes using their existing e-mail accounts without the need for any hardware or dedicated fax lines. Personal computers have also long been able to handle incoming and outgoing faxes using analogue modems or ISDN, eliminating the need for a stand-alone fax machine. These solutions are often ideally suited for users who only very occasionally need to use fax services. There are 17 million fax machines in the US, about one every 4.47 square miles.

Capabilities

There are several indicators of fax capabilities: Group, class, data transmission rate, and conformance with ITU-T (formerly CCITT) recommendations. Since the 1968 Carterphone decision, most fax machines have been designed to connect to standard PSTN lines and telephone numbers.

Group

Analog

Group 1 and 2 faxes are sent in the same manner as a frame of analog television, with each scanned line transmitted as a continuous analog signal. Horizontal resolution depended upon the quality of the scanner, transmission line, and the printer. Analog fax machines are obsolete and no longer manufactured. ITU-T Recommendations T.2 and T.3 were withdrawn as obsolete in July 1996.

- Group 1 faxes conform to the ITU-T Recommendation T.2. Group 1 faxes take six minutes to transmit a single page, with a vertical resolution of 96 scan lines per inch. Group 1 fax machines are obsolete and no longer manufactured.

- Group 2 faxes conform to the ITU-T Recommendations T.30 and T.3. Group 2 faxes take three minutes to transmit a single page, with a vertical resolution of 96 scan lines per inch. Group 2 fax machines are almost obsolete, and are no longer manufactured. Group 2 fax machines can interoperate with Group 3 fax machines.

Digital

A major breakthrough in the development of the modern facsimile system was the result of digital technology, where the analog signal from scanners was digitized and then compressed, resulting in the ability to transmit high rates of data across standard phone lines. The first digital fax machine was the Dacom Rapidfax first sold in late 1960s, which incorporated digital data compression technology developed by Lockheed for transmission of images from satellites.

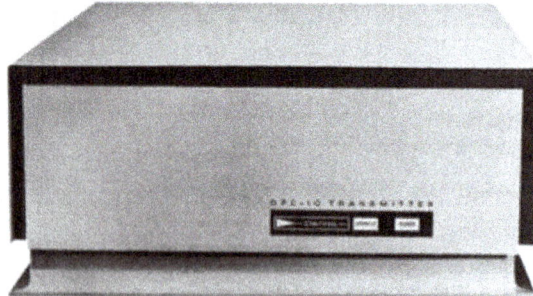

The Dacom DFC-10—the first digital fax machine.

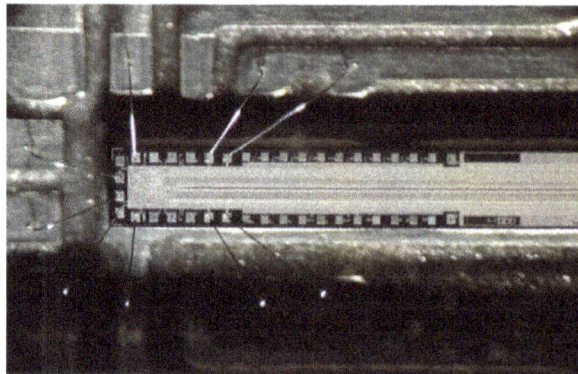

The chip in a fax machine. Only about one quarter of the length is shown. The thin line in the middle consists of photosensitive pixels. The read-out circuit is at left.

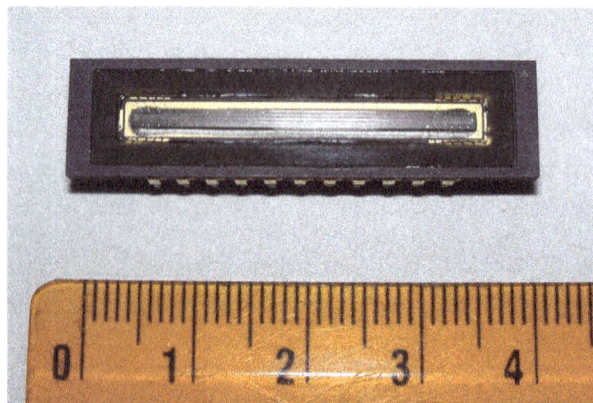

CCD MN8051 Matsushita CCD side, monochrome 2048-bit linear sensor +cm (centimeters)

Group 3 and 4 faxes are digital formats, and take advantage of digital compression methods to greatly reduce transmission times.

- Group 3 faxes conform to the ITU-T Recommendations T.30 and T.4. Group 3 faxes take between six and fifteen seconds to transmit a single page (not including the initial time for the fax machines to handshake and synchronize). The horizontal and vertical resolutions are allowed by the T.4 standard to vary among a set of fixed resolutions:

- o   Horizontal: 100 scan lines per inch

    - Vertical: 100 scan lines per inch ("Basic")

- o   Horizontal: 200 or 204 scan lines per inch

    - Vertical: 100 or 98 scan lines per inch ("Standard")

    - Vertical: 200 or 196 scan lines per inch ("Fine")

    - Vertical: 400 or 391 (note not 392) scan lines per inch ("Superfine")

- o   Horizontal: 300 scan lines per inch

    - Vertical: 300 scan lines per inch

- o   Horizontal: 400 or 408 scan lines per inch

    - Vertical: 400 or 391 scan lines per inch ("Ultrafine")

- Group 4 faxes conform to the ITU-T Recommendations T.563, T.503, T.521, T.6, T.62, T.70, T.411 to T.417. They are designed to operate over 64 kbit/s digital ISDN circuits. The allowed resolutions, a superset of those in the T.4 recommendation, are specified in the T.6 recommendation.

Fax Over IP (FoIP) can transmit and receive pre-digitized documents at near realtime speeds using ITU-T recommendation T.38 to send digitised images over an IP network using JPEG compression. T.38 is designed to work with VoIP services and often supported by analog telephone adapters used by legacy fax machines that need to connect through a VoIP service. Scanned documents are limited to the amount of time the user takes to load the document in a scanner and for the device to process a digital file. The resolution can vary from as little as 150 DPI to 9600 DPI or more. This type of faxing is not related to the e-mail to fax service that still uses fax modems at least one way.

Class

Computer modems are often designated by a particular fax class, which indicates how much processing is offloaded from the computer's CPU to the fax modem.

- Class 1 fax devices do fax data transfer where the T.4/T.6 data compression and T.30 session management are performed by software on a controlling computer. This is described in ITU-T recommendation T.31.

- Class 2 fax devices perform T.30 session management themselves, but the T.4/T.6 data compression is performed by software on a controlling computer. The relevant ITU-T recommendation is T.32.

- Class 2.0 is different from Class 2.

- Class 2.1 is an improvement of Class 2.0. Class 2.1 fax devices are referred to as "super G3"; they seem to be a little faster than Class 1/2/2.0.

- Class 3 fax devices are responsible for virtually the entire fax session, given little more than a phone number and the text to send (including rendering ASCII text as a raster image). These devices are not common.

Data Transmission Rate

Several different telephone line modulation techniques are used by fax machines. They are negotiated during the fax-modem handshake, and the fax devices will use the highest data rate that both fax devices support, usually a minimum of 14.4 kbit/s for Group 3 fax.

ITU Standard	Released Date	Data Rates (bit/s)	Modulation Method
V.27	1988	4800, 2400	PSK
V.29	1988	9600, 7200, 4800	QAM
V.17	1991	14,400; 12,000; 9600; 7200	TCM
V.34	1994	28,800	QAM
V.34bis	1998	33,600	QAM
ISDN	1986	64,000	digital

Note that "Super Group 3" faxes use V.34bis modulation that allows a data rate of up to 33.6 kbit/s.

Compression

As well as specifying the resolution (and allowable physical size of the image being faxed), the ITU-T T.4 recommendation specifies two compression methods for decreasing the amount of data that needs to be transmitted between the fax machines to transfer the image. The two methods defined in T.4 are:

- Modified Huffman (MH), and

- Modified READ (MR) (*Relative Element Address Designate*), optional

An additional method is specified in T.6:

- Modified Modified READ (MMR)

Later, other compression techniques were added as options to ITU-T recommendation T.30, such as the more efficient JBIG (T.82, T.85) for bi-level content, and JPEG (T.81), T.43, MRC (T.44), and T.45 for grayscale, palette, and colour content. Fax machines can negotiate at the start of the T.30 session to use the best technique implemented on both sides.

Modified Huffman

Modified Huffman (MH), specified in T.4 as the one-dimensional coding scheme, is a codebook-based run-length encoding scheme optimised to efficiently compress whitespace. As most faxes consist mostly of white space, this minimises the transmission time of most faxes. Each line scanned is compressed independently of its predecessor and successor.

Modified READ

Modified READ (MR), specified as an optional two-dimensional coding scheme in T.4, encodes the first scanned line using MH. The next line is compared to the first, the differences determined, and then the differences are encoded and transmitted. This is effective as most lines differ little from their predecessor. This is not continued to the end of the fax transmission, but only for a limited number of lines until the process is reset and a new 'first line' encoded with MH is produced. This limited number of lines is to prevent errors propagating throughout the whole fax, as the standard does not provide for error-correction. MR is an optional facility, and some fax machines do not use MR in order to minimise the amount of computation required by the machine. The limited number of lines is two for 'Standard' resolution faxes, and four for 'Fine' resolution faxes.

Modified Modified READ

The ITU-T T.6 recommendation adds a further compression type of Modified Modified READ (MMR), which simply allows for a greater number of lines to be coded by MR than in T.4. This is because T.6 makes the assumption that the transmission is over a circuit with a low number of line errors such as digital ISDN. In this case, there is no maximum number of lines for which the differences are encoded.

JBIG

In 1999, ITU-T recommendation T.30 added JBIG (ITU-T T.82) as another lossless bi-level compression algorithm, or more precisely a "fax profile" subset of JBIG (ITU-T T.85). JBIG-compressed pages result in 20% to 50% faster transmission than MMR-compressed pages, and up to 30-times faster transmission if the page includes halftone images.

JBIG performs adaptive compression, that is both the encoder and decoder collect statistical information about the transmitted image from the pixels transmitted so far, in order to predict the probability for each next pixel being either black or white. For each new pixel, JBIG looks at ten nearby, previously transmitted pixels. It counts, how often in the past the next pixel has been black or white in the same neighborhood, and estimates from that the probability distribution of the next pixel. This is fed into an arithmetic coder, which adds only a small fraction of a bit to the output sequence if the more probable pixel is then encountered.

The ITU-T T.85 "fax profile" constrains some optional features of the full JBIG standard, such that codecs do not have to keep data about more than the last three pixel rows of an image in memory at any time. This allows the streaming of "endless" images, where the height of the image may not be known until the last row is transmitted.

ITU-T T.30 allows fax machines to negotiate one of two options of the T.85 "fax profile":

- In "basic mode", the JBIG encoder must split the image into horizontal stripes of 128 lines (parameter L0=128), and restart the arithmetic encoder for each stripe.

- In "option mode", there is no such constraint.

Matsushita Whiteline Skip

A proprietary compression scheme employed on Panasonic fax machines is Matsushita Whiteline Skip (MWS). It can be overlaid on the other compression schemes, but is operative only when two

Panasonic machines are communicating with one another. This system detects the blank scanned areas between lines of text, and then compresses several blank scan lines into the data space of a single character. (JBIG implements a similar technique called "typical prediction", if header flag TPBON is set to 1.)

Typical Characteristics

Group 3 fax machines transfer one or a few printed or handwritten pages per minute in black-and-white (bitonal) at a resolution of 204×98 (normal) or 204×196 (fine) dots per square inch. The transfer rate is 14.4 kbit/s or higher for modems and some fax machines, but fax machines support speeds beginning with 2400 bit/s and typically operate at 9600 bit/s. The transferred image formats are called ITU-T (formerly CCITT) fax group 3 or 4. Group 3 faxes have the suffix .g3 and the MIME type image/g3fax.

The most basic fax mode transfers in black and white only. The original page is scanned in a resolution of 1728 pixels/line and 1145 lines/page (for A4). The resulting raw data is compressed using a modified Huffman code optimized for written text, achieving average compression factors of around 20. Typically a page needs 10 s for transmission, instead of about 3 minutes for the same uncompressed raw data of 1728×1145 bits at a speed of 9600 bit/s. The compression method uses a Huffman codebook for run lengths of black and white runs in a single scanned line, and it can also use the fact that two adjacent scanlines are usually quite similar, saving bandwidth by encoding only the differences.

Fax classes denote the way fax programs interact with fax hardware. Available classes include Class 1, Class 2, Class 2.0 and 2.1, and Intel CAS. Many modems support at least class 1 and often either Class 2 or Class 2.0. Which is preferable to use depends on factors such as hardware, software, modem firmware, and expected use.

Printing Process

Fax machines from the 1970s to the 1990s often used direct thermal printers with rolls of thermal paper as their printing technology, but since the mid-1990s there has been a transition towards plain-paper faxes:- thermal transfer printers, inkjet printers and laser printers.

One of the advantages of inkjet printing is that inkjets can affordably print in color; therefore, many of the inkjet-based fax machines claim to have color fax capability. There is a standard called ITU-T30e (formally ITU-T Recommendation T.30 Annex E) for faxing in color; unfortunately, it is not widely supported, so many of the color fax machines can only fax in color to machines from the same manufacturer.

Stroke Speed

Stroke speed in facsimile systems is the rate at which a fixed line perpendicular to the direction of scanning is crossed in one direction by a scanning or recording spot. Stroke speed is usually expressed as a number of strokes per minute. When the fax system scans in both directions, the stroke speed is twice this number. In most conventional 20th century mechanical systems, the stroke speed is equivalent to drum speed.

Fax Paper

Paper roll for direct thermal fax machine

As a precaution, thermal fax paper is typically not accepted in archives or as documentary evidence in some courts of law unless photocopied. This is because the image-forming coating is eradicable and brittle, and it tends to detach from the medium after a long time in storage.

Internet Fax

One popular alternative is to subscribe to an Internet fax service, allowing users to send and receive faxes from their personal computers using an existing email account. No software, fax server or fax machine is needed. Faxes are received as attached TIFF or PDF files, or in proprietary formats that require the use of the service provider's software. Faxes can be sent or retrieved from anywhere at any time that a user can get Internet access. Some services offer secure faxing to comply with stringent HIPAA and Gramm–Leach–Bliley Act requirements to keep medical information and financial information private and secure. Utilizing a fax service provider does not require paper, a dedicated fax line, or consumable resources.

Another alternative to a physical fax machine is to make use of computer software which allows people to send and receive faxes using their own computers, utilizing fax servers and unified messaging. A virtual (email) fax can be printed out and then signed and scanned back to computer before being emailed. Also the sender can attach a digital signature to the document file.

With the surging popularity of mobile phones, virtual fax machines can now be downloaded as applications for Android and iOS. These applications make use of the phone's internal camera to scan fax documents for upload or they can import from various cloud services.

Radio

Radio is the technology of using radio waves to carry information, such as sound, by systematically modulating some property of electromagnetic energy waves transmitted through space, such as their amplitude, frequency, phase, or pulse width. When radio waves strike an electrical conductor, the oscillating fields induce an alternating current in the conductor. The information in the waves can be extracted and transformed back into its original form.

The Alexandra Palace, here: mast of the broadcasting station

Radio systems need a transmitter to modulate (change) some property of the energy produced to impress a signal on it, for example using amplitude modulation or angle modulation (which can be frequency modulation or phase modulation). Radio systems also need an antenna to convert electric currents into radio waves, and vice versa. An antenna can be used for both transmitting and receiving. The electrical resonance of tuned circuits in radios allow individual stations to be selected. The electromagnetic wave is intercepted by a tuned receiving antenna. A radio receiver receives its input from an antenna and converts it into a form usable for the consumer, such as sound, pictures, digital data, measurement values, navigational positions, etc. Radio frequencies occupy the range from a 3 kHz to 300 GHz, although commercially important uses of radio use only a small part of this spectrum.

Classic radio receiver dial

A radio communication system sends signals by radio. The radio equipment involved in communication systems includes a transmitter and a receiver, each having an antenna and appropriate terminal equipment such as a microphone at the transmitter and a loudspeaker at the receiver in the case of a voice-communication system.

Etymology

The term "radio" is derived from the Latin word "radius", meaning "spoke of a wheel, beam of light, ray". It was first applied to communications in 1881 when, at the suggestion of French scientist Ernest Mercadier, Alexander Graham Bell adopted "radiophone" (meaning "radiated sound") as an alternate name for his photophone optical transmission system. However, this invention would not be widely adopted.

Following Heinrich Hertz's establishment of the existence of electromagnetic radiation in the late 1880s, a variety of terms were initially used for the phenomenon, with early descriptions of the radiation itself including "Hertzian waves", "electric waves", and "ether waves", while phrases describing its use in communications included "spark telegraphy", "space telegraphy", "aerography" and, eventually and most commonly, "wireless telegraphy". However, "wireless" included a broad variety of related electronic technologies, including electrostatic induction, electromagnetic induction and aquatic and earth conduction, so there was a need for a more precise term referring exclusively to electromagnetic radiation.

The first use of *radio-* in conjunction with electromagnetic radiation appears to have been by French physicist Édouard Branly, who in 1890 developed a version of a coherer receiver he called a *radio-conducteur*. The radio- prefix was later used to form additional descriptive compound and hyphenated words, especially in Europe, for example, in early 1898 the British publication *The Practical Engineer* included a reference to "the radiotelegraph" and "radiotelegraphy", while the French text of both the 1903 and 1906 Berlin Radiotelegraphic Conventions includes the phrases *radiotélégraphique* and *radiotélégrammes*.

The use of "radio" as a standalone word dates back to at least December 30, 1904, when instructions issued by the British Post Office for transmitting telegrams specified that "The word 'Radio'... is sent in the Service Instructions". This practice was universally adopted, and the word "radio" introduced internationally, by the 1906 Berlin Radiotelegraphic Convention, which included a Service Regulation specifying that "Radiotelegrams shall show in the preamble that the service is 'Radio'".

The switch to "radio" in place of "wireless" took place slowly and unevenly in the English-speaking world. Lee de Forest helped popularize the new word in the United States—in early 1907 he founded the DeForest Radio Telephone Company, and his letter in the June 22, 1907 *Electrical World* about the need for legal restrictions warned that "Radio chaos will certainly be the result until such stringent regulation is enforced". The United States Navy would also play a role. Although its translation of the 1906 Berlin Convention used the terms "wireless telegraph" and "wireless telegram", by 1912 it began to promote the use of "radio" instead. The term started to become preferred by the general public in the 1920s with the introduction of broadcasting. ("Broadcasting" is based upon an agricultural term meaning roughly "scattering seeds widely".) British Commonwealth countries continued to commonly use the term "wireless" until the mid-20th century, though the magazine of the British Broadcasting Corporation in the UK has been called Radio Times since its founding in the early 1920s.

In recent years the more general term "wireless" has gained renewed popularity, even for devices using electromagnetic radiation, through the rapid growth of short-range computer networking, e.g., Wireless Local Area Network (WLAN), Wi-Fi, and Bluetooth, as well as mobile telephony, e.g., GSM and UMTS cell phones. Today, the term "radio" specifies the transceiver device or chip, whereas "wireless" refers to the lack of physical connections; thus equipment employs embedded *radio* transceivers, but operates as *wireless* devices over *wireless* sensor networks.

Processes

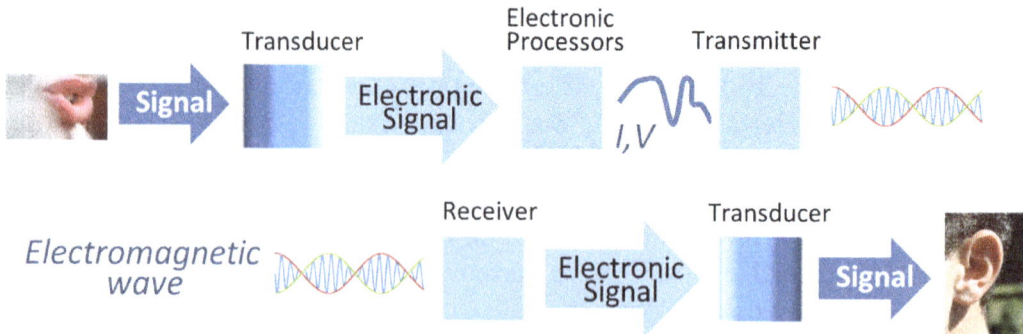

Radio communication. Information such as sound is converted by a transducer such as a microphone to an electrical signal, which modulates a radio wave sent from a transmitter. A receiver intercepts the radio wave and extracts the information-bearing electronic signal, which is converted back using another transducer such as a speaker.

Radio systems used for communication have the following elements. With more than 100 years of development, each process is implemented by a wide range of methods, specialised for different communications purposes.

Transmitter and Modulation

Each system contains a transmitter, This consists of a source of electrical energy, producing alternating current of a desired frequency of oscillation. The transmitter contains a system to modulate (change) some property of the energy produced to impress a signal on it. This modulation might be as simple as turning the energy on and off, or altering more subtle properties such as amplitude, frequency, phase, or combinations of these properties. The transmitter sends the modulated electrical energy to a tuned resonant antenna; this structure converts the rapidly changing alternating current into an electromagnetic wave that can move through free space (sometimes with a particular polarization).

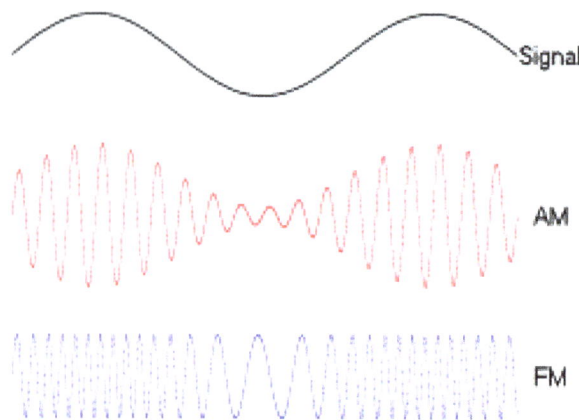

An audio signal (top) may be carried by an AM or FM radio wave.

Amplitude modulation of a carrier wave works by varying the strength of the transmitted signal in proportion to the information being sent. For example, changes in the signal strength can be used to reflect the sounds to be reproduced by a speaker, or to specify the light intensity of television pixels. It was the method used for the first audio radio transmissions, and remains in use today. "AM" is often used to refer to the medium wave broadcast band, but it is used in various radiotelephone services such as the Citizen Band, amateur radio and especially in aviation, due to its ability to be received under very weak signal conditions and its immunity to capture effect, allowing more than one signal to be heard simultaneously.

Frequency modulation varies the frequency of the carrier. The instantaneous frequency of the carrier is directly proportional to the instantaneous value of the input signal. FM has the "capture effect" whereby a receiver only receives the strongest signal, even when others are present. Digital data can be sent by shifting the carrier's frequency among a set of discrete values, a technique known as frequency-shift keying. FM is commonly used at Very high frequency (VHF) radio frequencies for high-fidelity broadcasts of music and speech. Analog TV sound is also broadcast using FM.

Angle modulation alters the instantaneous phase of the carrier wave to transmit a signal. It may be either FM or phase modulation (PM).

Antenna

Rooftop television antennas. Yagi-Uda antennas like these six are widely used at VHF and UHF frequencies.

An *antenna* (or *aerial*) is an electrical device which converts electric currents into radio waves, and vice versa. It is usually used with a radio transmitter or radio receiver. In transmission, a radio transmitter supplies an electric current oscillating at radio frequency (i.e. high frequency AC) to the antenna's terminals, and the antenna radiates the energy from the current as electromagnetic waves (radio waves). In reception, an antenna intercepts some of the power of an electromagnetic wave in order to produce a tiny voltage at its terminals, that is applied to a receiver to be amplified. Some antennas can be used for both transmitting and receiving, even simultaneously, depending on the connected equipment.

Propagation

Once generated, electromagnetic waves travel through space either directly, or have their path altered by reflection, refraction or diffraction. The intensity of the waves diminishes due to geometric dispersion (the inverse-square law); some energy may also be absorbed by the intervening medium in some cases. Noise will generally alter the desired signal; this electromagnetic interference comes from natural sources, as well as from artificial sources such as other transmitters and accidental radiators. Noise is also produced at every step due to the inherent properties of the devices used. If the magnitude of the noise is large enough, the desired signal will no longer be discernible; the signal-to-noise ratio is the fundamental limit to the range of radio communications.

Resonance

Electrical resonance of tuned circuits in radios allow individual stations to be selected. A resonant circuit will respond strongly to a particular frequency, and much less so to differing frequencies. This allows the radio receiver to discriminate between multiple signals differing in frequency.

Receiver and Demodulation

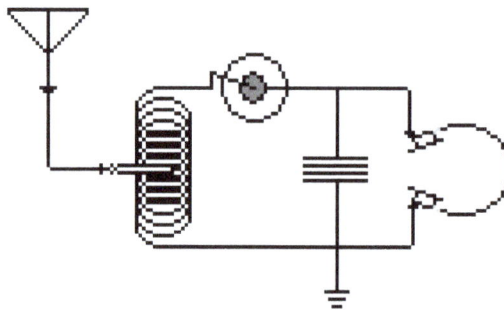

A crystal receiver, consisting of an antenna, adjustable electromagnetic coil, crystal rectifier, capacitor, headphones and ground connection.

The electromagnetic wave is intercepted by a tuned receiving antenna; this structure captures some of the energy of the wave and returns it to the form of oscillating electrical currents. At the receiver, these currents are demodulated, which is conversion to a usable signal form by a detector sub-system. The receiver is "tuned" to respond preferentially to the desired signals, and reject undesired signals.

Early radio systems relied entirely on the energy collected by an antenna to produce signals for the operator. Radio became more useful after the invention of electronic devices such as the vacuum tube and later the transistor, which made it possible to amplify weak signals. Today radio systems are used for applications from walkie-talkie children's toys to the control of space vehicles, as well as for broadcasting, and many other applications.

A *radio receiver* receives its input from an antenna, uses electronic filters to separate a wanted radio signal from all other signals picked up by this antenna, amplifies it to a level suitable for further processing, and finally converts through demodulation and decoding the signal into a form usable for the consumer, such as sound, pictures, digital data, measurement values, navigational positions, etc.

Radio Band

Light comparison		
Name	Frequency (Hz) (Wavelength)	Photon energy (eV)
Gamma ray	> 30 EHz (0.01 nm)	124 keV - 300+ GeV
X-Ray	30 EHz - 30 PHz (0.01 nm - 10 nm)	124 eV to 120 keV
Ultraviolet	30 PHz - 750 THz (10 nm - 400 nm)	3.1 eV to 124 eV
Visible	750 THz - 428.5 THz (400 nm - 700 nm)	1.7 eV - 3.1 eV
Infrared	428.5 THz - 300 GHz (700 nm - 1 mm)	1.24 meV - 1.7 eV
Microwave	300 GHz - 300 MHz (1 mm - 1 m)	1.24 μeV - 1.24 meV
Radio	300 MHz - 3 kHz (1 m - 100 km)	12.4 feV - 1.24 meV

Radio frequencies occupy the range from a 3 kHz to 300 GHz, although commercially important uses of radio use only a small part of this spectrum. Other types of electromagnetic radiation, with frequencies above the RF range, are infrared, visible light, ultraviolet, X-rays and gamma rays. Since the energy of an individual photon of radio frequency is too low to remove an electron from an atom, radio waves are classified as non-ionizing radiation.

Communication Systems

A *radio communication system* sends signals by radio. Types of radio communication systems deployed depend on technology, standards, regulations, radio spectrum allocation, user requirements, service positioning, and investment.

The radio equipment involved in communication systems includes a transmitter and a receiver, each having an antenna and appropriate terminal equipment such as a microphone at the transmitter and a loudspeaker at the receiver in the case of a voice-communication system.

The power consumed in a transmitting station varies depending on the distance of communication and the transmission conditions. The power received at the receiving station is usually only a tiny fraction of the transmitter's output, since communication depends on receiving the information, not the energy, that was transmitted.

Classical radio communications systems use frequency-division multiplexing (FDM) as a strategy to split up and share the available radio-frequency bandwidth for use by different parties communications concurrently. Modern radio communication systems include those that divide up a radio-frequency band by time-division multiplexing (TDM) and code-division multiplexing (CDM) as alternatives to the classical FDM strategy. These systems offer different tradeoffs in supporting multiple users, beyond the FDM strategy that was ideal for broadcast radio but less so for applications such as mobile telephony.

A radio communication system may send information only one way. For example, in broadcasting a single transmitter sends signals to many receivers. Two stations may take turns sending and receiving, using a single radio frequency; this is called "simplex." By using two radio frequencies, two stations may continuously and concurrently send and receive signals - this is called "duplex" operation.

History

In 1864 James Clerk Maxwell showed mathematically that electromagnetic waves could propagate through free space. The effects of electromagnetic waves (then-unexplained "action at a distance" sparking behavior) were actually observed before and after Maxwell's work by many inventors and experimenters including Luigi Galvani (1791), Peter Samuel Munk (1835), Joseph Henry (1842), Samuel Alfred Varley (1852), Edwin Houston, Elihu Thomson, Thomas Edison (1875) and David Edward Hughes (1878). Edison gave the effect the name "etheric force" and Hughes detected a spark impulse up to 500 yards (460 m) with a portable receiver, but none could identify what caused the phenomenon and it was usually written off as electromagnetic induction. In 1886 Heinrich Rudolf Hertz noticed the same sparking phenomenon and, in published experiments (1887-1888), was able to demonstrate the existence of electromagnetic waves in an experiment confirming Maxwell's theory of electromagnetism. The discovery of these "Hertzian waves" (radio waves) prompted many experiments by physicists. An August 1894 lecture by the British physicist Oliver Lodge, where he transmitted and received "Hertzian waves" at distances up to 50 meters, was followed up a year later with experiments by Indian physicist Jagadish Bose in radio microwave optics and construction of a radio based lightning detector by Russian physicist Alexander Stepanovich Popov. Starting in late 1894, Guglielmo Marconi began pursuing the idea of building a wireless telegraphy system based on Hertzian waves (radio). Marconi gained a patent on the system in 1896 and developed it into a commercial communication system over the next few years.

Early 20th century radio systems transmitted messages by continuous wave code only. Early attempts at developing a system of amplitude modulation for voice and music were demonstrated in 1900 and 1906, but had little success. World War I accelerated the development of radio for military communications, and in this era the first vacuum tubes were applied to radio transmitters and receivers. Electronic amplification was a key development in changing radio from an experimental practice by experts into a home appliance. After the war, commercial radio broadcasting began in the 1920s and became an important mass medium for entertainment and news.

World War II again accelerated development of radio for the wartime purposes of aircraft and land communication, radio navigation and radar. After the war, the experiments in television that had been interrupted were resumed, and it also became an important home entertainment medium.

Uses of Radio

Early uses were maritime, for sending telegraphic messages using Morse code between ships and land. The earliest users included the Japanese Navy scouting the Russian fleet during the Battle of Tsushima in 1905. One of the most memorable uses of marine telegraphy was during the sinking of the RMS *Titanic* in 1912, including communications between operators on the sinking ship and nearby vessels, and communications to shore stations listing the survivors.

Radio was used to pass on orders and communications between armies and navies on both sides in World War I; Germany used radio communications for diplomatic messages once it discovered that its submarine cables had been tapped by the British. The United States passed on President Woodrow Wilson's Fourteen Points to Germany via radio during the war. Broadcasting began from San Jose, California in 1909, and became feasible in the 1920s, with the widespread introduction of radio receivers, particularly in Europe and the United States. Besides broadcasting, point-to-point broadcasting, including telephone messages and relays of radio programs, became widespread in the 1920s and 1930s. Another use of radio in the pre-war years was the development of detection and locating of aircraft and ships by the use of radar (*RA*dio *D*etection *A*nd *R*anging).

Today, radio takes many forms, including wireless networks and mobile communications of all types, as well as radio broadcasting. Before the advent of television, commercial radio broadcasts included not only news and music, but dramas, comedies, variety shows, and many other forms of entertainment (the era from the late 1920s to the mid-1950s is commonly called radio's "Golden Age"). Radio was unique among methods of dramatic presentation in that it used only sound. F

Audio

One-way

Bakelite radio at the Bakelite Museum, Orchard Mill, Williton, Somerset, UK.

AM radio uses amplitude modulation, in which the amplitude of the transmitted signal is made proportional to the sound amplitude captured (transduced) by the microphone, while the transmitted frequency remains unchanged. Transmissions are affected by static and interference because lightning and other sources of radio emissions on the same frequency add their amplitudes to the original transmitted amplitude.

A Fisher 500 AM/FM hi-fi receiver from 1959.

In the early part of the 20th century, American AM radio stations broadcast with powers as high as 500 kW, and some could be heard worldwide; these stations' transmitters were commandeered for military use by the US Government during World War II. Currently, the maximum broadcast power for a civilian AM radio station in the United States and Canada is 50 kW, and the majority of stations that emit signals this powerful were grandfathered in. In 1986 KTNN received the last granted 50,000-watt class A license. These 50 kW stations are generally called "clear channel" stations, because within North America each of these stations has exclusive use of its broadcast frequency throughout part or all of the broadcast day.

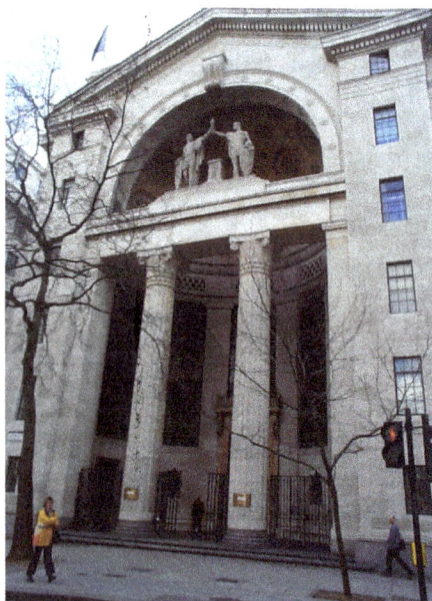

Bush House, old home of the BBC World Service.

FM broadcast radio sends music and voice with less noise than AM radio. It is often mistakenly thought that FM is higher fidelity than AM, but that is not true. AM is capable of the same audio bandwidth that FM employs. AM receivers typically use narrower filters in the receiver to recover the signal with less noise. AM stereo receivers can reproduce the same audio bandwidth that FM does due to the wider filter used in an AM stereo receiver, but today, AM radios limit the audio bandpass to 3–5 kHz. In frequency modulation, amplitude variation at the microphone causes the transmitter frequency to fluctuate. Because the audio signal modulates the frequency and not the amplitude, an FM signal is not subject to static and interference in the same way as AM signals. Due to its need for a wider bandwidth, FM is transmitted in the Very High Frequency (VHF, 30 MHz to 300 MHz) radio spectrum.

VHF radio waves act more like light, traveling in straight lines; hence the reception range is generally limited to about 50–200 miles (80–322 km). During unusual upper atmospheric conditions, FM signals are occasionally reflected back towards the Earth by the ionosphere, resulting in long distance FM reception. FM receivers are subject to the capture effect, which causes the radio to only receive the strongest signal when multiple signals appear on the same frequency. FM receivers are relatively immune to lightning and spark interference.

High power is useful in penetrating buildings, diffracting around hills, and refracting in the dense atmosphere near the horizon for some distance beyond the horizon. Consequently, 100,000-watt FM stations can regularly be heard up to 100 miles (160 km) away, and farther, 150 miles (240 km), if there are no competing signals.

A few old, "grandfathered" stations do not conform to these power rules. WBCT-FM (93.7) in Grand Rapids, Michigan, US, runs 320,000 watts ERP, and can increase to 500,000 watts ERP by the terms of its original license. Such a huge power level does not usually help to increase range as much as one might expect, because VHF frequencies travel in nearly straight lines over the horizon and off into space. Nevertheless, when there were fewer FM stations competing, this station could be heard near Bloomington, Illinois, US, almost 300 miles (480 km) away.

FM subcarrier services are secondary signals transmitted in a "piggyback" fashion along with the main program. Special receivers are required to utilize these services. Analog channels may contain alternative programming, such as reading services for the blind, background music or stereo sound signals. In some extremely crowded metropolitan areas, the sub-channel program might be an alternate foreign-language radio program for various ethnic groups. Sub-carriers can also transmit digital data, such as station identification, the current song's name, web addresses, or stock quotes. In some countries, FM radios automatically re-tune themselves to the same channel in a different district by using sub-bands.

Two-way

Aviation voice radios use Aircraft band VHF AM. AM is used so that multiple stations on the same channel can be received. (Use of FM would result in stronger stations blocking out reception of weaker stations due to FM's capture effect). Aircraft fly high enough that their transmitters can be received hundreds of miles away, even though they are using VHF.

Marine voice radios can use single sideband voice (SSB) in the shortwave High Frequency (HF—3 MHz to 30 MHz) radio spectrum for very long ranges or Marine VHF radio / *narrowband FM* in the VHF spectrum for much shorter ranges. Narrowband FM sacrifices fidelity to make more channels available within the radio spectrum, by using a smaller range of radio frequencies, usually with five kHz of deviation, versus the 75 kHz used by commercial FM broadcasts, and 25 kHz used for TV sound.

Degen DE1103, an advanced world mini-receiver with single sideband modulation and dual conversion

Government, police, fire and commercial voice services also use narrowband FM on special frequencies. Early police radios used AM receivers to receive one-way dispatches.

Civil and military HF (high frequency) voice services use shortwave radio to contact ships at sea, aircraft and isolated settlements. Most use single sideband voice (SSB), which uses less bandwidth than AM. On an AM radio SSB sounds like ducks quacking, or the adults in a Charlie Brown cartoon. Viewed as a graph of frequency versus power, an AM signal shows power where the frequencies of the voice add and subtract with the main radio frequency. SSB cuts the bandwidth in half by suppressing the carrier and one of the sidebands. This also makes the transmitter about three times more powerful, because it doesn't need to transmit the unused carrier and sideband.

TETRA, Terrestrial Trunked Radio is a digital cell phone system for military, police and ambulances. Commercial services such as XM, WorldSpace and Sirius offer encrypted digital satellite radio.

Telephony

Mobile phones transmit to a local cell site (transmitter/receiver) that ultimately connects to the public switched telephone network (PSTN) through an optic fiber or microwave radio and other network elements. When the mobile phone nears the edge of the cell site's radio coverage area, the central computer switches the phone to a new cell. Cell phones originally used FM, but now most use various digital modulation schemes. Recent developments in Sweden (such as DROPme) allow for the instant downloading of digital material from a radio broadcast (such as a song) to a mobile phone.

Satellite phones use satellites rather than cell towers to communicate.

Video

Analog television sends the picture as AM and the sound as AM or FM, with the sound carrier a fixed frequency (4.5 MHz in the NTSC system) away from the video carrier. Analog television also uses a vestigial sideband on the video carrier to reduce the bandwidth required.

Digital television uses 8VSB modulation in North America (under the ATSC digital television standard), and COFDM modulation elsewhere in the world (using the DVB-T standard). A Reed–Solomon error correction code adds redundant correction codes and allows reliable reception during moderate data loss. Although many current and future codecs can be sent in the MPEG transport stream container format, as of 2006 most systems use a standard-definition format almost identical to DVD: MPEG-2 video in Anamorphic widescreen and MPEG layer 2 (*MP2*) audio. High-definition television is possible simply by using a higher-resolution picture, but H.264/AVC is being considered as a replacement video codec in some regions for its improved compression. With the compression and improved modulation involved, a single "channel" can contain a high-definition program and several standard-definition programs.

Navigation

All satellite navigation systems use satellites with precision clocks. The satellite transmits its position, and the time of the transmission. The receiver listens to four satellites, and can figure its position as being on a line that is tangent to a spherical shell around each satellite, determined by the time-of-flight of the radio signals from the satellite. A computer in the receiver does the math.

Radio direction-finding is the oldest form of radio navigation. Before 1960 navigators used movable loop antennas to locate commercial AM stations near cities. In some cases they used marine radiolocation beacons, which share a range of frequencies just above AM radio with amateur radio operators. LORAN systems also used time-of-flight radio signals, but from radio stations on the ground.

Very High Frequency Omnidirectional Range (VOR), systems (used by aircraft), have an antenna array that transmits two signals simultaneously. A directional signal rotates like a lighthouse at a fixed rate. When the directional signal is facing north, an omnidirectional signal pulses. By measuring the difference in phase of these two signals, an aircraft can determine its bearing or radial from the station, thus establishing a line of position. An aircraft can get readings from two VORs and locate its position at the intersection of the two radials, known as a "fix."

When the VOR station is collocated with DME (Distance Measuring Equipment), the aircraft can determine its bearing and range from the station, thus providing a fix from only one ground station. Such stations are called VOR/DMEs. The military operates a similar system of navaids, called TACANs, which are often built into VOR stations. Such stations are called VORTACs. Because TACANs include distance measuring equipment, VOR/DME and VORTAC stations are identical in navigation potential to civil aircraft.

Radar

Radar (Radio Detection And Ranging) detects objects at a distance by bouncing radio waves off them. The delay caused by the echo measures the distance. The direction of the beam determines the direction of the reflection. The polarization and frequency of the return can sense the type of surface. Navigational radars scan a wide area two to four times per minute. They use very short waves that reflect from earth and stone. They are common on commercial ships and long-distance commercial aircraft.

General purpose radars generally use navigational radar frequencies, but modulate and polarize the pulse so the receiver can determine the type of surface of the reflector. The best general-purpose radars distinguish the rain of heavy storms, as well as land and vehicles. Some can superimpose sonar data and map data from GPS position.

Search radars scan a wide area with pulses of short radio waves. They usually scan the area two to four times a minute. Sometimes search radars use the Doppler effect to separate moving vehicles from clutter. Targeting radars use the same principle as search radar but scan a much smaller area far more often, usually several times a second or more. Weather radars resemble search radars, but use radio waves with circular polarization and a wavelength to reflect from water droplets. Some weather radar use the Doppler effect to measure wind speeds.

Data (Digital Radio)

2008 Pure One Classic digital radio

Most new radio systems are digital, including Digital TV, satellite radio, and Digital Audio Broadcasting. The oldest form of digital broadcast was spark gap telegraphy, used by pioneers such as Marconi. By pressing the key, the operator could send messages in Morse code by energizing a rotating commutating spark gap. The rotating commutator produced a tone in the receiver, where a simple spark gap would produce a hiss, indistinguishable from static. Spark-gap transmitters are now illegal, because their transmissions span several hundred megahertz. This is very wasteful of both radio frequencies and power.

The next advance was continuous wave telegraphy, or CW (Continuous Wave), in which a pure radio frequency, produced by a vacuum tube electronic oscillator was switched on and off by a key. A receiver with a local oscillator would "heterodyne" with the pure radio frequency, creating a whistle-like audio tone. CW uses less than 100 Hz of bandwidth. CW is still used, these days primarily by amateur radio operators (hams). Strictly, on-off keying of a carrier should be known as "Interrupted Continuous Wave" or ICW or on-off keying (OOK).

Radioteletype equipment usually operates on short-wave (HF) and is much loved by the military because they create written information without a skilled operator. They send a bit as one of two tones using frequency-shift keying. Groups of five or seven bits become a character printed by a teleprinter. From about 1925 to 1975, radioteletype was how most commercial messages were sent to less developed countries. These are still used by the military and weather services.

Aircraft use a 1200 Baud radioteletype service over VHF to send their ID, altitude and position, and get gate and connecting-flight data. Microwave dishes on satellites, telephone exchanges and TV stations usually use quadrature amplitude modulation (QAM). QAM sends data by changing both the phase and the amplitude of the radio signal. Engineers like QAM because it packs the most bits into a radio signal when given an exclusive (non-shared) fixed narrowband frequency range. Usually the bits are sent in "frames" that repeat. A special bit pattern is used to locate the beginning of a frame.

Modern GPS receivers.

Communication systems that limit themselves to a fixed narrowband frequency range are vulnerable to jamming. A variety of jamming-resistant spread spectrum techniques were initially developed for military use, most famously for Global Positioning System satellite transmissions. Commercial use of spread spectrum began in the 1980s. Bluetooth, most cell phones, and the 802.11b version of Wi-Fi each use various forms of spread spectrum.

Systems that need reliability, or that share their frequency with other services, may use "coded orthogonal frequency-division multiplexing" or COFDM. COFDM breaks a digital signal into as many as several hundred slower subchannels. The digital signal is often sent as QAM on the subchannels. Modern COFDM systems use a small computer to make and decode the signal with digital signal processing, which is more flexible and far less expensive than older systems that implemented separate electronic channels.

COFDM resists fading and ghosting because the narrow-channel QAM signals can be sent slowly. An adaptive system, or one that sends error-correction codes can also resist interference, because most interference can affect only a few of the QAM channels. COFDM is used for Wi-Fi, some cell phones, Digital Radio Mondiale, Eureka 147, and many other local area network, digital TV and radio standards.

Heating

Radio-frequency energy generated for heating of objects is generally not intended to radiate outside of the generating equipment, to prevent interference with other radio signals. Microwave ovens use intense radio waves to heat food. Diathermy equipment is used in surgery for sealing of blood vessels. Induction furnaces are used for melting metal for casting, and induction hobs for cooking.

Amateur Radio Service

Amateur radio station with multiple receivers and transceivers

Amateur radio, also known as "ham radio", is a hobby in which enthusiasts are licensed to communicate on a number of bands in the radio frequency spectrum non-commercially and for their own experiments. They may also provide emergency and service assistance in exceptional circumstances. This contribution has been very beneficial in saving lives in many instances.

Radio amateurs use a variety of modes, including efficient ones like Morse code and experimental ones like Low-Frequency Experimental Radio. Several forms of radio were pioneered by radio amateurs and later became commercially important, including FM, single-sideband (SSB), AM, digital packet radio and satellite repeaters. Some amateur frequencies may be disrupted illegally by power-line internet service.

Unlicensed Radio Services

Unlicensed, government-authorized personal radio services such as Citizens' band radio in Australia, most of the Americas, and Europe, and Family Radio Service and Multi-Use Radio Service in North America exist to provide simple, usually short range communication for individuals and small groups, without the overhead of licensing. Similar services exist in other parts of the world. These radio services involve the use of handheld units.

Wi-Fi also operates in unlicensed radio bands and is very widely used to network computers.

Free radio stations, sometimes called pirate radio or "clandestine" stations, are unauthorized, unlicensed, illegal broadcasting stations. These are often low power transmitters operated on sporadic schedules by hobbyists, community activists, or political and cultural dissidents. Some pirate stations operating offshore in parts of Europe and the United Kingdom more closely resembled legal stations, maintaining regular schedules, using high power, and selling commercial advertising time.

Radio Control (RC)

Radio remote controls use radio waves to transmit control data to a remote object as in some early forms of guided missile, some early TV remotes and a range of model boats, cars and airplanes.

Large industrial remote-controlled equipment such as cranes and switching locomotives now usually use digital radio techniques to ensure safety and reliability.

In Madison Square Garden, at the Electrical Exhibition of 1898, Nikola Tesla successfully demonstrated a radio-controlled boat. He was awarded U.S. patent No. 613,809 for a "Method of and Apparatus for Controlling Mechanism of Moving Vessels or Vehicles."

Terrestrial Television

Over the Air antenna

Terrestrial television or broadcast television is a type of television broadcasting in which the television signal is transmitted by radio waves from the terrestrial (Earth based) transmitter of a television station to a TV receiver having an antenna. The term is more common in Europe, while in North America it is referred to as broadcast television or sometimes over-the-air television (OTA). The term "terrestrial" is used to distinguish this type from the newer technologies of satellite television (direct broadcast satellite or DBS television), in which the television signal is transmitted to the receiver from an overhead satellite, and cable television, in which the signal is carried to the receiver through a cable.

Terrestrial television was the first technology used for television broadcasting, with the first long-distance public television broadcast from Washington, D.C., on 7 April 1927. The BBC began broadcasting in 1929, and had a regular schedule of television programmes in 1930. However these early experimental systems had insufficient picture quality to attract the public, due to their mechanical scan technology, and television didn't become widespread until after World War II with the advent of electronic scan technology. The television broadcasting business followed the model of radio networks, with local television stations in cities and towns affiliated with television networks, either commercial (in USA) or government-controlled (in Europe), which provided content. Television broadcasts were in black and white until the 1960s, when color television broadcasting began.

There was no other method of television delivery until the 1950s with the beginnings of cable television and *community antenna television* (CATV). CATV was, initially, only a re-broadcast of over-the-air signals. With the widespread adoption of cable across the United States in the 1970s and 1980s, viewing of terrestrial television broadcasts has been in decline; in 2013 it was estimated that about 7% of US households used an antenna. A slight increase in use began after the 2009 final conversion to digital terrestrial television broadcasts, which offer HDTV image quality as an alternative to CATV for cord cutters.

Europe

Following the ST61 conference, UHF frequencies were first used in the UK in 1964 with the introduction of BBC2. In UK, VHF channels were kept on the old 405-line system, while UHF was used solely for 625-line broadcasts (which later used PAL colour). Television broadcasting in the 405-line system continued after the introduction of four analogue programmes in the UHF bands until the last 405-line transmitters were switched off on January 6, 1985. VHF Band III was used in other countries around Europe for PAL broadcasts until planned phase out and switchover to digital television.

The success of analogue terrestrial television across Europe varied from country to country. Although each country had rights to a certain number of frequencies by virtue of the ST61 plan, not all of them were brought into service.

North America

In North America terrestrial television underwent a revolutionary transformation with the acceptance of the NTSC standard for color television broadcasts in 1953. Later, Europe and the rest of the world either chose between the later PAL and SECAM color television standards, or adopted NTSC. Japan also uses a version of NTSC.

North American terrestrial broadcast television operates on television channels 2 through 6 (VHF-low band, 54 to 88 MHz, known as band I in Europe), 7 through 13 (VHF-high band, 174 to 216 MHz, known as band III elsewhere), and 14 through 51 (UHF television band, 470 to 698 MHz, elsewhere bands IV and V). Channel numbers represent actual frequencies used to broadcast the television signal. Additionally, television translators and signal boosters can be used to rebroadcast a terrestrial television signal using an otherwise unused channel to cover areas with marginal reception.

Analog television channels 2 through 6, 7 through 13 and 14 through 51 are only used for LPTV translator stations in the U.S. Channels 52 through 69 are still used in exceptional circumstances for LPTV translators in some remote regions of Canada. The Canadian Broadcasting Corporation received permission to continue analog broadcasting at severely reduced power levels in some mandatory markets until January 8, 2012. It was expected that these transmitters would be shut down rather than converted to digital. Some older cathode ray tube television sets have a built-in transmitter.

The rise of digital terrestrial television, especially high-definition television (HDTV), may mark an end to the decline of broadcast television reception by traditional receiving antennas, which can receive over-the-air HDTV signals.

Asia

Terrestrial television broadcast in Asia started as early as 1939 in Japan through a series of experiments done by NHK Broadcasting Institute of Technology. However, these experiments were interrupted by the beginning of the World War II in the Pacific. On February 1, 1953, NHK (Japan Broadcasting Corporation) began broadcasting. On August 28, 1953, Nippon TV (Nippon Television Network Corporation), the first commercial television broadcaster in Asia was launched. Meanwhile, in the Philippines, Alto Broadcasting System (now ABS-CBN), the first commercial television broadcaster in Southeast Asia, launched its first commercial terrestrial television station DZAQ-TV on October 23, 1953 with the help of Radio Corporation of America (RCA).

Digital Terrestrial Television

By the mid 1990s, the interest in digital television across Europe was such the CEPT convened the "Chester '97" conference to agree means by which digital television could be inserted into the ST61 frequency plan.

The introduction of digital television in the late 1990s and early years of the 21st century led the ITU to call a Regional Radiocommunication Conference to abrogate the ST61 plan and to put a new plan for digital broadcasting only in its place.

In December 2005 the European Union decided to cease all analog audio and analog video television transmissions by 2012 and switch all terrestrial television broadcasting to digital audio and digital video (all EU countries have agreed on using DVB-T). The Netherlands completed the transition in December 2006, and some EU member states decided to complete their switchover as early as 2008 (Sweden), and (Denmark) in 2009. While the UK began the switch in late 2007, it was not completed until 24 October 2012. Norway ceased all analogue television transmissions on December 1, 2009. Two member states (not specified in the announcement) have expressed concerns that they might not be able to proceed to the switchover by 2012 due to technical limitations; the rest of the EU member states had stopped analog television transmissions by the end 2012.

Many countries are developing and evaluating digital terrestrial television systems.

Australia has adopted the DVB-T standard and the government's industry regulator, the Australian Communications and Media Authority, has mandated that all analogue transmissions will cease by 2012. Mandated digital conversion commenced early in 2009 with a graduated program. The first centre to experience analog switch-off will be the remote Victorian regional town of Mildura, in 2010. The government will supply underprivileged houses across the nation with free digital set-top DTV converter boxes in order to minimise any conversion disruption. Australia's major free-to-air television networks have all been granted digital transmission licences and are each required to broadcast at least one high-definition and one standard-definition channel into all of their markets.

In North America a specification laid out by the ATSC has become the standard for digital terrestrial television. In the United States the Federal Communications Commission (FCC) set the final deadline for the switchoff of analog service for June 12, 2009. All television receivers must now include a digital tuner. In Canada, the Canadian Radio-television and Telecommunications Commission (CRTC), set August 31, 2011 as the date that over-the-air analog transmission service ceased in most parts of the country except in Northern Canada.   In Mexico the Federal Telecommunications Institute (IFT) set the final deadline for the end of Analog Television for December 31, 2015.

Competition for Radio Spectrum

In late 2009, US competition for the limited available radio spectrum led to debate over the possible re-allocation of frequencies currently occupied by television, and the FCC began asking for comments on how to increase the bandwidth available for wireless broadband. Some have proposed mixing the two together, on different channels that are already open (like White Spaces), while others have proposed "repacking" some stations and forcing them off certain channels, just a few years after the same thing was done (without compensation to the broadcasters) in the DTV transition in the United States.

Some US commentators have proposed the closing down of over-the-air TV broadcasting, on the grounds that available spectrum might be better used, and requiring viewers to shift to satellite or cable reception. This would eliminate mobile TV, which has been delayed several years by the FCC's decision to choose ATSC and its proprietary 8VSB modulation, instead of the worldwide COFDM standard used for all other digital terrestrial broadcasting around the world. Compared to Europe and Asia, this has hamstrung mobile TV in the US, because ATSC cannot be received while in motion (or often even while stationary) without ATSC-M/H as terrestrial DVB-T or ISDB-T can even without DVB-H or 1seg.

The National Association of Broadcasters has organized to fight such proposals, and public comments are also being taken by the FCC through mid-December 2009, in preparation for a plan to be released in mid-February 2010.

Telegraphy

Telegraphy is the long-distance transmission of textual or symbolic (as opposed to verbal or audio) messages without the physical exchange of an object bearing the message. Thus semaphore is a method of telegraphy, whereas pigeon post is not.

Telegraphy requires that the method used for encoding the message be known to both sender and receiver. Such methods are designed according to the limits of the signalling medium used. The use of smoke signals, beacons, reflected light signals, and flag semaphore signals are early examples. In the 19th century, the harnessing of electricity led to the invention of electrical telegraphy. The advent of radio in the early 20th century brought about radiotelegraphy and other forms of wireless telegraphy. In the Internet age, telegraphic means developed greatly in sophistication and ease of use, with natural language interfaces that hide the underlying code, allowing such technologies as electronic mail and instant messaging.

Terminology

The word "telegraph" was first coined by the French inventor of the Semaphore line, Claude Chappe, who also coined the word "semaphore".

A "telegraph" is a device for transmitting and receiving messages over long distances, i.e., for telegraphy. The word "telegraph" alone now generally refers to an electrical telegraph.

Wireless telegraphy is also known as "CW", for continuous wave (a carrier modulated by on-off keying), as opposed to the earlier radio technique of using a spark gap.

Contrary to the extensive definition used by Chappe, Morse argued that the term *telegraph* can strictly be applied only to systems that transmit *and* record messages at a distance. This is to be distinguished from *semaphore*, which merely transmits messages. Smoke signals, for instance, are to be considered semaphore, not telegraph. According to Morse, telegraph dates only from 1832, when Pavel Schilling invented one of the earliest electrical telegraphs.

A telegraph message sent by an electrical telegraph operator or telegrapher using Morse code (or a printing telegraph operator using plain text) was known as a *telegram*. A *cablegram* was a message sent by a submarine telegraph cable, often shortened to a *cable* or a *wire*. Later, a *Telex* was a message sent by a Telex network, a switched network of teleprinters similar to a telephone network.

A *wire picture* or *wire photo* was a newspaper picture that was sent from a remote location by a facsimile telegraph. A *diplomatic telegram*, also known as a diplomatic cable, is the term given to a confidential communication between a diplomatic mission and the foreign ministry of its parent country. These continue to be called telegrams or cables regardless of the method used for transmission.

History

Even though early telegraphic precedents, such as signalling through the lighting of pyres, have existed since ancient times, long-distance telegraphy (transmission of complex messages) started in 1792 in the form of semaphore lines, or optical telegraphs, that sent messages to a distant observer through line-of-sight signals. Commercial electrical telegraphs were introduced from 1837.

Optical Telegraph

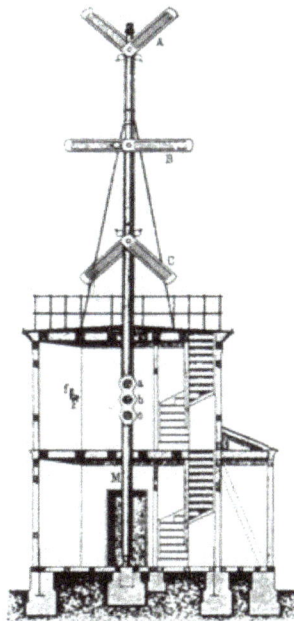

Construction schematic of a Prussian optical telegraph (or semaphore) tower, C. 1835

The first telegraphs came in the form of optical telegraph, including the use of smoke signals, beacons or reflected light, which have existed since ancient times. Early proposals for an optical telegraph system were made to the Royal Society by Robert Hooke in 1684 and were first implemented on an experimental level by Sir Richard Lovell Edgeworth in 1767.

Demonstration of the semaphore

The first successful semaphore network was invented by Claude Chappe and operated in France from 1793 through 1846.

During 1790–1795, at the height of the French Revolution, France needed a swift and reliable communication system to thwart the war efforts of its enemies. In 1790, the Chappe brothers set about devising a system of communication that would allow the central government to receive intelligence and to transmit orders in the shortest possible time. On 2 March 1791 at 11 am, they sent the message "si vous réussissez, vous serez bientôt couverts de gloire" (If you succeed, you will soon bask in glory) between Brulon and Parce, a distance of 16 kilometres (9.9 mi). The first means used a combination of black and white panels, clocks, telescopes, and codebooks to send their message.

In 1792 Claude was appointed *Ingénieur-Télégraphiste* and charged with establishing a line of stations between Paris and Lille, a distance of 230 kilometres (about 143 miles). It was used to carry dispatches for the war between France and Austria. In 1794, it brought news of a French capture of Condé-sur-l'Escaut from the Austrians less than an hour after it occurred.

The Prussian system was put into effect in the 1830s. However they were highly dependent on good weather and daylight to work and even then could accommodate only about two words per minute. . The last commercial semaphore link ceased operation in Sweden in 1880. As of 1895, France still operated coastal commercial semaphore telegraph stations, for ship-to-shore communication.

Electrical Telegraphs

Early Developments

The first suggestion for using electricity as a means of communication appeared in the "Scots Magazine" in 1753. Using one wire for each letter of the alphabet, a message could be transmitted by connecting the wire terminals in turn to an electrostatic machine, and observing the deflection of pith balls at the far end. Telegraphs employing electrostatic attraction were the basis of early experiments in electrical telegraphy in Europe, but were abandoned as being impractical and were never developed into a useful communication system.

One very early experiment in electrical telegraphy was an *electrochemical telegraph* created by the German physician, anatomist, and inventor Samuel Thomas von Sömmering in 1809, based on an earlier, less robust design of 1804 by Spanish polymath and scientist Francisco Salva Campillo. Both their designs employed multiple wires (up to 35) in order to visually represent most Latin letters and numerals. Thus, messages could be conveyed electrically up to a few kilometers (in von Sömmering's design), with each of the telegraph receiver's wires immersed in a separate glass tube of acid. As an electric current was applied by the sender representing each digit of a message, it would at the recipient's end electrolyse the acid in its corresponding tube, releasing a stream of hydrogen bubbles next to its associated letter or numeral. The telegraph receiver's operator would visually observe the bubbles and could then record the transmitted message, albeit at a very low baud rate.

The first working telegraph was built by the English inventor Francis Ronalds in 1816 and used static electricity. At the family home on Hammersmith Mall, he set up a complete subterranean system in a 175 yard long trench as well as an eight mile long overhead telegraph. The lines were connected at both ends to clocks marked with the letters of the alphabet and electrical impulses sent along the wire were used to transmit messages. Offering his invention to the Admiralty in July 1816, it was rejected as "wholly unnecessary". His account of the scheme and the possibilities of rapid global communication in *Descriptions of an Electrical Telegraph and of some other Electrical Apparatus* was the first published work on electric telegraphy and even described the risk of signal retardation due to induction. Elements of Ronalds' design were utilised in the subsequent commercialisation of the telegraph over 20 years later.

Pavel Schilling, an early pioneer of electrical telegraphy

An early electromagnetic telegraph design was created by Russian diplomat Pavel Schilling in 1832. He set it up in his apartment in St. Petersburg and demonstrated the long-distance transmission of signals by positioning two telegraphs of his invention in two different rooms of his apartment. Schilling was the first to put into practice the idea of a binary system of signal transmissions.

Carl Friedrich Gauss and Wilhelm Weber built the first electromagnetic telegraph used for *regular* communication in 1833 in Göttingen, connecting Göttingen Observatory and the Institute of Physics, covering a distance of about 1 km. The setup consisted of a coil that could be moved up and down over the end of two magnetic steel bars. The resulting induction current was transmitted through two wires to the receiver, consisting of a galvanometer. The direction of the current could be reversed by commuting the two wires in a special switch. Therefore, Gauss and Weber chose to encode the alphabet in a binary code, using positive and negative currents as the two states.

Commercial Telegraphy

Telegraph networks were expensive to build, but financing was readily available, especially from London bankers. By 1852, National systems were in operation major countries as follows:

- United States, 20 companies with 23,000 miles of wire.
- Great Britain, Cooke-Wheatstone company and minor companies, with 2200 miles of wire.
- Prussia, 1400 miles of wire, Siemens system.
- Austria, 1000 miles of wire, Siemens system.
- Canada, 900 miles of wire.
- France, 700 miles of wire; optical systems dominant.
- Australia, 0. First line opened in 1854.

Cooke and Wheatstone System

Cooke and Wheatstone's five-needle, six-wire telegraph

The first commercial electrical telegraph was co-developed by Sir William Fothergill Cooke and Charles Wheatstone. In May 1837 they patented the Cooke and Wheatstone system, which used a number of needles on a board that could be moved to point to letters of the alphabet. The patent recommended a five-needle system, but any number of needles could be used depending on the number of characters it was required to code. A four-needle system was installed between Euston and Camden Town in London on a rail line being constructed by Robert Stephenson between London and Birmingham. It was successfully demonstrated on 25 July 1837. Euston needed to signal to an engine house at Camden Town to start hauling the locomotive up the incline. As at Liverpool, the electric telegraph was in the end rejected in favour of a pneumatic system with whistles.

Cooke and Wheatstone had their first commercial success with a system installed on the Great Western Railway over the 13 miles (21 km) from Paddington station to West Drayton in 1838, the first commercial telegraph in the world. This was a five-needle, six-wire system. The cables were originally installed underground in a steel conduit. However, the cables soon began to fail as a result of deteriorating insulation and were replaced with uninsulated wires on poles. As an interim measure, a two-needle system was used with three of the remaining working underground wires, which despite using only two needles had a greater number of codes. But when the line was extended to Slough in 1843, a one-needle, two-wire system was installed.

From this point the use of the electric telegraph started to grow on the new railways being built from London. The London and Blackwall Railway (another rope-hauled application) was equipped with the Cooke and Wheatstone telegraph when it opened in 1840, and many others followed. The one-needle telegraph proved highly successful on British railways, and 15,000 sets were still in use at the end of the nineteenth century. Some remained in service in the 1930s. In September 1845 the financier John Lewis Ricardo and Cooke formed the Electric Telegraph Company, the first public telegraphy company in the world. This company bought out the Cooke and Wheatstone patents and solidly established the telegraph business.

As well as the rapid expansion of the use of the telegraphs along the railways, they soon spread into the field of mass communication with the instruments being installed in post offices across the country. The era of mass personal communication had begun.

Morse System

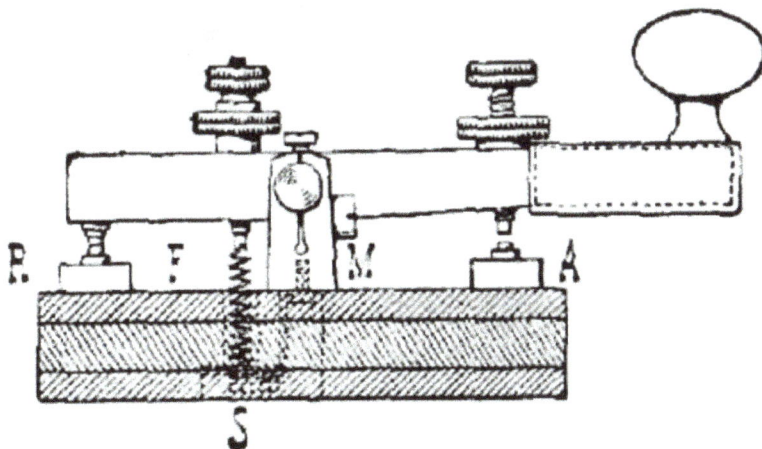

A Morse key

An electrical telegraph was independently developed and patented in the United States in 1837 by Samuel Morse. His assistant, Alfred Vail, developed the Morse code signalling alphabet with Morse. The first telegram in the United States was sent by Morse on 11 January 1838, across two miles (3 km) of wire at Speedwell Ironworks near Morristown, New Jersey, although it was only later, in 1844, that he sent the message "WHAT HATH GOD WROUGHT" from the Capitol in Washington to the old Mt. Clare Depot in Baltimore. From then on, commercial telegraphy took off in America with lines linking all the major metropolitan centres on the East Coast within the next decade. The overland telegraph connected the west coast of the continent to the east coast by 24 October 1861, bringing an end to the Pony Express.

The Morse telegraphic apparatus was officially adopted as the standard for European telegraphy in 1851. Only Great Britain with its extensive overseas empire kept the needle telegraph of Cooke and Wheatstone. In 1858, Morse introduced wired communication to Latin America when he established a telegraph system in Puerto Rico, then a Spanish Colony. The line was inaugurated on March 1, 1859, in a ceremony flanked by the Spanish and American flags.

Another early system was that of Edward Davy, who demonstrated his in Regent's Park in 1837 and was granted a patent on 4 July 1838. He also developed an electric relay.

Telegraphic Improvements

Telegraphy was driven by the need to reduce sending costs, either in hand-work per message or by increasing the sending rate. While many experimental systems employing moving pointers and various electrical encodings proved too complicated and unreliable, a successful advance in the sending rate was achieved through the development of telegraphese.

The first message is received by the Submarine Telegraph Company in London from Paris on the Foy-Breguet instrument in 1851

The first system that didn't require skilled technicians to operate, was Sir Charles Wheatstone's ABC system in 1840 where the letters of the alphabet were arranged around a clock-face, and the signal caused a needle to indicate the letter. This early system required the receiver to be present in real time to record the message and it reached speeds of up to 15 words a minute.

Before Telegraphy, a letter by post from London took	
days	to reach
12	New York in USA
13	Alexandria in Egypt
19	Constantinople in Ottoman Turkey
33	Bombay in India
44	Calcutta in Bengal
45	Singapore
57	Shanghai in China
73	Sydney in Australia

In 1846, Alexander Bain patented a chemical telegraph in Edinburgh. The signal current made a readable mark on a moving paper tape soaked in a mixture of ammonium nitrate and potassium ferrocyanide, which gave a blue mark when a current was passed through it.

David Edward Hughes invented the printing telegraph in 1855; it used a keyboard of 26 keys for the alphabet and a spinning type wheel that determined the letter being transmitted by the length of time that had elapsed since the previous transmission. The system allowed for automatic recording on the receiving end. The system was very stable and accurate and became the accepted around the world.

The next improvement was the Baudot code of 1874. French engineer Émile Baudot patented a printing telegraph in which the signals were translated automatically into typographic characters. Each character was assigned a unique code based on the sequence of just five contacts. Operators had to maintain a steady rhythm, and the usual speed of operation was 30 words per minute.

A Baudot keyboard, 1884

By this point reception had been automated, but the speed and accuracy of the transmission was still limited to the skill of the human operator. The first practical automated system was patented by Charles Wheatstone, the original inventor of the telegraph. The message (in Morse code) was typed onto a piece of perforated tape using a keyboard-like device called the 'Stick Punch'. The transmitter automatically ran the tape through and transmitted the message at the then exceptionally high speed of 70 words per minute.

Teleprinters

Phelps' Electro-motor Printing Telegraph from circa 1880, the last and most advanced telegraphy mechanism designed by George May Phelps

Teleprinters were invented in order to send and receive messages without the need for operators trained in the use of Morse code. A system of two teleprinters, with one operator trained to use a typewriter, replaced two trained Morse code operators. The teleprinter system improved message speed and delivery time, making it possible for messages to be flashed across a country with little manual intervention.

Early teleprinters used the ITA-1 Baudot code, a five-bit code. This yielded only thirty-two codes, so it was over-defined into two "shifts", "letters" and "figures". An explicit, unshared shift code prefaced each set of letters and figures. In 1901 Baudot's code was modified by Donald Murray and Around 1930, the CCITT introduced the International Telegraph Alphabet No. 2 (ITA2) code as an international standard.

A Siemens T100 Telex machine

By 1935, message routing was the last great barrier to full automation. Large telegraphy providers began to develop systems that used telephone-like rotary dialling to connect teletypewriters. These machines were called "Telex" (TELegraph EXchange). Telex machines first performed rotary-telephone-style pulse dialling for circuit switching, and then sent data by Baudot code. This "type A" Telex routing functionally automated message routing.

Telex began in Germany as a research and development program in 1926 that became an operational teleprinter service in 1933. The service was operated by the Reichspost (Reich postal service) and had a speed of 50 baud - approximately 66 words-per-minute.

At the rate of 45.45 (±0.5%) baud—considered speedy at the time—up to 25 telex channels could share a single long-distance telephone channel by using *voice frequency telegraphy multiplexing*, making telex the least expensive method of reliable long-distance communication.

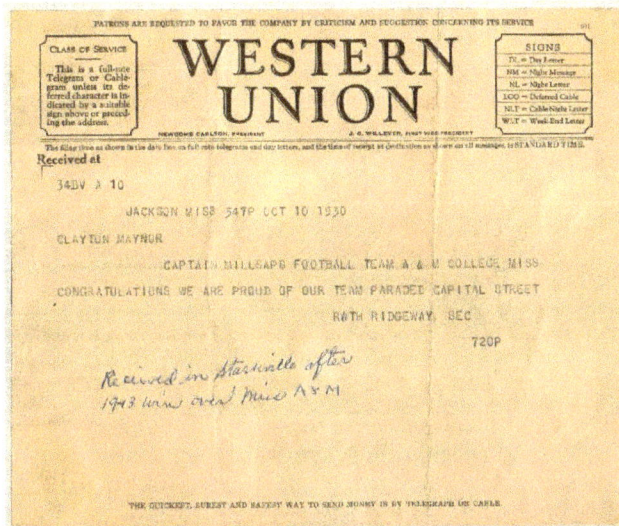

Western Union telegram circa 1930

Automatic teleprinter exchange service was introduced into Canada by CPR Telegraphs and CN Telegraph in July 1957 and in 1958, Western Union started to build a Telex network in the United States.

Beginning in 1956 telegrams begun to be transmitted over the Telex network using the ITU F.20 standard named Gentex in order to lower the costs for some European telecommunications companies by allowing the sending telegraph station to connect directly to the receiving station.

Oceanic Telegraph Cables

Soon after the first successful telegraph systems were operational, the possibility of transmitting messages across the sea by way of submarine communications cables was first mooted. One of the primary technical challenges was to insulate the submarine cable sufficiently to prevent the current from leaking out into the water. In 1842, a Scottish surgeon William Montgomerie introduced Gutta-percha, the adhesive juice of the *Palaquium gutta* tree, to Europe. Michael Faraday and Wheatstone soon discovered the merits of gutta-percha as an insulator, and in 1845, the latter suggested that it should be employed to cover the wire which was proposed to be laid from Dover to Calais. It was tried on a wire laid across the Rhine between Deutz and Cologne. In 1849, C.V. Walker, electrician to the South Eastern Railway, submerged a two-mile wire coated with gutta-percha off the coast from Folkestone, which was tested successfully.

John Watkins Brett, an engineer from Bristol, sought and obtained permission from Louis-Philippe in 1847 to establish telegraphic communication between France and England. The first undersea cable was laid in 1850 and connected London with Paris. After an initial exchange of greetings between Queen Victoria and President Napoleon, it was almost immediately severed by a French fishing vessel. The line was relaid the next year and then followed by connections to Ireland and the Low Countries.

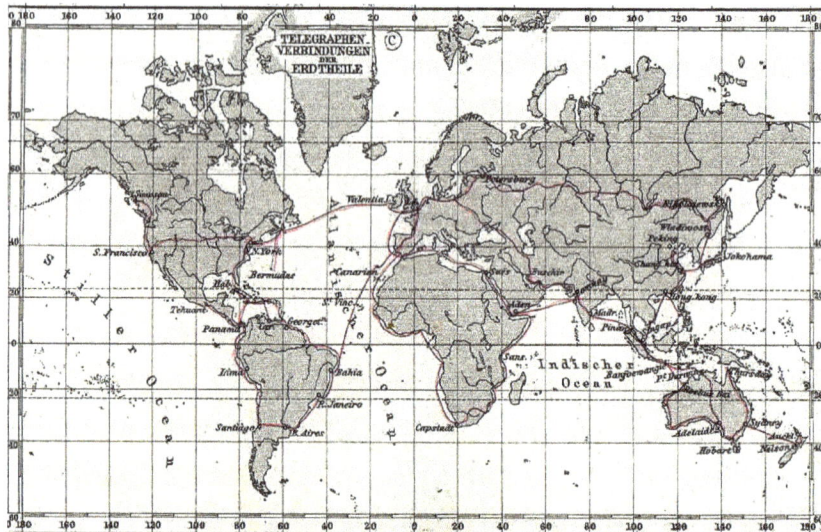

Major telegraph lines across the Earth in 1891

The Atlantic Telegraph Company was formed in London in 1856 to undertake to construct a commercial telegraph cable across the Atlantic Ocean. It was successfully completed on 18 July 1866 by the ship SS *Great Eastern*, captained by Sir James Anderson after many mishaps along the way. Earlier transatlantic submarine cables installations were attempted in 1857, 1858 and 1865. The 1857 cable only operated intermittently for a few days or weeks before it failed. The study of underwater telegraph cables accelerated interest in mathematical analysis of very long transmission lines. An overland telegraph from Britain to India was first connected in 1866 but was unreliable so a submarine telegraph cable was connected in 1870. Several telegraph companies were combined to form the *Eastern Telegraph Company* in 1872.

Australia was first linked to the rest of the world in October 1872 by a submarine telegraph cable at Darwin. This brought news reportage from the rest of the world. The telegraph across the Pacific was completed in 1902, finally encircling the world.

From the 1850s until well into the 20th century, British submarine cable systems dominated the world system. This was set out as a formal strategic goal, which became known as the All Red Line. In 1896, there were thirty cable laying ships in the world and twenty-four of them were owned by British companies. In 1892, British companies owned and operated two-thirds of the world's cables and by 1923, their share was still 42.7 percent. During World War I, Britain's telegraph communications were almost completely uninterrupted, while it was able to quickly cut Germany's cables worldwide.

Later Technology

Facsimile

Alexander Bain's facsimile machine, 1850

In 1843 Scottish inventor Alexander Bain invented a device that could be considered the first facsimile machine. He called his invention a "recording telegraph". Bain's telegraph was able to transmit images by electrical wires. Frederick Bakewell made several improvements on Bain's design and demonstrated a telefax machine. In 1855 an Italian abbot, Giovanni Caselli, also created an electric telegraph that could transmit images. Caselli called his invention "Pantelegraph". Pantelegraph was successfully tested and approved for a telegraph line between Paris and Lyon.

In 1881, English inventor Shelford Bidwell constructed the *scanning phototelegraph* that was the first telefax machine to scan any two-dimensional original, not requiring manual plotting or drawing. Around 1900, German physicist Arthur Korn invented the *Bildtelegraph*, widespread in continental Europe especially, since a widely noticed transmission of a wanted-person photograph from Paris to London in 1908, used until the wider distribution of the radiofax. Its main competitors were the *Bélinographe* by Édouard Belin first, then since the 1930s the *Hellschreiber*, invented in 1929 by German inventor Rudolf Hell, a pioneer in mechanical image scanning and transmission.

Wireless Telegraphy

Post Office Engineers inspect Marconi's equipment on Flat Holm, May 1897

The late 1880s through the 1890s saw the discovery and then development of a newly understood phenomenon into a form of wireless telegraphy, called *Hertzian wave* wireless telegraphy, radio-telegraphy, or (later) simply "radio". Between 1886 and 1888 Heinrich Rudolf Hertz published the results of his experiments where he was able to transmit electromagnetic waves (radio waves) through the air, proving James Clerk Maxwell's 1873 theory of electromagnetic radiation. Many scientists and inventors experimented with this new phenomenon but the general consensus was that these new waves (similar to light) would be just as short range as light, and therefore useless for long range communication.

At the end of 1894 the young Italian inventor Guglielmo Marconi began working on the idea of building a commercial wireless telegraphy system based on the use of Hertzian waves (radio waves), a line of inquiry that he noted other inventors did not seem to be pursuing. Building on the ideas of previous scientists and inventors Marconi re-engineered their apparatus by trial and error attempting to build a radio based wireless telegraphic system that would function the same as wired telegraphy. He would work on the system through 1895 in his lab and then in field tests making improvements to extend its range. After many breakthroughs, including applying the wired telegraphy concept of grounding the transmitter and receiver, Marconi was able, by early 1896, to transmit radio far beyond the short ranges that had been predicted. Having failed to interest the Italian government, the 22-year-old inventor brought his telegraphy system to Britain in 1896 and met William Preece, a Welshman, who was a major figure in the field and Chief Engineer of the General Post Office. A series of demonstrations for the British government followed—by March 1897, Marconi had transmitted Morse code signals over a distance of about 6 kilometres (3.7 mi) across Salisbury Plain.

Marconi watching associates raising the kite (a "Levitor" by B.F.S. Baden-Powell) used to lift the antenna at St. John's, Newfoundland, December 1901

On 13 May 1897, Marconi, assisted by George Kemp, a Cardiff Post Office engineer, transmitted the first wireless signals over water to Lavernock (near Penarth in Wales) from Flat Holm. The message sent was "ARE YOU READY". From his Fraserburgh base, he transmitted the first long-distance, cross-country wireless signal to Poldhu in Cornwall. His star rising, he was soon sending signals across The English channel (1899), from shore to ship (1899) and finally across the Atlantic (1901). A study of these demonstrations of radio, with scientists trying to figure out how a phenomenon predicted to have a short range could transmit "over the horizon", led to the discovery of a radio reflecting layer in the Earth's atmosphere in 1902, later called the ionosphere.

Radiotelegraphy proved effective for rescue work in sea disasters by enabling effective communication between ships and from ship to shore. In 1904 Marconi began the first commercial service to transmit nightly news summaries to subscribing ships, which could incorporate them into their on-board newspapers. A regular transatlantic radio-telegraph service was finally begun on 17 October 1907. Notably, Marconi's apparatus was used to help rescue efforts after the sinking of *Titanic*. Britain's postmaster-general summed up, referring to the *Titanic* disaster, "Those who have been saved, have been saved through one man, Mr. Marconi…and his marvellous invention."

Internet

Around 1965, DARPA commissioned a study of decentralized switching systems. Some of the ideas developed in this study provided inspiration for the development of the ARPANET packet switching research network, which later grew to become the public Internet.

As the PSTN became a digital network, T-carrier "synchronous" networks became commonplace in the U.S. A T1 line has a "frame" of 193 bits that repeats 8000 times per second. The first bit, called the "sync" bit, alternates between 1 and 0 to identify the start of the frames. The rest of the frame provides 8 bits for each of 24 separate voice or data channels. Customarily, a T-1 link is sent over a balanced twisted pair, isolated with transformers to prevent current flow. Europeans adopted a similar system (E-1) of 32 channels (with one channel for frame synchronisation).

Later, SONET and SDH were adapted to combine carrier channels into groups that could be sent over optic fiber. The capacity of an optic fiber is often extended with wavelength division multiplexing, rather than rerigging new fibre. Rigging several fibres in the same structures as the first fibre is usually easy and inexpensive, and many fibre installations include unused spare "dark fibre", "dark wavelengths", and unused parts of the SONET frame, so-called "virtual channels".

In 2002, the Internet was used by Kevin Warwick at the University of Reading to communicate neural signals, in purely electronic form, telegraphically between the nervous systems of two humans, potentially opening up a new form of communication combining the Internet and telegraphy.

In 2006, a well-defined communication channel used for telegraphy was established by the SONET standard OC-768, which sent about 40 gigabits per second.

The theoretical maximum capacity of an optic fiber is more than 10^{12} bits (one terabit or one trillion bits) per second. In 2006, no existing encoding system approached this theoretical limit, even with wavelength division multiplexing.

Since the Internet operates over any digital transmission medium, further evolution of telegraphic technology will be effectively concealed from users.

E-mail

E-mail was first invented for CTSS and similar time sharing systems of the era in the mid-1960s. At first, e-mail was possible only between different accounts on the same computer (typically a mainframe). ARPANET allowed different computers to be connected to allow e-mails to be relayed from computer to computer, with the first ARPANET e-mail being sent in 1971. Multics also pioneered instant messaging between computer users in the mid-1970s. With the growth of the Internet, e-mail began to be possible between any two computers with access to the Internet.

Various private networks like UUNET (founded 1987), the Well (1985), and GEnie (1985) had e-mail from the 1970s, but subscriptions were quite expensive for an individual, US$25 to US$50 per month, just for e-mail. Internet use was then largely limited to government, academia and other government contractors until the net was opened to commercial use in the 1980s.

By the early 1990s, modems made e-mail a viable alternative to Telex systems in a business environment. But individual e-mail accounts were not widely available until local Internet service providers were in place, although demand grew rapidly, as e-mail was seen as the Internet's killer app. It allowed anyone to email anyone, whereas previously, different system had been walled off from each other, such that America Online subscribers could email only other America Online subscribers, Compuserve subscribers could email only other Compuserve subscribers, etc. The broad user base created by the demand for e-mail smoothed the way for the rapid acceptance of the World Wide Web in the mid-1990s. Fax machines were another technology that helped displace the telegram.

On Monday, 12 July 1999, a final telegram was sent from the National Liberty Ship Memorial, the SS Jeremiah O'Brien, in San Francisco Bay to President Bill Clinton in the White House. Officials of Globe Wireless reported that "The message was 95 words, and it took six or eight minutes to copy it." They then transmitted the message to the White House via e-mail. That event was also used to mark the final commercial U.S. ship-to-shore telegraph message transmitted from North America by Globe Wireless, a company founded in 1911. Sent from its wireless station at Half Moon Bay, California, the sign-off message was a repeat of Samuel F. B. Morse's message 155 years earlier, "What hath God wrought?"

21st Century Decline

- In Australia, Australia Post closed its telegram service on 7 March 2011. In the Victorian town of Beechworth, visitors can send telegrams to family members or friends from the Beechworth Telegraph Station.

- In Bahrain, Batelco still offers telegram services. They are thought to be more formal than an email or a fax, but less so than a letter. So should a death or anything of importance occur, telegrams would be sent.

- In Belgium, Belgacom still offers telegram services within the country and internationally. It sent 63.000 telegrams in 2010.

- In Canada, Telegrams Canada still offers telegram services. AT&T Canada (previously CNCP Telecommunications) had discontinued its telegram service in 2001 and later became MTS Allstream.

- In France, Orange S.A. still offers a telegram service, although not transmitted by telegraph any more.

- In Germany, Deutsche Post delivers telegrams the next day as ordinary mail. Deutsche Post discontinued service to foreign countries on 31 December 2000. A private firm, TelegrammDirekt.de, offers delivery in Germany and service to a number of foreign countries.

- In Hungary, Magyar Posta still offers (national only) telegram services.

- In India, state-owned BSNL discontinued telegram services from 15 July 2013. Telegrams to foreign countries had been discontinued in May 2013.

- In Iran, telex services are still provided by Telecommunication Infrastructure Company of I.R.Iran.

- In Ireland, Eircom – the country's largest telecommunication company and former PTT – formally discontinued telegram service on 30 July 2002.

- In Israel, the Israel Postal Company still offers telegram services. Telegrams may be sent via the internet or by a telephone operator. Illustrated telegrams are available for special occasions.

- In Italy, Poste Italiane still offers telegram services. Around 2.5 million telegrams are sent annually, primarily for births, weddings, and funerals.

- In Japan, NTT provides a telegram (*denpou*) service used mainly for special occasions such as weddings, funerals, graduations, etc. Local offices offer telegrams printed on special decorated paper and envelopes.

- In Lithuania, telegram service was closed by the only provider Teo LT on 15 October 2007.

- In Malaysia, Telekom Malaysia has ceased its telegram service effective 1 July 2012.

- In Mexico, telegrams are still used as a low-cost service for people who cannot afford or do not have access to e-mail.

- In Nepal, Nepal Telecom closed its telegram service on 1 January 2009.

- In the Netherlands, the telegram service was sold by KPN to the Swiss-based company Unitel Telegram Services in 2001.

- In New Zealand, New Zealand Post closed its telegram service in 1999. It later reinstated the service in 2003 for use only by business customers, primarily for debt collection or other important business notices.

- In Pakistan, the Pakistan Telecommunication Company Ltd ceased telegram services on 27 January 2006.

- In the Philippines, telegram services by the government's Telecommunications Office or *Tanggapan ng Telekomunikasyon* ceased on 20 September 2013. The last telegram was sent on that day at 3:15 PM.

- In Russia, Central Telegraph (subsidiary of national operator Rostelecom) still offers telegram service. "Regular" or "Urgent" telegrams can be sent to any address in Russia and other countries. So called "Stylish" telegrams printed on an artistic postcards are also available.

- In Serbia, JP Pošta Srbije Beograd, the state-owned post, provides a telegram service. It is commonly used to express condolences, official notifications of death or to congratulate anniversaries, births, graduations etc. Telegrams may be sent by using special telephone number or directly at the post office. Telegrams are delivered on the same day for recipients in territories covered by post offices with telegram delivery service and are delivered as regular mail for post offices which do not have telegram delivery service. In internal traffic, length of message is limited to 800 characters and is charged at flat rate, while in international traffic telegrams are charged by word. International delivery is possible for recipients in Croatia, Slovenia, Montenegro, Bosnia and Herzegovina, and Macedonia.

- In Slovakia, the Slovak post closed its telegram service on 1 January 2007.

- In Slovenia, Pošta Slovenije d.o.o. (Slovenian Post) provides a telegram service still commonly used for special occasions such as births, anniversaries, condolences, graduations, etc. It is considered more formal than email or SMS. Telegrams are usually printed in a typewriter font on greeting or condolences cards delivered in a specific yellow envelope. It is also possible to send gifts (e.g. chocolates, wine, plush toys, flowers) together with a message. The telegrams can be sent from local post offices, over the phone or online to addresses in Slovenia only.

- In Sweden, Telia ceased telegraph services in 2002.

- In Switzerland, Unitel Telegram Services took over telegram services from the national PTTs. Telegrams can still be sent to and from most countries.

- In Thailand, Thailand Post ceased its telegram service on 30 April 2008, at 20.00 local time.

- In the United Kingdom, the international telegram service formerly provided by British Telecom was sold in 2003 to an independent company, Telegrams Online, which promotes the use of telegrams as a retro greeting card or invitation.

- In the United States, Western Union sent its last telegram on 27 January 2006. iTelegram bought Western Union's telex service and offers a range of telegram-like services.

Social Implications

Prior to the electrical telegraph, nearly all information was limited to traveling at the speed of a human or animal. The telegraph freed communication from the constraints of space and time and revolutionized the global economy and society. By the end of the 19th century, the telegraph was becoming an increasingly common medium of communication for ordinary people. The telegraph isolated the message (information) from the physical movement of objects or the process.

Telegraphy facilitated the growth of organizations "in the railroads, consolidated financial and commodity markets, and reduced information costs within and between firms". This immense growth in the business sectors influenced society to embrace the use of telegrams.

Worldwide telegraphy changed the gathering of information for news reporting. Messages and information would now travel far and wide, and the telegraph demanded a language "stripped of the local, the regional; and colloquial", to better facilitate a worldwide media language. Media language had to be standardized, which led to the gradual disappearance of different forms of speech and styles of journalism and storytelling.

Newspaper Names

Numerous newspapers and news outlets in various countries, such as *The Daily Telegraph* in Britain, *The Telegraph* in India, *De Telegraaf* in the Netherlands and the Jewish Telegraphic Agency in the US, were given names which include the word "telegraph" due to their having received news by means of electric telegraphy. Some of these names are retained even though more sophisticated means are now used.

Telegram Length

The average length of a telegram in the 1900s in the US was 11.93 words, more than half of the messages were 10 words or fewer.

According to another study the mean length of the telegrams sent in the UK before 1950 was 14.6 words or 78.8 characters.

For German telegrams the mean length is 11.5 words or 72.4 characters. At the end of the 19th century the average length of a German telegram was calculated as 14.2 words.

Telephony

Telephony is the field of technology involving the development, application, and deployment of telecommunication services for the purpose of electronic transmission of voice, fax, or data, between distant parties. The history of telephony is intimately linked to the invention and development of the telephone.

Telephony is commonly referred to as the construction or operation of telephones and telephonic systems and as a system of telecommunications in which telephonic equipment is employed in the transmission of speech or other sound between points, with or without the use of wires. The term is also used frequently to refer to computer hardware, software, and computer network systems, that perform functions traditionally performed by telephone equipment. In this context the technology is specifically referred to as Internet telephony, or voice over Internet Protocol (VoIP).

Overview

The first telephones were connected directly in pairs. Each user had a separate telephone wired to the locations he might wish to reach. This quickly became inconvenient and unmanageable when people wanted to communicate with more than a few people. The inventions of the telephone exchange provided the solution for establishing telephone connections with any other telephone in service in the local area. Each telephone was connected to the exchange via one wire pair, the local loop. Nearby exchanges in other service areas were connected with trunk lines and long distance service could be established by relaying the calls through multiple exchanges.

Initially the switchboards were manually operated by an attendant, a *switchboard operator*. When a customer cranked a handle on the telephone, it turned on an indicator on the board in front of the operator who would plug the operator headset into that jack and offer service. The caller had to ask for the called party by name, later by number, and the operator connected one end of a circuit into the called party jack to alert them. If the called station answered the operator disconnected their headset and completed the station-to-station circuit. Trunk calls were made with the assistance of other operators at other exchangers in the network.

In modern times, most telephones are plugged into telephone jacks. The jacks are connected by inside wiring to a drop wire which connects the building to a cable. Cables usually bring a large number of drop wires from all over a district access network to one wire center or telephone exchange. When a telephone user wants to make a telephone call, equipment at the exchange examines the dialed telephone number and connects that telephone line to another in the same wire center, or to a trunk to a distant exchange. Most of the exchanges in the world are interconnected through a system of larger switching systems, forming the public switched telephone network (PSTN).

After the middle of the 20th century, fax and data became important secondary users of the network created to carry voices, and late in the century, parts of the network were upgraded with ISDN and DSL to improve handling of such traffic.

Today, telephony uses digital technology (digital telephony) in the provisioning of telephone services and systems. Telephone calls can be provided digitally, but may be restricted to cases in which the last mile is digital, or where the conversion between digital and analog signals takes place inside the telephone. This advancement has reduced costs in communication, and improved the quality of voice services. The first implementation of this, ISDN, permitted all data transport from end-to-end speedily over telephone lines. This service was later made much less important due to the ability to provide digital services based on the IP protocol.

Since the advent of personal computer technology in the 1980s, computer telephony integration (CTI) has progressively provided more sophisticated telephony services, initiated and controlled by the computer, such as making and receiving voice, fax, and data calls with telephone directory services and caller identification. The integration of telephony software and computer systems is a major development in the evolution of the automated office. The term is used in describing the computerized services of call centers, such as those that direct your phone call to the right department at a business you're calling. It's also sometimes used to describe the ability to use your personal computer to initiate and manage phone calls (in which case you can think of your computer as your personal call center). CTI is not a new concept and has been used in the past in large telephone networks, but only dedicated call centers could justify the costs of the required equipment installation. Primary telephone service providers are offering information services such as automatic number identification, which is a telephone service architecture that separates CTI services from call switching and will make it easier to add new services. Dialed Number Identification Service (DNIS) on a scale is wide enough for its implementation to bring real value to business or residential telephone usage. A new generation of applications (middleware) is being developed as a result of standardization and availability of low cost computer telephony links.

Recent Developments

The term's scope has been broadened with the advent of the different new communication technologies. In its broadest sense, the terms encompasses phone communication, Internet calling, mobile communication, faxing, voicemail and video conferencing. Telephony's initial idea returns to POTS, (an acronym for "plain old telephone service") technically called the PSTN (public-switched telephone network).

This system is being fiercely challenged by and to a great extent yielding to Voice over IP (VoIP) technology, which is also commonly referred to as IP Telephony and Internet Telephony. IP telephony is a modern form of telephony which uses the TCP/IP protocol popularized by the Internet to transmit digitized voice data. Also, unlike traditional phone service, IP telephony service is relatively unregulated by government. In the United States, the Federal Communications Commission (FCC) regulates phone-to-phone connections, but says they do not plan to regulate connections between a phone user and an IP telephony service provider.Using the Internet, calls travel as packets of data on shared lines, avoiding the tolls of the PSTN. The challenge in IP telephony is to deliver the voice, fax, or video packets in a dependable flow to the user. Much of IP telephony focuses on that challenge.

Digital Telephony

Starting with the introduction of the transistor, invented in 1947 by Bell Laboratories, to amplification and switching circuits in the 1950s, and through development of computer-based electronic switching systems, the public switched telephone network (PSTN) has gradually evolved towards automation and digitization of signaling and audio transmissions.

Digital telephony is the use of digital electronics in the operation and provisioning of telephony systems and services. Since the 1960s a digital core network has replaced the traditional analog transmission and signaling systems, and much of the access network has also been digitized.

Digital telephony has dramatically improved the capacity, quality, and cost of the network. End-to-end analog telephone networks were first modified in the early 1960s by upgrading transmission networks with Digital Signal 1 (DS1/T1) carrier systems, designed to support the basic 3 kHz voice channel by sampling the bandwidth-limited analog voice signal and encoding using PCM. While digitization allows wideband voice on the same channel, the improved quality of a wider analog voice channel did not find a large market in the PSTN.

Later transmission methods such as SONET and fiber optic transmission further advanced digital transmission. Although analog carrier systems existed that multiplexed multiple analog voice channels onto a single transmission medium, digital transmission allowed lower cost and more channels multiplexed on the transmission medium. Today the end instrument often remains analog but the analog signals are typically converted to digital signals at the serving area interface (SAI), central office (CO), or other aggregation point. Digital loop carriers (DLC) place the digital network ever closer to the customer premises, relegating the analog local loop to legacy status.

Milestones in Digital Telephony

- early experiments with pulse code modulation in telephony

- the 8-bit, 8 kHz standard is developed; Nyquist's theorem and the standard 3.5 kHz telephony bandwidth

- DS0 as the basic digital telephony bitstream standard

- non-linear quantization: A-law vs. μ-law, and transcoding between the two

- bit error rate and intelligibility

- first practical digital telephone systems put into service

- the U.S. T-carrier system and the European E-carrier system developed to carry digital telephony

- introduction of space-time switching in fully digital electronic switching systems

- replacement of tone signaling with digital signaling for trunks

- in-band signaling vs. out-of-band signaling

- the problem of bit-robbing

- development of SS7

- emergence of fiber optic networking allows greater reliability and call capacity

- transition from plesiochronous transmission to synchronous systems like SONET/SDH

- optical self-healing ring networks further increase reliability

- digital/optical systems revolutionize international long-distance networks, particularly undersea cables

- digital telephone exchanges eliminate moving parts, make exchange equipment much smaller and more reliable

- separation of exchange and concentrator functions

- roll-out of digital systems throughout the PSTN

- provision of intelligent network services

- digital speech coding and compression

- speech compression on international digital trunks

- phone tapping in the digital environment

- introduction of digital mobile telephony, specialized compression algorithms for high bit error rates

- direct digital termination to customers via ISDN; PRI catches on, BRI mostly does not, except in Germany

- the effects of digital telephony, and digital termination at the ISP, on modem performance

- voice over IP as a carrier strategy

- emergence of ADSL leads to voice over IP becoming a consumer product, and the slow demise of dial-up Internet access

- expected convergence of VoIP, mobile telephony, etc.

- flattening of telephony tariffs, increasing moves towards flat rate pricing as the marginal cost of telephony drops further and further.

IP Telephony

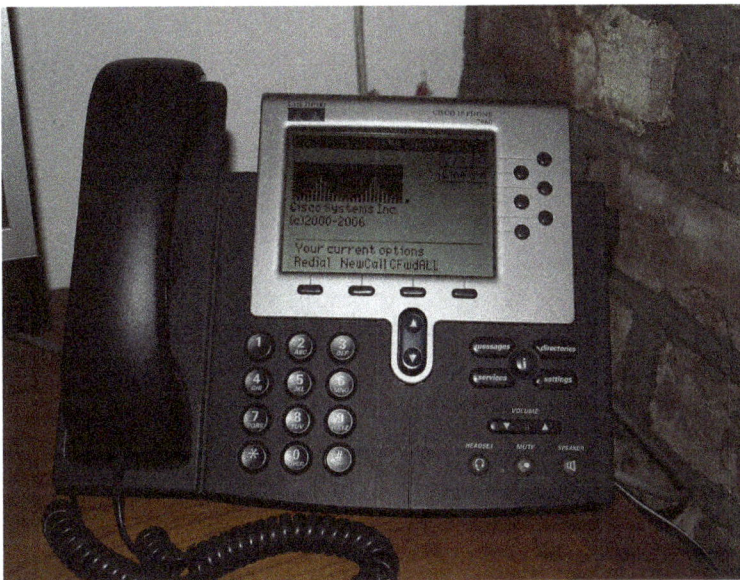

A commercial IP telephone, with keypad, control keys, and screen functions
to perform configuration and user features.

A specialization of digital telephony, Internet Protocol (IP) telephony involves the application of digital networking technology that was the foundation to the Internet to create, transmit, and receive telecommunications sessions over computer networks. Internet telephony is commonly known as voice over Internet Protocol (VoIP), reflecting the principle, but it has been referred with many other terms. VoIP has proven to be a disruptive technology that is rapidly replacing traditional telephone infrastructure technologies. As of January 2005, up to 10% of telephone subscribers in Japan and South Korea have switched to this digital telephone service. A January 2005 *Newsweek* article suggested that Internet telephony may be "the next big thing". As of 2006, many VoIP companies offer service to consumers and businesses.

IP telephony uses an Internet connection and hardware IP phones, analog telephone adapters, or softphone computer applications to transmit conversations encoded as data packets. In addition to replacing plain old telephone service (POTS), IP telephony services compete with mobile phone services by offering free or lower cost connections via WiFi hotspots. VoIP is also used on private networks which may or may not have a connection to the global telephone network.

Fixed telephone lines per 100 inhabitants 1997-2007 (ITU)

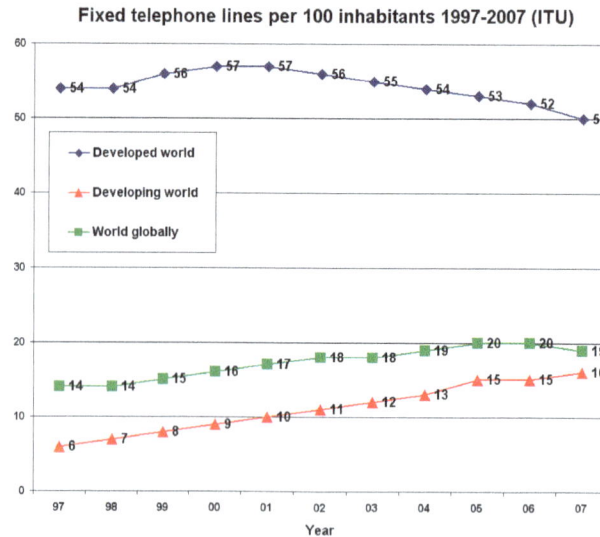

Fixed telephone lines per 100 inhabitants 1997–2007

Social Impact Research

Direct person-to-person communication includes non-verbal cues expressed in facial and other bodily articulation, that cannot be transmitted in traditional voice telephony. Video telephony restores such interactions to varying degrees. Social Context Cues Theory is a model to measure the success of different types of communication in maintaining the non-verbal cues present in face-to-face interactions. The research examines many different cues, such as the physical context, different facial expressions, body movements, tone of voice, touch and smell.

Various communication cues are lost with the usage of the telephone. The communicating parties are not able to identify the body movements, and lack touch and smell. Although this diminished ability to identify social cues is well known, Wiesenfeld, Raghuram, and Garud point out that there is a value and efficiency to the type of communication for different tasks. They examine work places in which different types of communication, such as the telephone, are more useful than face-to-face interaction.

The expansion of communication to mobile telephone service has created a different filter of the social cues than the land-line telephone. The use of instant messaging, such as *texting*, on mobile telephones has created a sense of community. In *The Social Construction of Mobile Telephony* it is suggested that each phone call and text message is more than an attempt to converse. Instead, it is a gesture which maintains the social network between family and friends. Although there is a loss of certain social cues through telephones, mobile phones bring new forms of expression of different cues that are understood by different audiences. New language additives attempt to compensate for the inherent lack of non-physical interaction.

Another social theory supported through telephony is the Media Dependency Theory. This theory concludes that people use media or a resource to attain certain goals. This theory states that there is a link between the media, audience, and the large social system. Telephones, depending on the person, help attain certain goals like accessing information, keeping in contact with others, sending quick communication, entertainment, etc.

Mobile Telephony

Mobile telephony is the provision of telephone services to phones which may move around freely rather than stay fixed in one location. Mobile phones connect to a terrestrial cellular network of base stations (cell sites), whereas satellite phones connect to orbiting satellites. Both networks are interconnected to the public switched telephone network (PSTN) to allow any phone in the world to be dialed.

Mobile phone tower

In 2010 there were estimated to be five billion mobile cellular subscriptions in the world.

History

According to internal memos, American Telephone & Telegraph discussed developing a wireless phone in 1915, but were afraid that deployment of the technology could undermine its monopoly on wired service in the U.S.

Public mobile phone systems were first introduced in the years after the Second World War and made use of technology developed before and during the conflict. The first system opened in St Louis, Missouri, USA in 1946 whilst other countries followed in the succeeding decades. The UK introduced its 'System 1' manual radiotelephone service as the South Lancashire Radiophone Service in 1958. Calls were made via an operator using handsets identical to ordinary phone handsets. The phone itself was a large box located in the boot (trunk) of the vehicle containing valves and other early electronic components. Although an uprated manual service ('System 3') was extended to cover most of the UK, automation did not arrive until 1981 with 'System 4'. Although this non-cellular service, based on German B-Netz technology, was expanded rapidly throughout the UK between 1982 and 1985 and continued in operation for several years before finally closing in Scotland, it was overtaken by the introduction in January 1985 of two cellular systems - the British Telecom/Securicor 'Cellnet' service and the Racal/Millicom/Barclays 'Vodafone' (from voice + data + phone) service. These cellular systems were based on US Advanced Mobile Phone Service

(AMPS) technology, the modified technology being named Total Access Communication System (TACS).

In 1947 Bell Labs was the first to propose a cellular radio telephone network. The primary innovation was the development of a network of small overlapping cell sites supported by a call switching infrastructure that tracks users as they move through a network and passes their calls from one site to another without dropping the connection. In 1956 the MTA system was launched in Sweden. The early efforts to develop mobile telephony faced two significant challenges: allowing a great number of callers to use the comparatively few available frequencies simultaneously and allowing users to seamlessly move from one area to another without having their calls dropped. Both problems were solved by Bell Labs employee Amos Joel who, in 1970 applied for a patent for a mobile communications system. However, a business consulting firm calculated the entire U.S. market for mobile telephones at 100,000 units and the entire worldwide market at no more than 200,000 units based on the ready availability of pay telephones and the high cost of constructing cell towers. As a consequence, Bell Labs concluded that the invention was "of little or no consequence," leading it not to attempt to commercialize the invention. The invention earned Joel induction into the National Inventors Hall of Fame in 2008. The first call on a handheld mobile phone was made on April 3, 1973 by Martin Cooper, then of Motorola to his opposite number in Bell Labs who were also racing to be first. Bell Labs went on to install the first trial cellular network in Chicago in 1978. This trial system was licensed by the FCC to ATT for commercial use in 1982 and, as part of the divestiture arrangements for the breakup of ATT, the AMPS technology was distributed to local telcos. The first commercial system opened in Chicago in October 1983. A system designed by Motorola also operated in the Washington D.C./Baltimore area from summer 1982 and became a full public service later the following year. Japan's first commercial radiotelephony service was launched by NTT in 1978.

The first fully automatic first generation cellular system was the Nordic Mobile Telephone (NMT) system, simultaneously launched in 1981 in Denmark, Finland, Norway and Sweden. NMT was the first mobile phone network featuring international roaming. The Swedish electrical engineer Östen Mäkitalo started to work on this vision in 1966, and is considered as the father of the NMT system and some consider him also the father of the cellular phone.

The advent of cellular technology encouraged European countries to co-operate in the development of a pan-European cellular technology to rival those of the US and Japan. This resulted in the GSM system, the initials originally from the *Groupe Spécial Mobile* that was charged with the specification and development tasks but latterly as the 'Global System for Mobile Communications'. The GSM standard eventually spread outside Europe and is now the most widely used cellular technology in the world and the de facto standard. The industry association, the GSMA, now represents 219 countries and nearly 800 mobile network operators. There are now estimated to be over 5 billion phone subscriptions according to the "List of countries by number of mobile phones in use" (although some users have multiple subscriptions, or inactive subscriptions), which also makes the mobile phone the most widely spread technology and the most common electronic device in the world.

The first mobile phone to enable internet connectivity and wireless email, the Nokia Communicator, was released in 1996, creating a new category of multi-use devices called smartphones. In 1999 the first mobile internet service was launched by NTT DoCoMo in Japan under the i-Mode service.

By 2007 over 798 million people around the world accessed the internet or equivalent mobile internet services such as WAP and i-Mode at least occasionally using a mobile phone rather than a personal computer.

Cellular Systems

Mobile phone subscribers per 100 inhabitants 1997-2007

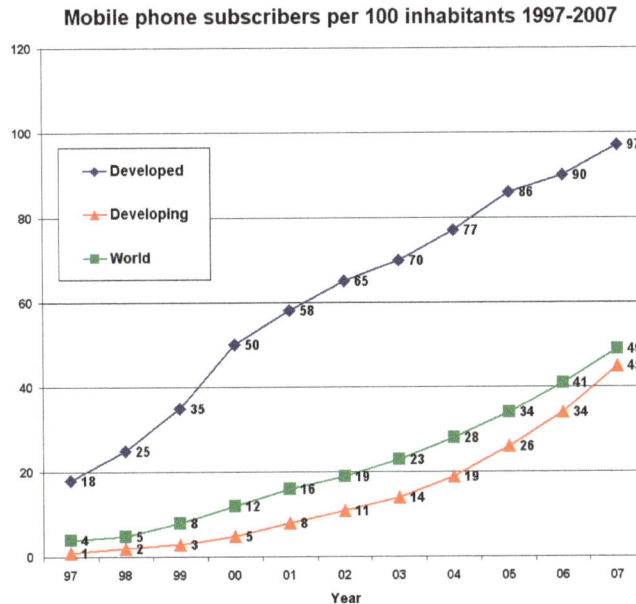

Mobile phone subscriptions, not subscribers, per 100 inhabitants 1997-2007

Mobile phones send and receive radio signals with any number of cell site base stations fitted with microwave antennas. These sites are usually mounted on a tower, pole or building, located throughout populated areas, then connected to a cabled communication network and switching system. The phones have a low-power transceiver that transmits voice and data to the nearest cell sites, normally not more than 8 to 13 km (approximately 5 to 8 miles) away. In areas of low coverage, a cellular repeater may be used, which uses a long distance high-gain dish antenna or yagi antenna to communicate with a cell tower far outside of normal range, and a repeater to rebroadcast on a small short-range local antenna that allows any cellphone within a few meters to function properly.

When the mobile phone or data device is turned on, it registers with the mobile telephone exchange, or switch, with its unique identifiers, and can then be alerted by the mobile switch when there is an incoming telephone call. The handset constantly listens for the strongest signal being received from the surrounding base stations, and is able to switch seamlessly between sites. As the user moves around the network, the "handoffs" are performed to allow the device to switch sites without interrupting the call.

Cell sites have relatively low-power (often only one or two watts) radio transmitters which broadcast their presence and relay communications between the mobile handsets and the switch. The switch in turn connects the call to another subscriber of the same wireless service provider or to the public telephone network, which includes the networks of other wireless carriers. Many of these sites are camouflaged to blend with existing environments, particularly in scenic areas.

The dialogue between the handset and the cell site is a stream of digital data that includes digitised audio (except for the first generation analog networks). The technology that achieves this depends on the system which the mobile phone operator has adopted. The technologies are grouped by generation. The first-generation systems started in 1979 with Japan, are all analog and include AMPS and NMT. Second-generation systems, started in 1991 in Finland, are all digital and include GSM, CDMA and TDMA.

The nature of cellular technology renders many phones vulnerable to 'cloning': anytime a cell phone moves out of coverage (for example, in a road tunnel), when the signal is re-established, the phone sends out a 're-connect' signal to the nearest cell-tower, identifying itself and signalling that it is again ready to transmit. With the proper equipment, it's possible to intercept the re-connect signal and encode the data it contains into a 'blank' phone—in all respects, the 'blank' is then an exact duplicate of the real phone and any calls made on the 'clone' will be charged to the original account. This problem was widespread with the first generation analogue technology, however the modern digital standards such as GSM greatly improve security and make cloning hard to achieve.

In an effort to limit the potential harm from having a transmitter close to the user's body, the first fixed/mobile cellular phones that had a separate transmitter, vehicle-mounted antenna, and handset (known as *car phones* and *bag phones*) were limited to a maximum 3 watts Effective Radiated Power. Modern *handheld* cellphones which must have the transmission antenna held inches from the user's skull are limited to a maximum transmission power of 0.6 watts ERP. Regardless of the potential biological effects, the reduced transmission range of modern handheld phones limits their usefulness in rural locations as compared to car/bag phones, and handhelds require that cell towers are spaced much closer together to compensate for their lack of transmission power.

Usage

By Civilians

This Railfone found on some Amtrak trains in North America uses cellular technology.

An increasing number of countries, particularly in Europe, now have more mobile phones than people. According to the figures from Eurostat, the European Union's in-house statistical office, Luxembourg had the highest mobile phone penetration rate at 158 mobile subscriptions per 100 people, closely followed by Lithuania and Italy. In Hong Kong the penetration rate reached 139.8% of the population in July 2007. Over 50 countries have mobile phone subscription penetration rates higher than that of the population and the Western European average penetration rate was 110% in 2007 (source Informa 2007). Canada currently has the lowest rates of mobile phone penetrations in the industrialised world at 58%.

There are over five hundred million active mobile phone accounts in China, as of 2007, but the total penetration rate there still stands below 50%. The total number of mobile phone subscribers in the world was estimated at 2.14 billion in 2005. The subscriber count reached 2.7 billion by end of 2006 according to Informa, and 3.3 billion by November, 2007, thus reaching an equivalent of over half the planet's population. Around 80% of the world's population has access to mobile phone coverage, as of 2006. This figure is expected to increase to 90% by the year 2010.

In some developing countries with little "landline" telephone infrastructure, mobile phone use has quadrupled in the last decade. The rise of mobile phone technology in developing countries is often cited as an example of the leapfrog effect. Many remote regions in the third world went from having no telecommunications infrastructure to having satellite based communications systems. At present, Africa has the largest growth rate of cellular subscribers in the world, its markets expanding nearly twice as fast as Asian markets. The availability of prepaid or 'pay-as-you-go' services, where the subscriber is not committed to a long term contract, has helped fuel this growth in Africa as well as in other continents.

On a numerical basis, India is the largest growth market, adding about 6 million mobile phones every month. It currently has a mobile subscriber base of 937.06 million mobile phones.

Traffic

Since the world is operating quickly to 3G and 4G networks, mobile traffic through video is heading high. It is expected that by end of 2018, the global traffic will reach an annual rate of 190 exabytes/ year. This is the result of people shifting to smart phones now-a-days. It is predicted by 2018, mobile traffic will reach by 10 billion connections with 94% traffic comes from Smartphones, laptops and tablets. Also 69% of mobile traffic from Videos since we have high definition screens available in smart phones and 176.9 wearable devices to be at use. Apparently, 4G will be dominating the traffic by 51% of total mobile data by 2018.

By Government Agencies

Law Enforcement

Law enforcement have used mobile phone evidence in a number of different ways. Evidence about the physical location of an individual at a given time can be obtained by triangulating the individual's cellphone between several cellphone towers. This triangulation technique can be used to show that an individual's cellphone was at a certain location at a certain time. The concerns over terrorism and terrorist use of technology prompted an inquiry by the British House of Commons

Home Affairs Select Committee into the use of evidence from mobile phone devices, prompting leading mobile telephone forensic specialists to identify forensic techniques available in this area. NIST have published guidelines and procedures for the preservation, acquisition, examination, analysis, and reporting of digital information present on mobile phones can be found under the NIST Publication SP800-101.

In the UK in 2000 it was claimed that recordings of mobile phone conversations made on the day of the Omagh bombing were crucial to the police investigation. In particular, calls made on two mobile phones which were tracked from south of the Irish border to Omagh and back on the day of the bombing, were considered of vital importance.

Further example of criminal investigations using mobile phones is the initial location and ultimate identification of the terrorists of the 2004 Madrid train bombings. In the attacks, mobile phones had been used to detonate the bombs. However, one of the bombs failed to detonate, and the SIM card in the corresponding mobile phone gave the first serious lead about the terrorists to investigators. By tracking the whereabouts of the SIM card and correlating other mobile phones that had been registered in those areas, police were able to locate the terrorists.

Disaster Response

The Finnish government decided in 2005 that the fastest way to warn citizens of disasters was the mobile phone network. In Japan, mobile phone companies provide immediate notification of earthquakes and other natural disasters to their customers free of charge. In the event of an emergency, disaster response crews can locate trapped or injured people using the signals from their mobile phones. An interactive menu accessible through the phone's Internet browser notifies the company if the user is safe or in distress. In Finland rescue services suggest hikers carry mobile phones in case of emergency even when deep in the forests beyond cellular coverage, as the radio signal of a cellphone attempting to connect to a base station can be detected by overflying rescue aircraft with special detection gear. Also, users in the United States can sign up through their provider for free text messages when an AMBER Alert goes out for a missing person in their area.

However, most mobile phone networks operate close to capacity during normal times, and spikes in call volumes caused by widespread emergencies often overload the system just when it is needed the most. Examples reported in the media where this has occurred include the September 11, 2001 attacks, the 2003 Northeast blackouts, the 2005 London Tube bombings, Hurricane Katrina, the 2006 Hawaii earthquake, and the 2007 Minnesota bridge collapse.

Under FCC regulations, all mobile telephones must be capable of dialing emergency telephone numbers, regardless of the presence of a SIM card or the payment status of the account.

Impact on Society

Human Health

Since the introduction of mobile phones, concerns (both scientific and public) have been raised about the potential health impacts from regular use. But by 2008, American mobile phones transmitted and received more text messages than phone calls. Numerous studies have reported no significant relationship between mobile phone use and health, but the effect of mobile phone usage on health continues to be an area of public concern.

For example, at the request of some of their customers, Verizon created usage controls that meter service and can switch phones off, so that children could get some sleep. There have also been attempts to limit use by persons operating moving trains or automobiles, coaches when writing to potential players on their teams, and movie theater audiences. By one measure, nearly 40% of automobile drivers aged 16 to 30 years old text while driving, and by another, 40% of teenagers said they could text blindfolded.

18 studies have been conducted on the link between cell phones and brain cancer; A review of these studies found that cell phone use of 10 years or more "give a consistent pattern of an increased risk for acoustic neuroma and glioma". The tumors are found mostly on the side of the head that the mobile phone is in contact with. In July 2008, Dr. Ronald Herberman, director of the University of Pittsburgh Cancer Institute, warned about the radiation from mobile phones. He stated that there was no definitive proof of the link between mobile phones and brain tumors but there was enough studies that mobile phone usage should be reduced as a precaution. To reduce the amount of radiation being absorbed hands free devices can be used or texting could supplement calls. Calls could also be shortened or limit mobile phone usage in rural areas. Radiation is found to be higher in areas that are located away from mobile phone towers.

According to Reuters, The British Association of Dermatologists is warning of a rash occurring on people's ears or cheeks caused by an allergic reaction from the nickel surface commonly found on mobile devices' exteriors. There is also a theory it could even occur on the fingers if someone spends a lot of time text messaging on metal menu buttons. In 2008, Lionel Bercovitch of Brown University in Providence, Rhode Island, and his colleagues tested 22 popular handsets from eight different manufacturers and found nickel on 10 of the devices.

Human Behaviour

Culture and Customs

Cellular phones allow people to communicate from almost anywhere at their leisure.

Between the 1980s and the 2000s, the mobile phone has gone from being an expensive item used by the business elite to a pervasive, personal communications tool for the general population. In most countries, mobile phones outnumber land-line phones, with fixed landlines numbering 1.3 billion but mobile subscriptions 3.3 billion at the end of 2007.

In many markets from Japan and South Korea, to Europe, to Malaysia, Singapore, Taiwan and Hong Kong, most children age 8-9 have mobile phones and the new accounts are now opened for customers aged 6 and 7. Where mostly parents tend to give hand-me-down used phones to their youngest children, in Japan already new cameraphones are on the market whose target age group is under 10 years of age, introduced by KDDI in February 2007. The USA also lags on this measure, as in the US so far, about half of all children have mobile phones. In many young adults' households it has supplanted the land-line phone. Mobile phone usage is banned in some countries, such as North Korea and restricted in some other countries such as Burma.

Given the high levels of societal mobile phone service penetration, it is a key means for people to communicate with each other. The SMS feature spawned the "texting" sub-culture amongst younger users. In December 1993, the first person-to-person SMS text message was transmitted in Finland. Currently, texting is the most widely used data service; 1.8 billion users generated $80 billion of revenue in 2006 (source ITU). Many phones offer Instant Messenger services for simple, easy texting. Mobile phones have Internet service (e.g. NTT DoCoMo's i-mode), offering text messaging via e-mail in Japan, South Korea, China, and India. Most mobile internet access is much different from computer access, featuring alerts, weather data, e-mail, search engines, instant messages, and game and music downloading; most mobile internet access is hurried and short.

Because mobile phones are often used publicly, social norms have been shown to play a major role in the usage of mobile phones. Furthermore, the mobile phone can be a fashion totem custom-decorated to reflect the owner's personality and may be a part of their self-identity. This aspect of the mobile telephony business is, in itself, an industry, e.g. ringtone sales amounted to $3.5 billion in 2005. Mobile phone use on aircraft is starting to be allowed with several airlines already offering the ability to use phones during flights. Mobile phone use during flights used to be prohibited and many airlines still claim in their in-plane announcements that this prohibition is due to possible interference with aircraft radio communications. Shut-off mobile phones do not interfere with aircraft avionics. The recommendation why phones should not be used during take-off and landing, even on planes that allow calls or messaging, is so that passengers pay attention to the crew for any possible accident situations, as most aircraft accidents happen on take-off and landing.

Etiquette

Mobile phone use can be an important matter of social discourtesy: phones ringing during funerals or weddings; in toilets, cinemas and theatres. Some book shops, libraries, bathrooms, cinemas, doctors' offices and places of worship prohibit their use, so that other patrons will not be disturbed by conversations. Some facilities install signal-jamming equipment to prevent their use, although in many countries, including the US, such equipment is illegal.

Many US cities with subway transit systems underground are studying or have implemented mobile phone reception in their underground tunnels for their riders, and trains, particularly those involving long-distance services, often offer a "quiet carriage" where phone use is prohibited, much like the designated non-smoking carriage of the past. Most schools in the United States and Europe and Canada have prohibited mobile phones in the classroom, or in school in an effort to limit class disruptions.

A working group made up of Finnish telephone companies, public transport operators and communications authorities has launched a campaign to remind mobile phone users of courtesy, especially when using mass transit—what to talk about on the phone, and how to. In particular, the campaign wants to impact loud mobile phone usage as well as calls regarding sensitive matters.

Use by Drivers

The use of mobile phones by people who are driving has become increasingly common, for example as part of their job, as in the case of delivery drivers who are calling a client, or socially as for commuters who are chatting with a friend. While many drivers have embraced the convenience of using their cellphone while driving, some jurisdictions have made the practice against the law, such as Australia, the Canadian provinces of British Columbia, Quebec, Ontario, Nova Scotia, and Newfoundland and Labrador as well as the United Kingdom, consisting of a zero-tolerance system operated in Scotland and a warning system operated in England, Wales, and Northern Ireland. Officials from these jurisdictions argue that using a mobile phone while driving is an impediment to vehicle operation that can increase the risk of road traffic accidents.

Studies have found vastly different relative risks (RR). Two separate studies using case-crossover analysis each calculated RR at 4, while an epidemiological cohort study found RR, when adjusted for crash-risk exposure, of 1.11 for men and 1.21 for women.

A simulation study from the University of Utah Professor David Strayer compared drivers with a blood alcohol content of 0.08% to those conversing on a cell phone, and after controlling for driving difficulty and time on task, the study concluded that cell phone drivers exhibited greater impairment than intoxicated drivers. Meta-analysis by The Canadian Automobile Association and The University of Illinois found that response time while using both hands-free and handheld phones was approximately 0.5 standard deviations higher than normal driving (i.e., an average driver, while talking on a cell phone, has response times of a driver in roughly the 40th percentile).

Driving while using a hands-free device is not safer than driving while using a hand-held phone, as concluded by case-crossover studies. epidemiological studies, simulation studies, and meta-analysis. Even with this information, California initiated new Wireless Communications Device Law (effective January 1, 2009) makes it an infraction to write, send, or read text-based communication on an electronic wireless communications device, such as a cell phone, while driving a motor vehicle. Two additional laws dealing with the use of wireless telephones while driving went into effect July 1, 2008. The first law prohibits all drivers from using a handheld wireless telephone while operating a motor vehicle. The law allows a driver to use a wireless telephone to make emergency calls to a law enforcement agency, a medical provider, the fire department, or other emergency services agency. The base fine for the FIRST offense is $20 and $50 for subsequent convictions. With penalty assessments, the fine can be more than triple the base fine amount. videos about California cellular phone laws; with captions (California Vehicle Code [VC] §23123). Motorists 18 and over may use a "hands-free device. The second law effective July 1, 2008, prohibits drivers under the age of 18 from using a wireless telephone or hands-free device while operating a motor vehicle (VC §23124)The consistency of increased crash risk between hands-free and hand-held phone use is at odds with legislation in over 30 countries that prohibit hand-held phone use but allow hands-free. Scientific literature is mixed on the dangers of talking on a phone versus those of talking with a passenger, with the Accident Research Unit at the University of Nottingham finding that the number of utterances was usually higher for mobile calls when compared to blindfolded and non-blindfolded passengers, but the University of Illinois meta-analysis concluding that passenger conversations were just as costly to driving performance as cell phone ones.

Use on Aircraft

As of 2007, several airlines are experimenting with base station and antenna systems installed on the airplane, allowing low power, short-range connection of any phones aboard to remain connected to the aircraft's base station. Thus, they would not attempt connection to the ground base stations as during take off and landing. Simultaneously, airlines may offer phone services to their travelling passengers either as full voice and data services, or initially only as SMS text messaging and similar services. The Australian airline Qantas is the first airline to run a test aeroplane in this configuration in the autumn of 2007. Emirates has announced plans to allow limited mobile phone usage on some flights. However, in the past, commercial airlines have prevented the use of cell phones and laptops, due to the assertion that the frequencies emitted from these devices may disturb the radio waves contact of the airplane.

On March 20, 2008, an Emirates flight was the first time voice calls have been allowed in-flight on commercial airline flights. The breakthrough came after the European Aviation Safety Agency (EASA) and the United Arab Emirates-based General Civil Aviation Authority (GCAA) granted full approval for the AeroMobile system to be used on Emirates. Passengers were able to make and receive voice calls as well as use text messaging. The system automatically came into operation as the Airbus A340-300 reached cruise altitude. Passengers wanting to use the service received a text message welcoming them to the AeroMobile system when they first switched their phones on. The approval by EASA has established that GSM phones are safe to use on airplanes, as the AeroMobile system does not require the modification of aircraft components deemed "sensitive," nor does it require the use of modified phones.

In any case, there are inconsistencies between practices allowed by different airlines and even on the same airline in different countries. For example, Delta Air Lines may allow the use of mobile phones immediately after landing on a domestic flight within the US, whereas they may state "not until the doors are open" on an international flight arriving in the Netherlands. In April 2007 the US Federal Communications Commission officially prohibited passengers' use of cell phones during a flight.

In a similar vein, signs are put up in many countries, such as Canada, the UK and the U.S., at petrol stations prohibiting the use of mobile phones, due to possible safety issues. However, it is unlikely that mobile phone use can cause any problems, and in fact "petrol station employees have themselves spread the rumour about alleged incidents."

Environmental Impacts

Cellular antenna disguised to look like a tree

Like all high structures, cellular antenna masts pose a hazard to low flying aircraft. Towers over a certain height or towers that are close to airports or heliports are normally required to have warning lights. There have been reports that warning lights on cellular masts, TV-towers and other high structures can attract and confuse birds. US authorities estimate that millions of birds are killed near communication towers in the country each year.

Some cellular antenna towers have been camouflaged to make them less obvious on the horizon, and make them look more like a tree.

An example of the way mobile phones and mobile networks have sometimes been perceived as a threat is the widely reported and later discredited claim that mobile phone masts are associated with the Colony Collapse Disorder (CCD) which has reduced bee hive numbers by up to 75% in many areas, especially near cities in the US. The Independent newspaper cited a scientific study claiming it provided evidence for the theory that mobile phone masts *are* a major cause in the collapse of bee populations, with controlled experiments demonstrating a rapid and catastrophic effect on individual hives near masts. Mobile phones were in fact not covered in the study, and the original researchers have since emphatically disavowed any connection between their research, mobile phones, and CCD, specifically indicating that the Independent article had misinterpreted their results and created "a horror story". While the initial claim of damage to bees was widely reported, the corrections to the story were almost non-existent in the media.

There are more than 500 million used mobile phones in the US sitting on shelves or in landfills, and it is estimated that over 125 million will be discarded this year alone. The problem is growing at a rate of more than two million phones per week, putting tons of toxic waste into landfills daily. Several companies offer to buy back and recycle mobile phones from users. In the United States many unwanted but working mobile phones are donated to women's shelters to allow emergency communication.

Tariff Models

Mobile phone shop in Uganda

Payment Methods

There are two principal ways to pay for mobile telephony: the 'pay-as-you-go' model where conversation time is purchased and added to a phone unit via an Internet account or in shops or ATMs, or the contract model where bills are paid by regular intervals after the service has been consumed. It is increasingly common for a consumer to purchase a basic package and then bolt-on services and functionality to create a subscription customised to the users needs.

Pay as you go (also known as "pre-pay" or "prepaid") accounts were invented simultaneously in Portugal and Italy and today form more than half of all mobile phone subscriptions. USA, Canada, Costa Rica, Japan, Israel and Finland are among the rare countries left where most phones are still contract-based.

Incoming Call Charges

In the early days of mobile telephony, the operators (carriers) charged for all air time consumed by the mobile phone user, which included both outbound and inbound telephone calls. As mobile phone adoption rates increased, competition between operators meant that some decided not to charge for incoming calls in some markets (also called "calling party pays").

The European market adopted a calling party pays model throughout the GSM environment and soon various other GSM markets also started to emulate this model.

In Hong Kong, Singapore, Canada, and the United States, it is common for the party receiving the call to be charged per minute, although a few carriers are beginning to offer unlimited received phone calls. This is called the "Receiving Party Pays" model. In China, it was reported that both of its two operators will adopt the caller-pays approach as early as January 2007.

One disadvantage of the receiving party pays systems is that phone owners keep their phones turned off to avoid receiving unwanted calls, which results in the total voice usage rates (and profits) in Calling Party Pays countries outperform those in Receiving Party Pays countries. To avoid the problem of users keeping their phone turned off, most Receiving Party Pays countries have either switched to Calling Party Pays, or their carriers offer additional incentives such as a large number of monthly minutes at a sufficiently discounted rate to compensate for the inconvenience.

Note that when a user roaming in another country, international roaming tariffs apply to all calls received, regardless of the model adopted in the home country.

Instant Messaging

A buddy list in Pidgin 2.0

Instant messaging (IM) is a type of online chat that offers real-time text transmission over the Internet. A LAN messenger operates in a similar way over a local area network. Short messages are typically transmitted bi-directionally between two parties, when each user chooses to complete a thought and select "send". Some IM applications can use push technology to provide real-time text, which transmits messages character by character, as they are composed. More advanced instant messaging can add file transfer, clickable hyperlinks, Voice over IP, or video chat.

Non-IM types of chat include multicast transmission, usually referred to as "chat rooms", where participants might be anonymous or might be previously known to each other (for example collaborators on a project that is using chat to facilitate communication). Instant messaging systems tend to facilitate connections between specified known users (often using a contact list also known as a "buddy list" or "friend list"). Depending on the IM protocol, the technical architecture can be peer-to-peer (direct point-to-point transmission) or client-server (a central server retransmits messages from the sender to the communication device).

Overview

Instant messaging is a set of communication technologies used for text-based communication between two or more participants over the Internet or other types of networks. IM–chat happens in real-time. Of importance is that online chat and instant messaging differ from other technologies such as email due to the perceived quasi-synchrony of the communications by the users. Some systems permit messages to be sent to users not then 'logged on' (*offline messages*), thus removing some differences between IM and email (often done by sending the message to the associated email account).

IM allows effective and efficient communication, allowing immediate receipt of acknowledgment or reply. However IM is basically not necessarily supported by transaction control. In many cases, instant messaging includes added features which can make it even more popular. For example, users may see each other via webcams, or talk directly for free over the Internet using a microphone and headphones or loudspeakers. Many applications allow file transfers, although they are usually limited in the permissible file-size.

It is usually possible to save a text conversation for later reference. Instant messages are often logged in a local message history, making it similar to the persistent nature of emails.

History

Command-line Unix "talk", using a split screen user interface, was popular in the 1980s and early 1990s.

Though the term dates from the 1990s, instant messaging predates the Internet, first appearing on multi-user operating systems like Compatible Time-Sharing System (CTSS) and Multiplexed Information and Computing Service (Multics) in the mid-1960s. Initially, some of these systems were used as notification systems for services like printing, but quickly were used to facilitate communication with other users logged into the same machine. As networks developed, the protocols spread with the networks. Some of these used a peer-to-peer protocol (e.g. talk, ntalk and ytalk), while others required peers to connect to a server. The Zephyr Notification Service (still in use at some institutions) was invented at MIT's Project Athena in the 1980s to allow service providers to locate and send messages to users.

Parallel to instant messaging were early online chat facilities, the earliest of which was Talkomatic (1973) on the PLATO system. During the bulletin board system (BBS) phenomenon that peaked during the 1980s, some systems incorporated chat features which were similar to instant messaging; Freelancin' Roundtable was one prime example. The first such general-availability commercial online chat service (as opposed to PLATO, which was educational) was the CompuServe CB Simulator in 1980, created by CompuServe executive Alexander "Sandy" Trevor in Columbus, Ohio.

Early instant messaging programs were primarily real-time text, where characters appeared as they were typed. This includes the Unix "talk" command line program, which was popular in the 1980s and early 1990s. Some BBS chat programs (i.e. Celerity BBS) also used a similar interface. Modern implementations of real-time text also exist in instant messengers, such as AOL's Real-Time IM as an optional feature.

In the latter half of the 1980s and into the early 1990s, the Quantum Link online service for Commodore 64 computers offered user-to-user messages between concurrently connected customers, which they called "On-Line Messages" (or OLM for short), and later "FlashMail." (Quantum Link later became America Online and made AOL Instant Messenger (AIM), discussed later). While the Quantum Link client software ran on a Commodore 64, using only the Commodore's PETSCII text-graphics, the screen was visually divided into sections and OLMs would appear as a yellow bar saying "Message From:" and the name of the sender along with the message across the top of whatever the user was already doing, and presented a list of options for responding. As such, it could be considered a type of graphical user interface (GUI), albeit much more primitive than the later Unix, Windows and Macintosh based GUI IM software. OLMs were what Q-Link called "Plus Services" meaning they charged an extra per-minute fee on top of the monthly Q-Link access costs.

Modern, Internet-wide, GUI-based messaging clients as they are known today, began to take off in the mid-1990s with PowWow, ICQ, and AOL Instant Messenger. Similar functionality was offered by CU-SeeMe in 1992; though primarily an audio/video chat link, users could also send textual messages to each other. AOL later acquired Mirabilis, the authors of ICQ; a few years later ICQ (now owned by AOL) was awarded two patents for instant messaging by the U.S. patent office. Meanwhile, other companies developed their own software; (Excite, MSN, Ubique, and Yahoo!), each with its own proprietary protocol and client; users therefore had to run multiple client applications if they wished to use more than one of these networks. In 1998, IBM released IBM Lotus Sametime, a product based on technology acquired when IBM bought Haifa-based Ubique and Lexington-based Databeam.

In 2000, an open source application and open standards-based protocol called Jabber was launched. The protocol was standardized under the name Extensible Messaging and Presence Protocol (XMPP). XMPP servers could act as gateways to other IM protocols, reducing the need to run multiple clients. Multi-protocol clients can use any of the popular IM protocols by using additional local libraries for each protocol. IBM Lotus Sametime's November 2007 release added IBM Lotus Sametime Gateway support for XMPP.

As of 2010, social networking providers often offer IM abilities. Facebook Chat is a form of instant messaging, and Twitter can be thought of as a Web 2.0 instant messaging system. Similar server-side chat features are part of most dating websites, such as OKCupid or PlentyofFish. The spread of smartphones and similar devices in the late 2000s also caused increased competition with conventional instant messaging, by making text messaging services still more ubiquitous.

Many instant messaging services offer video calling features, voice over IP and web conferencing services. Web conferencing services can integrate both video calling and instant messaging abilities. Some instant messaging companies are also offering desktop sharing, IP radio, and IPTV to the voice and video features.

The term "Instant Messenger" is a service mark of Time Warner and may not be used in software not affiliated with AOL in the United States. For this reason, in April 2007, the instant messaging client formerly named Gaim (or gaim) announced that they would be renamed "Pidgin".

Clients

Each modern IM service generally provides its own client, either a separately installed piece of software, or a browser-based client. These usually only work with the supplier company's service, although some allow limited function with other services. Third party client software applications exist that will connect with most of the major IM services.

Interoperability

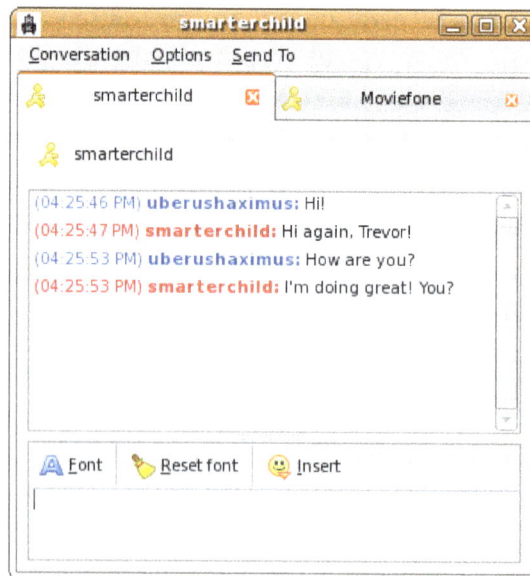

Pidgin's tabbed chat window in Linux

Standard complementary instant messaging applications offer functions like file transfer, contact list(s), the ability to hold several simultaneous conversations, etc. These may be all the functions that a small business needs, but larger organizations will require more sophisticated applications that can work together. The solution to finding applications capable of this is to use enterprise versions of instant messaging applications. These include titles like XMPP, Lotus Sametime, Microsoft Office Communicator, etc., which are often integrated with other enterprise applications such as workflow systems. These enterprise applications, or enterprise application integration (EAI), are built to certain constraints, namely storing data in a common format.

There have been several attempts to create a unified standard for instant messaging: IETF's Session Initiation Protocol (SIP) and SIP for Instant Messaging and Presence Leveraging Extensions (SIMPLE), Application Exchange (APEX), Instant Messaging and Presence Protocol (IMPP), the open XML-based Extensible Messaging and Presence Protocol (XMPP), and Open Mobile Alliance's Instant Messaging and Presence Service developed specifically for mobile devices.

Most attempts at producing a unified standard for the major IM providers (AOL, Yahoo! and Microsoft) have failed, and each continues to use its own proprietary protocol.

However, while discussions at IETF were stalled, Reuters signed the first inter-service provider connectivity agreement on September 2003. This agreement enabled AIM, ICQ and MSN Messenger users to talk with Reuters Messaging counterparts and vice versa. Following this, Microsoft, Yahoo! and AOL agreed to a deal in which Microsoft's Live Communications Server 2005 users would also have the possibility to talk to public instant messaging users. This deal established SIP/SIMPLE as a standard for protocol interoperability and established a connectivity fee for accessing public instant messaging groups or services. Separately, on October 13, 2005, Microsoft and Yahoo! announced that by the 3rd quarter of 2006 they would interoperate using SIP/SIMPLE, which was followed, in December 2005, by the AOL and Google strategic partnership deal in which Google Talk users would be able to communicate with AIM and ICQ users provided they have an AIM account.

There are two ways to combine the many disparate protocols:

- Combine the many disparate protocols inside the IM *client application.*

- Combine the many disparate protocols inside the IM *server* application. This approach moves the task of communicating with the other services to the server. Clients need not know or care about other IM protocols. For example, LCS 2005 Public IM Connectivity. This approach is popular in XMPP servers; however, the so-called transport projects suffer the same reverse engineering difficulties as any other project involved with closed protocols or formats.

Some approaches allow organizations to deploy their own, private instant messaging network by enabling them to restrict access to the server (often with the IM network entirely behind their firewall) and administer user permissions. Other corporate messaging systems allow registered users to also connect from outside the corporation LAN, by using an encrypted, firewall-friendly, HTTPS-based protocol. Usually, a dedicated corporate IM server has several advantages, such as pre-populated contact lists, integrated authentication, and better security and privacy.

Certain networks have made changes to prevent them from being used by such multi-network IM clients. For example, Trillian had to release several revisions and patches to allow its users to access the MSN, AOL, and Yahoo! networks, after changes were made to these networks. The major IM providers usually cite the need for formal agreements, and security concerns as reasons for making these changes.

The use of proprietary protocols has meant that many instant messaging networks have been incompatible and users have been unable to reach users on other networks. This may have allowed social networking with IM-like features and text messaging an opportunity to gain market share at the expense of IM.

IM Language

Users sometimes make use of internet slang or text speak to abbreviate common words or expressions to quicken conversations or reduce keystrokes. The language has become widespread, with well-known expressions such as 'lol' translated over to face-to-face language.

Emotions are often expressed in shorthand, such as the abbreviation LOL, BRB and TTYL; respectively laugh(ing) out loud, be right back, and talk to you later.

Some, however, attempt to be more accurate with emotional expression over IM. Real time reactions such as (*chortle*) (*snort*) (*guffaw*) or (*eye-roll*) are becoming more popular. Also there are certain standards that are being introduced into mainstream conversations including, '#' indicates the use of sarcasm in a statement and '*' which indicates a spelling mistake and/or grammatical error in the prior message, followed by a correction.

Business Application

Instant messaging has proven to be similar to personal computers, email, and the World Wide Web, in that its adoption for use as a business communications medium was driven primarily by individual employees using consumer software at work, rather than by formal mandate or provisioning by corporate information technology departments. Tens of millions of the consumer IM accounts in use are being used for business purposes by employees of companies and other organizations.

In response to the demand for business-grade IM and the need to ensure security and legal compliance, a new type of instant messaging, called "Enterprise Instant Messaging" ("EIM") was created when Lotus Software launched IBM Lotus Sametime in 1998. Microsoft followed suit shortly thereafter with Microsoft Exchange Instant Messaging, later created a new platform called Microsoft Office Live Communications Server, and released Office Communications Server 2007 in October 2007. Oracle Corporation has also jumped into the market recently with its Oracle Beehive unified collaboration software. Both IBM Lotus and Microsoft have introduced federation between their EIM systems and some of the public IM networks so that employees may use one interface to both their internal EIM system and their contacts on AOL, MSN, and Yahoo. As of 2010, leading EIM platforms include IBM Lotus Sametime, Microsoft Office Communications Server, Jabber XCP and Cisco Unified Presence.Industry-focused EIM platforms as Reuters Messaging and Bloomberg Messaging also provide IM abilities to financial services companies.

The adoption of IM across corporate networks outside of the control of IT organizations creates risks and liabilities for companies who do not effectively manage and support IM use. Companies implement specialized IM archiving and security products and services to mitigate these risks and provide safe, secure, productive instant messaging abilities to their employees. IM is increasingly becoming a feature of enterprise software rather than a stand-alone application.

Types of Products

IM products can usually be categorised into two types: Enterprise Instant Messaging (EIM) and Consumer Instant Messaging (CIM). Enterprise solutions use an internal IM server, however this isn't always feasible, particularly for smaller businesses with limited budgets. The second option, using a CIM provides the advantage of being inexpensive to implement and has little need for investing in new hardware or server software.

For corporate use, encryption and conversation archiving are usually regarded as important features due to security concerns. Sometimes the use of different operating systems in organizations requires use of software that supports more than one platform. For example, many software companies use Windows in administration departments but have software developers who use Linux.

Security Risks

Crackers (malicious or black hat hackers) have consistently used IM networks as vectors for delivering phishing attempts, "poison URLs", and virus-laden file attachments from 2004 to the present, with over 1100 discrete attacks listed by the IM Security Center in 2004–2007. Hackers use two methods of delivering malicious code through IM: delivery of viruses, trojan horses, or spyware within an infected file, and the use of "socially engineered" text with a web address that entices the recipient to click on a URL connecting him or her to a website that then downloads malicious code.

Viruses, computer worms, and trojans usually propagate by sending themselves rapidly through the infected user's contact list. An effective attack using a poisoned URL may reach tens of thousands of users in a short period when each user's contact list receives messages appearing to be from a trusted friend. The recipients click on the web address, and the entire cycle starts again. Infections may range from nuisance to criminal, and are becoming more sophisticated each year.

IM connections usually occur in plain text, making them vulnerable to eavesdropping. Also, IM client software often requires the user to expose open UDP ports to the world, raising the threat posed by potential security vulnerabilities.

Compliance Risks

In addition to the malicious code threat, the use of instant messaging at work also creates a risk of non-compliance to laws and regulations governing use of electronic communications in businesses.

In the United States

In the United States alone there are over 10,000 laws and regulations related to electronic messaging and records retention. The better-known of these include the Sarbanes–Oxley Act, HIPAA, and SEC 17a-3.

Clarification from the Financial Industry Regulatory Authority (FINRA) was issued to member firms in the financial services industry in December, 2007, noting that "electronic communications", "email", and "electronic correspondence" may be used interchangeably and can include such forms of electronic messaging as *instant messaging* and text messaging. Changes to Federal Rules of Civil Procedure, effective December 1, 2006, created a new category for electronic records which may be requested during discovery in legal proceedings.

World-wide

Most nations also regulate use of electronic messaging and electronic records retention in similar fashion as the United States. The most common regulations related to IM at work involve the need to produce archived business communications to satisfy government or judicial requests under law. Many instant messaging communications fall into the category of business communications that must be archived and retrievable.

Security and Archiving

In the early 2000s, a new class of IT security provider emerged to provide remedies for the risks and liabilities faced by corporations who chose to use IM for business communications. The IM security providers created new products to be installed in corporate networks for the purpose of archiving, content-scanning, and security-scanning IM traffic moving in and out of the corporation. Similar to the e-mail filtering vendors, the IM security providers focus on the risks and liabilities described above.

With rapid adoption of IM in the workplace, demand for IM security products began to grow in the mid-2000s. By 2007, the preferred platform for the purchase of security software had become the "computer appliance", according to IDC, who estimate that by 2008, 80% of network security products will be delivered via an appliance.

By 2014 however, the level of safety offered by instant messengers was still extremely poor. According to a scorecard made by the Electronic Frontier Foundation, only 7 out of 39 instant messengers received a perfect score, whereas the most popular instant messengers at the time only attained a score of 2 out of 7. A number of studies have shown that IM services are quite vulnerable for providing user privacy.

User Base

While some numbers are given by the owners of a complete instant messaging system, others are provided by commercial vendors of a part of a distributed system. Some companies may be motivated to inflate their numbers to raise advertising earnings or attract partners, clients, or customers. Importantly, some numbers are reported as the number of *active* users (with no shared standard of that activity), others indicate total user accounts, while others indicate only the users logged in during an instance of peak use.

Since the acquisitions of 2010 and later and with the wide availability of smartphones, the virtual communities of those conglomerates are becoming the user base of most instant messaging services:

- Facebook (through Facebook Messenger, Instagram, WhatsApp)
- Tencent Holdings Limited (WeChat, Tencent QQ, Qzone)
- Google (Google+ and YouTube users (merging))
- Microsoft (Skype and Windows Live / MSN messenger (merged))
- Twitter (API)
- LinkedIn (API)

Statistics

Instant messenger client	Company	Usage
BlackBerry Messenger	BlackBerry	91 million total users (October 2014)
AIM	AOL, Inc	53 million active users (September 2006)
XMPP	XMPP Standards Foundation	1200+ million (September 2011)
eBuddy	eBuddy	35 million users (October2006)
iMessage	Apple Inc.	140 million users (June 2012)
Facebook Messenger	Facebook	900 million active users (April 2016)
Windows Live Messenger	Microsoft Corporation	330 million active monthly (June 2009)
Yahoo! Messenger	Yahoo!, Inc.	22 million users (Unknown)
QQ	Tencent Holdings Limited	176+ million peak online users, 840+ million "active" (Q2 2009)
IBM Sametime	IBM Corp.	15 million (enterprise) users (Unknown)
Skype	Microsoft Corporation	34 million peak online (February 2012), 560 million total (April 2010)
MXit	MXit Lifestyle (Pty) Ltd.	7.4 million monthly subscribers (majority in South Africa (July 2013)
Xfire	Xfire, Inc.	24 million registered users (January 2014)
Gadu-Gadu	GG Network S.A.	6.5 million users active daily (majority in Poland) (June 2010)
ICQ	ICQ LLC.	4 million active (September 2006)
Paltalk	Paltalk.com	5.5 million monthly unique users (August 2013)
IMVU	IMVU, inc.	1 million users (June 2007)

Voice Over IP

Voice over Internet Protocol (Voice over IP, VoIP and IP telephony) is a methodology and group of technologies for the delivery of voice communications and multimedia sessions over Internet Protocol (IP) networks, such as the Internet. The terms Internet telephony, broadband telephony, and broadband phone service specifically refer to the provisioning of communications services (voice, fax, SMS, voice-messaging) over the public Internet, rather than via the public switched telephone network (PSTN).

The steps and principles involved in originating VoIP telephone calls are similar to traditional digital telephony and involve signaling, channel setup, digitization of the analog voice signals, and encoding. Instead of being transmitted over a circuit-switched network; however, the digital information is packetized, and transmission occurs as IP packets over a packet-switched network. They transport audio streams using special media delivery protocols that encode audio and video with audio codecs, and video codecs. Various codecs exist that optimize the media stream based on application requirements and network bandwidth; some implementations rely on narrowband

and compressed speech, while others support high fidelity stereo codecs. Some popular codecs include μ-law and a-law versions of G.711, G.722, a popular open source voice codec known as iLBC, a codec that only uses 8 kbit/s each way called G.729, and many others.

Early providers of voice-over-IP services offered business models and technical solutions that mirrored the architecture of the legacy telephone network. Second-generation providers, such as Skype, have built closed networks for private user bases, offering the benefit of free calls and convenience while potentially charging for access to other communication networks, such as the PSTN. This has limited the freedom of users to mix-and-match third-party hardware and software. Third-generation providers, such as Google Talk, have adopted the concept of federated VoIP—which is a departure from the architecture of the legacy networks. These solutions typically allow dynamic interconnection between users on any two domains on the Internet when a user wishes to place a call.

In addition to VoIP phones, VoIP is available on many smartphones, personal computers, and on Internet access devices. Calls and SMS text messages may be sent over 3G/4G or Wi-Fi.

Protocols

Voice over IP has been implemented in various ways using both proprietary protocols and protocols based on open standards. VoIP protocols include:

- Session Initiation Protocol (SIP)
- H.323
- Media Gateway Control Protocol (MGCP)
- Gateway Control Protocol (Megaco, H.248)
- Real-time Transport Protocol (RTP)
- Real-time Transport Control Protocol (RTCP)
- Secure Real-time Transport Protocol (SRTP)
- Session Description Protocol (SDP)
- Inter-Asterisk eXchange (IAX)
- Jingle XMPP VoIP extensions
- Skype protocol

The H.323 protocol was one of the first VoIP protocols that found widespread implementation for long-distance traffic, as well as local area network services. However, since the development of newer, less complex protocols such as MGCP and SIP, H.323 deployments are increasingly limited to carrying existing long-haul network traffic.

These protocols can be used by special-purpose software, such as Jitsi, or integrated into a web page (web-based VoIP), like Google Talk.

Adoption

Consumer Market

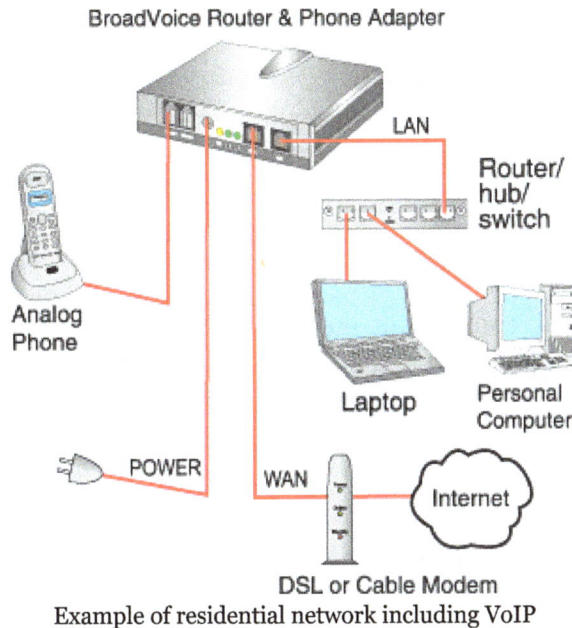

Example of residential network including VoIP

A major development that started in 2004 was the introduction of mass-market VoIP services that utilize existing broadband Internet access, by which subscribers place and receive telephone calls in much the same manner as they would via the public switched telephone network (PSTN). Full-service VoIP phone companies provide inbound and outbound service with direct inbound dialing. Many offer unlimited domestic calling for a flat monthly subscription fee. This sometimes includes international calls to certain countries. Phone calls between subscribers of the same provider are usually free when flat-fee service is not available. A VoIP phone is necessary to connect to a VoIP service provider. This can be implemented in several ways:

- Dedicated VoIP phones connect directly to the IP network using technologies such as wired Ethernet or Wi-Fi. They are typically designed in the style of traditional digital business telephones.

- An analog telephone adapter is a device that connects to the network and implements the electronics and firmware to operate a conventional analog telephone attached through a modular phone jack. Some residential Internet gateways and cablemodems have this function built in.

- A softphone is application software installed on a networked computer that is equipped with a microphone and speaker, or headset. The application typically presents a dial pad and display field to the user to operate the application by mouse clicks or keyboard input.

PSTN and Mobile Network Providers

It is becoming increasingly common for telecommunications providers to use VoIP telephony over dedicated and public IP networks to connect switching centers and to interconnect with other telephony network providers; this is often referred to as "IP backhaul".

Smartphones and Wi-Fi-enabled mobile phones may have SIP clients built into the firmware or available as an application download.

Corporate Use

Because of the bandwidth efficiency and low costs that VoIP technology can provide, businesses are migrating from traditional copper-wire telephone systems to VoIP systems to reduce their monthly phone costs. In 2008, 80% of all new Private branch exchange (PBX) lines installed internationally were VoIP.

VoIP solutions aimed at businesses have evolved into unified communications services that treat all communications—phone calls, faxes, voice mail, e-mail, Web conferences, and more—as discrete units that can all be delivered via any means and to any handset, including cellphones. Two kinds of competitors are competing in this space: one set is focused on VoIP for medium to large enterprises, while another is targeting the small-to-medium business (SMB) market.

VoIP allows both voice and data communications to be run over a single network, which can significantly reduce infrastructure costs.

The prices of extensions on VoIP are lower than for PBX and key systems. VoIP switches may run on commodity hardware, such as personal computers. Rather than closed architectures, these devices rely on standard interfaces.

VoIP devices have simple, intuitive user interfaces, so users can often make simple system configuration changes. Dual-mode phones enable users to continue their conversations as they move between an outside cellular service and an internal Wi-Fi network, so that it is no longer necessary to carry both a desktop phone and a cellphone. Maintenance becomes simpler as there are fewer devices to oversee.

Skype, which originally marketed itself as a service among friends, has begun to cater to businesses, providing free-of-charge connections between any users on the Skype network and connecting to and from ordinary PSTN telephones for a charge.

In the United States the Social Security Administration (SSA) is converting its field offices of 63,000 workers from traditional phone installations to a VoIP infrastructure carried over its existing data network.

Quality of Service

Communication on the IP network is perceived as less reliable in contrast to the circuit-switched public telephone network because it does not provide a network-based mechanism to ensure that data packets are not lost, and are delivered in sequential order. It is a best-effort network without fundamental Quality of Service (QoS) guarantees. Voice, and all other data, travels in packets over IP networks with fixed maximum capacity. This system may be more prone to congestion and DoS attacks than traditional circuit switched systems; a circuit switched system of insufficient capacity will refuse new connections while carrying the remainder without impairment, while the quality of real-time data such as telephone conversations on packet-switched networks degrades dramatically. Therefore, VoIP implementations may face problems with latency, packet loss, and jitter.

By default, network routers handle traffic on a first-come, first-served basis. Fixed delays cannot be controlled as they are caused by the physical distance the packets travel. They are especially problematic when satellite circuits are involved because of the long distance to a geostationary satellite and back; delays of 400–600 ms are typical. Latency can be minimized by marking voice packets as being delay-sensitive with QoS methods such as DiffServ.

Network routers on high volume traffic links may introduce latency that exceeds permissible thresholds for VoIP. When the load on a link grows so quickly that its switches experience queue overflows, congestion results and data packets are lost. This signals a transport protocol like TCP to reduce its transmission rate to alleviate the congestion. But VoIP usually uses UDP not TCP because recovering from congestion through retransmission usually entails too much latency. So QoS mechanisms can avoid the undesirable loss of VoIP packets by immediately transmitting them ahead of any queued bulk traffic on the same link, even when that bulk traffic queue is overflowing.

VoIP endpoints usually have to wait for completion of transmission of previous packets before new data may be sent. Although it is possible to preempt (abort) a less important packet in mid-trans-mission, this is not commonly done, especially on high-speed links where transmission times are short even for maximum-sized packets. An alternative to preemption on slower links, such as di-alup and digital subscriber line (DSL), is to reduce the maximum transmission time by reducing the maximum transmission unit. But every packet must contain protocol headers, so this increases relative header overhead on every link traversed, not just the bottleneck (usually Internet access) link.

The receiver must resequence IP packets that arrive out of order and recover gracefully when packets arrive too late or not at all. Jitter results from the rapid and random (i.e. unpredict-able) changes in queue lengths along a given Internet path due to competition from other users for the same transmission links. VoIP receivers counter jitter by storing incoming pack-ets briefly in a "de-jitter" or "playout" buffer, deliberately increasing latency to improve the chance that each packet will be on hand when it is time for the voice engine to play it. The added delay is thus a compromise between excessive latency and excessive dropout, i.e. mo-mentary audio interruptions.

Although jitter is a random variable, it is the sum of several other random variables that are at least somewhat independent: the individual queuing delays of the routers along the Internet path in question. Thus according to the central limit theorem, we can model jitter as a gaussian random variable. This suggests continually estimating the mean delay and its standard deviation and set-ting the playout delay so that only packets delayed more than several standard deviations above the mean will arrive too late to be useful. In practice, however, the variance in latency of many Internet paths is dominated by a small number (often one) of relatively slow and congested "bot-tleneck" links. Most Internet backbone links are now so fast (e.g. 10 Gbit/s) that their delays are dominated by the transmission medium (e.g. optical fiber) and the routers driving them do not have enough buffering for queuing delays to be significant.

It has been suggested to rely on the packetized nature of media in VoIP communications and transmit the stream of packets from the source phone to the destination phone simultaneously across different routes (multi-path routing). In such a way, temporary failures have less impact on the communication quality. In capillary routing it has been suggested to use at the packet level

Fountain codes or particularly raptor codes for transmitting extra redundant packets making the communication more reliable.

A number of protocols have been defined to support the reporting of quality of service (QoS) and quality of experience (QoE) for VoIP calls. These include RTCP Extended Report (RFC 3611), SIP RTCP Summary Reports, H.460.9 Annex B (for H.323), H.248.30 and MGCP extensions. The RFC 3611 VoIP Metrics block is generated by an IP phone or gateway during a live call and contains information on packet loss rate, packet discard rate (because of jitter), packet loss/discard burst metrics (burst length/density, gap length/density), network delay, end system delay, signal / noise / echo level, Mean Opinion Scores (MOS) and R factors and configuration information related to the jitter buffer.

RFC 3611 VoIP metrics reports are exchanged between IP endpoints on an occasional basis during a call, and an end of call message sent via SIP RTCP Summary Report or one of the other signaling protocol extensions. RFC 3611 VoIP metrics reports are intended to support real time feedback related to QoS problems, the exchange of information between the endpoints for improved call quality calculation and a variety of other applications.

Rural areas in particular are greatly hindered in their ability to choose a VoIP system over PBX. This is generally down to the poor access to superfast broadband in rural country areas. With the release of 4G data, there is a potential for corporate users based outside of populated areas to switch their internet connection to 4G data, which is comparatively as fast as a regular superfast broadband connection. This greatly enhances the overall quality and user experience of a VoIP system in these areas. This method was already trialled in rural Germany, surpassing all expectations.

DSL and ATM

DSL modems provide Ethernet (or Ethernet over USB) connections to local equipment, but inside they are actually Asynchronous Transfer Mode (ATM) modems. They use ATM Adaptation Layer 5 (AAL5) to segment each Ethernet packet into a series of 53-byte ATM cells for transmission, re-assembling them back into Ethernet frames at the receiving end. A virtual circuit identifier (VCI) is part of the 5-byte header on every ATM cell, so the transmitter can multiplex the active virtual circuits (VCs) in any arbitrary order. Cells from the *same* VC are always sent sequentially.

However, a majority of DSL providers use only one VC for each customer, even those with bundled VoIP service. Every Ethernet frame must be completely transmitted before another can begin. If a second VC were established, given high priority and reserved for VoIP, then a low priority data packet could be suspended in mid-transmission and a VoIP packet sent right away on the high priority VC. Then the link would pick up the low priority VC where it left off. Because ATM links are multiplexed on a cell-by-cell basis, a high priority packet would have to wait at most 53 byte times to begin transmission. There would be no need to reduce the interface MTU and accept the resulting increase in higher layer protocol overhead, and no need to abort a low priority packet and resend it later.

ATM has substantial header overhead: $5/53 = 9.4\%$, roughly twice the total header overhead of a 1500 byte Ethernet frame. This "ATM tax" is incurred by every DSL user whether or not they take advantage of multiple virtual circuits - and few can.

ATM's potential for latency reduction is greatest on slow links, because worst-case latency decreases with increasing link speed. A full-size (1500 byte) Ethernet frame takes 94 ms to transmit at 128 kbit/s but only 8 ms at 1.5 Mbit/s. If this is the bottleneck link, this latency is probably small enough to ensure good VoIP performance without MTU reductions or multiple ATM VCs. The latest generations of DSL, VDSL and VDSL2, carry Ethernet without intermediate ATM/AAL5 layers, and they generally support IEEE 802.1p priority tagging so that VoIP can be queued ahead of less time-critical traffic.

Layer 2

A number of protocols that deal with the data link layer and physical layer include quality-of-service mechanisms that can be used to ensure that applications like VoIP work well even in congested scenarios. Some examples include:

- IEEE 802.11e is an approved amendment to the IEEE 802.11 standard that defines a set of quality-of-service enhancements for wireless LAN applications through modifications to the Media Access Control (MAC) layer. The standard is considered of critical importance for delay-sensitive applications, such as voice over wireless IP.

- IEEE 802.1p defines 8 different classes of service (including one dedicated to voice) for traffic on layer-2 wired Ethernet.

- The ITU-T G.hn standard, which provides a way to create a high-speed (up to 1 gigabit per second) Local area network (LAN) using existing home wiring (power lines, phone lines and coaxial cables). G.hn provides QoS by means of "Contention-Free Transmission Opportunities" (CFTXOPs) which are allocated to flows (such as a VoIP call) which require QoS and which have negotiated a "contract" with the network controllers.

VoIP Performance Metrics

The quality of voice transmission is characterized by several metrics that may be monitored by network elements, by the user agent hardware or software. Such metrics include network packet loss, packet jitter, packet latency (delay), post-dial delay, and echo. The metrics are determined by VoIP performance testing and monitoring.

PSTN Integration

The Media VoIP Gateway connects the digital media stream, so as to complete creating the path for voice as well as data media. It includes the interface for connecting the standard PSTN networks with the ATM and Inter Protocol networks. The Ethernet interfaces are also included in the modern systems, which are specially designed to link calls that are passed via the VoIP.

E.164 is a global FGFnumbering standard for both the PSTN and PLMN. Most VoIP implementations support E.164 to allow calls to be routed to and from VoIP subscribers and the PSTN/PLMN. VoIP implementations can also allow other identification techniques to be used. For example, Skype allows subscribers to choose "Skype names" (usernames) whereas SIP implementations can use URIs similar to email addresses. Often VoIP implementations employ methods of translating non-E.164 identifiers to E.164 numbers and vice versa, such as the Skype-In service provided by Skype and the ENUM service in IMS and SIP.

Echo can also be an issue for PSTN integration. Common causes of echo include impedance mismatches in analog circuitry and acoustic coupling of the transmit and receive signal at the receiving end.

Number Portability

Local number portability (LNP) and Mobile number portability (MNP) also impact VoIP business. In November 2007, the Federal Communications Commission in the United States released an order extending number portability obligations to interconnected VoIP providers and carriers that support VoIP providers. Number portability is a service that allows a subscriber to select a new telephone carrier without requiring a new number to be issued. Typically, it is the responsibility of the former carrier to "map" the old number to the undisclosed number assigned by the new carrier. This is achieved by maintaining a database of numbers. A dialed number is initially received by the original carrier and quickly rerouted to the new carrier. Multiple porting references must be maintained even if the subscriber returns to the original carrier. The FCC mandates carrier compliance with these consumer-protection stipulations.

A voice call originating in the VoIP environment also faces challenges to reach its destination if the number is routed to a mobile phone number on a traditional mobile carrier. VoIP has been identified in the past as a Least Cost Routing (LCR) system, which is based on checking the destination of each telephone call as it is made, and then sending the call via the network that will cost the customer the least. This rating is subject to some debate given the complexity of call routing created by number portability. With GSM number portability now in place, LCR providers can no longer rely on using the network root prefix to determine how to route a call. Instead, they must now determine the actual network of every number before routing the call.

Therefore, VoIP solutions also need to handle MNP when routing a voice call. In countries without a central database, like the UK, it might be necessary to query the GSM network about which home network a mobile phone number belongs to. As the popularity of VoIP increases in the enterprise markets because of least cost routing options, it needs to provide a certain level of reliability when handling calls.

MNP checks are important to assure that this quality of service is met. Handling MNP lookups before routing a call provides some assurance that the voice call will actually work.

Emergency Calls

A telephone connected to a land line has a direct relationship between a telephone number and a physical location, which is maintained by the telephone company and available to emergency responders via the national emergency response service centers in form of emergency subscriber lists. When an emergency call is received by a center the location is automatically determined from its databases and displayed on the operator console.

In IP telephony, no such direct link between location and communications end point exists. Even a provider having hardware infrastructure, such as a DSL provider, may only know the approximate location of the device, based on the IP address allocated to the network router and the known service address. However, some ISPs do not track the automatic assignment of IP addresses to customer equipment.

IP communication provides for device mobility. For example, a residential broadband connection may be used as a link to a virtual private network of a corporate entity, in which case the IP address being used for customer communications may belong to the enterprise, not being the network address of the residential ISP. Such off-premises extensions may appear as part of an upstream IP PBX. On mobile devices, e.g., a 3G handset or USB wireless broadband adapter, the IP address has no relationship with any physical location known to the telephony service provider, since a mobile user could be anywhere in a region with network coverage, even roaming via another cellular company.

At the VoIP level, a phone or gateway may identify itself with a Session Initiation Protocol (SIP) registrar by its account credentials. In such cases, the Internet telephony service provider (ITSP) only knows that a particular user's equipment is active. Service providers often provide emergency response services by agreement with the user who registers a physical location and agrees that emergency services are only provided to that address if an emergency number is called from the IP device.

Such emergency services are provided by VoIP vendors in the United States by a system called Enhanced 911 (E911), based on the Wireless Communications and Public Safety Act of 1999. The VoIP E911 emergency-calling system associates a physical address with the calling party's telephone number. All VoIP providers that provide access to the public switched telephone network are required to implement E911, a service for which the subscriber may be charged. However, end-customer participation in E911 is not mandatory and customers may opt out of the service.

The VoIP E911 system is based on a static table lookup. Unlike in cellular phones, where the location of an E911 call can be traced using assisted GPS or other methods, the VoIP E911 information is only accurate so long as subscribers, who have the legal responsibility, are diligent in keeping their emergency address information current.

Fax Support

Support for fax has been problematic in many VoIP implementations, as most voice digitization and compression codecs are optimized for the representation of the human voice and the proper timing of the modem signals cannot be guaranteed in a packet-based, connection-less network. An alternative IP-based solution for delivering fax-over-IP called T.38 is available. Sending faxes using VoIP is sometimes referred to as FoIP, or Fax over IP.

The T.38 protocol is designed to compensate for the differences between traditional packet-less communications over analog lines and packet-based transmissions which are the basis for IP communications. The fax machine could be a traditional fax machine connected to the PSTN, or an ATA box (or similar). It could be a fax machine with an RJ-45 connector plugged straight into an IP network, or it could be a computer pretending to be a fax machine. Originally, T.38 was designed to use UDP and TCP transmission methods across an IP network. TCP is better suited for use between two IP devices. However, older fax machines, connected to an analog system, benefit from UDP near real-time characteristics due to the "no recovery rule" when a UDP packet is lost or an error occurs during transmission. UDP transmissions are preferred as they do not require testing for dropped packets and as such since each T.38 packet transmission includes a majority of the data sent in the prior packet, a T.38 termination point has a higher degree of success in re-assembling the fax transmission back into its original form for interpretation by the end device. This in an attempt to overcome the obstacles of simulating real time transmissions using packet based protocol.

There have been updated versions of T.30 to resolve the fax over IP issues, which is the core fax protocol. Some newer high end fax machines have T.38 built-in capabilities which allow the user to plug right into the network and transmit/receive faxes in native T.38 like the Ricoh 4410NF Fax Machine. A unique feature of T.38 is that each packet contains a portion of the main data sent in the previous packet. With T.38, two successive lost packets are needed to actually lose any data. The data one will lose will only be a small piece, but with the right settings and error correction mode, there is an increased likelihood that they will receive enough of the transmission to satisfy the requirements of the fax machine for output of the sent document.

While many late-model analog telephone adapters (ATAs) support T.38, uptake has been limited as many voice-over-IP providers perform least-cost routing which selects the least expensive PSTN gateway in the called city for an outbound message. There is typically no means to ensure that that gateway is T.38 capable. Providers often place their own equipment (such as an Asterisk PBX installation) in the signal path, which creates additional issues as every link in the chain must be T.38 aware for the protocol to work. Similar issues arise if a provider is purchasing local direct in-ward dial numbers from the lowest bidder in each city, as many of these may not be T.38 enabled.

Power Requirements

Telephones for traditional residential analog service are usually connected directly to telephone company phone lines which provide direct current to power most basic analog handsets independently of locally available electrical power.

IP Phones and VoIP telephone adapters connect to routers or cable modems which typically depend on the availability of mains electricity or locally generated power. Some VoIP service providers use customer premises equipment (e.g., cablemodems) with battery-backed power supplies to assure uninterrupted service for up to several hours in case of local power failures. Such battery-backed devices typically are designed for use with analog handsets.

Some VoIP service providers implement services to route calls to other telephone services of the subscriber, such a cellular phone, in the event that the customer's network device is inaccessible to terminate the call.

The susceptibility of phone service to power failures is a common problem even with traditional analog service in areas where many customers purchase modern telephone units that operate with wireless handsets to a base station, or that have other modern phone features, such as built-in voicemail or phone book features.

Security

The security concerns of VoIP telephone systems are similar to those of any Internet-connected device. This means that hackers who know about these vulnerabilities can institute denial-of-service attacks, harvest customer data, record conversations and compromise voicemail messages. The quality of internet connection determines the quality of the calls. VoIP phone service also will not work if there is power outage and when the internet connection is down. The 9-1-1 or 112 service provided by VoIP phone service is also different from analog phone which is associated with a fixed address. The emergency center may not be able to determine your location based on your virtual phone number. Compromised VoIP user account or session credentials may enable an attacker to incur substantial charges from third-party services, such as long-distance or international telephone calling.

The technical details of many VoIP protocols create challenges in routing VoIP traffic through firewalls and network address translators, used to interconnect to transit networks or the Internet. Private session border controllers are often employed to enable VoIP calls to and from protected networks. Other methods to traverse NAT devices involve assistive protocols such as STUN and Interactive Connectivity Establishment (ICE).

Many consumer VoIP solutions do not support encryption of the signaling path or the media, however securing a VoIP phone is conceptually easier to implement than on traditional telephone circuits. A result of the lack of encryption is a relative easy to eavesdrop on VoIP calls when access to the data network is possible. Free open-source solutions, such as Wireshark, facilitate capturing VoIP conversations.

Standards for securing VoIP are available in the Secure Real-time Transport Protocol (SRTP) and the ZRTP protocol for analog telephony adapters as well as for some softphones. IPsec is available to secure point-to-point VoIP at the transport level by using opportunistic encryption.

Government and military organizations use various security measures to protect VoIP traffic, such as voice over secure IP (VoSIP), secure voice over IP (SVoIP), and secure voice over secure IP (SVoSIP). The distinction lies in whether encryption is applied in the telephone or in the network or both. Secure voice over secure IP is accomplished by encrypting VoIP with protocols such as SRTP or ZRTP. Secure voice over IP is accomplished by using Type 1 encryption on a classified network, like SIPRNet. Public Secure VoIP is also available with free GNU programs and in many popular commercial VoIP programs via libraries such as ZRTP.

Caller ID

Voice over IP protocols and equipment provide caller ID support that is compatible with the facility provided in the public switched telephone network (PSTN). Many VoIP service providers also allow callers to configure arbitrary caller ID information.

Compatibility with Traditional Analog Telephone Sets

Most analog telephone adapters do not decode dial pulses generated by rotary dial telephones, supporting only touch-tone signaling, but pulse-to-tone converters are commercially available.

Support for Other Telephony Devices

Some special telephony services, such as those that operate in conjunction with digital video recorders, satellite television receivers, alarm systems, conventional modems over PSTN lines, may be impaired when operated over VoIP services, because of incompatibilities in design.

Operational Cost

VoIP has drastically reduced the cost of communication by sharing network infrastructure between data and voice. A single broad-band connection has the ability to transmit more than one telephone call. Secure calls using standardized protocols, such as Secure Real-time Transport Protocol, as most of the facilities of creating a secure telephone connection over traditional phone lines, such as digitizing and digital transmission, are already in place with VoIP. It is only necessary to encrypt and authenticate the existing data stream. Automated software, such as a virtual PBX, may eliminate the need of personnel to greet and switch incoming calls.

Regulatory and Legal Issues

As the popularity of VoIP grows, governments are becoming more interested in regulating VoIP in a manner similar to PSTN services.

Throughout the developing world, countries where regulation is weak or captured by the dominant operator, restrictions on the use of VoIP are imposed, including in Panama where VoIP is taxed, Guyana where VoIP is prohibited and India where its retail commercial sales is allowed but only for long distance service. In Ethiopia, where the government is nationalising telecommunication service, it is a criminal offence to offer services using VoIP. The country has installed firewalls to prevent international calls being made using VoIP. These measures were taken after the popularity of VoIP reduced the income generated by the state owned telecommunication company.

European Union

In the European Union, the treatment of VoIP service providers is a decision for each national telecommunications regulator, which must use competition law to define relevant national markets and then determine whether any service provider on those national markets has "significant market power" (and so should be subject to certain obligations). A general distinction is usually made between VoIP services that function over managed networks (via broadband connections) and VoIP services that function over unmanaged networks (essentially, the Internet).

The relevant EU Directive is not clearly drafted concerning obligations which can exist independently of market power (e.g., the obligation to offer access to emergency calls), and it is impossible to say definitively whether VoIP service providers of either type are bound by them. A review of the EU Directive is under way and should be complete by 2007.

Middle East

In the UAE and Oman it is illegal to use any form of VoIP, to the extent that Web sites of Gizmo5 are blocked. Providing or using VoIP services is illegal in Oman. Those who violate the law stand to be fined 50,000 Omani Rial (about 130,317 US dollars) or spend two years in jail or both. In 2009, police in Oman have raided 121 Internet cafes throughout the country and arrested 212 people for using/providing VoIP services.

India

In India, it is legal to use VoIP, but it is illegal to have VoIP gateways inside India. This effectively means that people who have PCs can use them to make a VoIP call to any number, but if the remote side is a normal phone, the gateway that converts the VoIP call to a POTS call is not permitted by law to be inside India. Foreign based Voip server services are illegal to use in India.

In the interest of the Access Service Providers and International Long Distance Operators the Internet telephony was permitted to the ISP with restrictions. Internet Telephony is considered to be different service in its scope, nature and kind from real time voice as offered by other Access Service Providers and Long Distance Carriers. Hence the following type of Internet Telephony are permitted in India:

(a) PC to PC; within or outside India.

(b) PC / a device / Adapter conforming to standard of any international agencies like- ITU or IETF etc. in India to PSTN/PLMN abroad.

(c) Any device / Adapter conforming to standards of International agencies like ITU, IETF etc. connected to ISP node with static IP address to similar device / Adapter; within or outside India.

(d) Except whatever is described in condition (ii) above, no other form of Internet Telephony is permitted.

(e) In India no Separate Numbering Scheme is provided to the Internet Telephony. Presently the 10 digit Numbering allocation based on E.164 is permitted to the Fixed Telephony, GSM, CDMA wireless service. For Internet Telephony the numbering scheme shall only conform to IP addressing Scheme of Internet Assigned Numbers Authority (IANA). Translation of E.164 number / private number to IP address allotted to any device and vice versa, by ISP to show compliance with IANA numbering scheme is not permitted.

(f) The Internet Service Licensee is not permitted to have PSTN/PLMN connectivity. Voice communication to and from a telephone connected to PSTN/PLMN and following E.164 numbering is prohibited in India.

South Korea

In South Korea, only providers registered with the government are authorized to offer VoIP services. Unlike many VoIP providers, most of whom offer flat rates, Korean VoIP services are generally metered and charged at rates similar to terrestrial calling. Foreign VoIP providers encounter high barriers to government registration. This issue came to a head in 2006 when Internet service providers providing personal Internet services by contract to United States Forces Korea members residing on USFK bases threatened to block off access to VoIP services used by USFK members as an economical way to keep in contact with their families in the United States, on the grounds that the service members' VoIP providers were not registered. A compromise was reached between USFK and Korean telecommunications officials in January 2007, wherein USFK service members arriving in Korea before June 1, 2007, and subscribing to the ISP services provided on base may continue to use their US-based VoIP subscription, but later arrivals must use a Korean-based VoIP provider, which by contract will offer pricing similar to the flat rates offered by US VoIP providers.

United States

In the United States, the Federal Communications Commission requires all interconnected VoIP service providers to comply with requirements comparable to those for traditional telecommunications service providers. VoIP operators in the US are required to support local number portability; make service accessible to people with disabilities; pay regulatory fees, universal service contributions, and other mandated payments; and enable law enforcement authorities to conduct surveillance pursuant to the Communications Assistance for Law Enforcement Act (CALEA).

Operators of "Interconnected" VoIP (fully connected to the PSTN) are mandated to provide Enhanced 911 service without special request, provide for customer location updates, clearly disclose any limitations on their E-911 functionality to their consumers, obtain affirmative acknowledgements of these disclosures from all consumers, and 'may not allow their customers to "opt-out" of 911 service.' VoIP operators also receive the benefit of certain US telecommunications regulations, including an entitlement to interconnection and exchange of traffic with incumbent local exchange carriers via wholesale carriers. Providers of "nomadic" VoIP service—those who are unable to determine the location of their users—are exempt from state telecommunications regulation.

Another legal issue that the US Congress is debating concerns changes to the Foreign Intelligence Surveillance Act. The issue in question is calls between Americans and foreigners. The National Security Agency (NSA) is not authorized to tap Americans' conversations without a warrant—but the Internet, and specifically VoIP does not draw as clear a line to the location of a caller or a call's recipient as the traditional phone system does. As VoIP's low cost and flexibility convinces more and more organizations to adopt the technology, the surveillance for law enforcement agencies becomes more difficult. VoIP technology has also increased security concerns because VoIP and similar technologies have made it more difficult for the government to determine where a target is physically located when communications are being intercepted, and that creates a whole set of new legal challenges.

History

The early developments of packet network designs by Paul Baran and other researchers were motivated by a desire for a higher degree of circuit redundancy and network avalability in face of infrastructure failures than was possible in the circuit-switched networks in telecommunications in the mid-twentieth century. In 1973, Danny Cohen first demonstrated a form of packet voice as part of a flight simulator application, which operated across the early ARPANET. In the following time span of about two decades, various forms of packet telephony were developed and industry interest groups formed to support the new technologies. Following the termination of the ARPANET project, and expansion of the Internet for commercial traffic, IP telephony became an established area of interest in commercial labs of the major IT concerns, such Microsoft and Intel, and open-source software, such as VocalTec, became available by the mid-1990s. By the late 1990s, the first softswitches became available, and new protocols, such as H.323, the Media Gateway Control Protocol (MGCP) and the Session Initiation Protocol (SIP) gained widespread attention. In the early 2000s, the proliferation of high-bandwidth always-on Internet connections to residential dwellings and businesses, spawned an industry of Internet telephony service providers (ITSPs). The development of open-source telephony software, such as Asterisk PBX, fueled widespread interest and entrepreneurship in voiceover-IP services, applying new Internet technology paradigms, such as cloud services to telephony.

Milestones

- 1973: Packet voice application by Danny Cohen.

- 1974: The Institute of Electrical and Electronic Engineers (IEEE) publishes a paper entitled "A Protocol for Packet Network Interconnection".

- 1974: Network Voice Protocol (NVP) tested over ARPANET in August 1974, carrying 16k CVSD encoded voice.

- 1977: Danny Cohen and Jon Postel of the USC Information Sciences Institute, and Vint Cerf of the Defense Advanced Research Projects Agency (DARPA), agree to separate IP from TCP, and create UDP for carrying real-time traffic.

- 1981: IPv4 is described in RFC 791.

- 1985: The National Science Foundation commissions the creation of NSFNET.

- 1986: Proposals from various standards organizations for Voice over ATM, in addition to commercial packet voice products from companies such as StrataCom.

- 1991: First Voice-over-IP application, Speak Freely, is released into the public domain. It was originally written by John Walker and further developed by Brian C. Wiles.

- 1992: The Frame Relay Forum conducts development of standards for Voice over Frame Relay.

- 1994: MTALK, a freeware VoIP application for Linux.

- 1995: VocalTec releases the first commercial Internet phone software.

  o Beginning in 1995, Intel, Microsoft and Radvision initiated standardization activities for VoIP communications system.

- 1996:

  o ITU-T begins development of standards for the transmission and signaling of voice communications over Internet Protocol networks with the H.323 standard.

  o US telecommunication companies petition the US Congress to ban Internet phone technology.

- 1997: Level 3 began development of its first softswitch, a term they coined in 1998.

- 1999:

  o The Session Initiation Protocol (SIP) specification RFC 2543 is released.

  o Mark Spencer of Digium develops the first open source private branch exchange (PBX) software (Asterisk).

- 2004: Commercial VoIP service providers proliferate.

- 2007: VOIP device manufacturers and sellers boom in Asia, specifically in the Philippines where many families of overseas workers reside.

- 2011: Raise of WebRTC technology which allows VoIP directly in browsers.

- 2015: Trend of using VoIP services in cloud: PBXes and contact centers, it means higher requirements to IP network to achieve good quality of service and reliability.

Videoconferencing

A Tandberg T3 high resolution telepresence room in use (2008).

Videoconferencing (VC) is the conduct of a videoconference (also known as a video conference or videoteleconference) by a set of telecommunication technologies which allow two or more locations to communicate by simultaneous two-way video and audio transmissions. It has also been called 'visual collaboration' and is a type of groupware.

Videoconferencing differs from videophone calls in that it's designed to serve a conference or multiple locations rather than individuals. It is an intermediate form of videotelephony, first used commercially in Germany during the late-1930s and later in the United States during the early 1970s as part of AT&T's development of Picturephone technology.

Indonesian and U.S. students participating in an educational videoconference (2010).

With the introduction of relatively low cost, high capacity broadband telecommunication services in the late 1990s, coupled with powerful computing processors and video compression techniques, videoconferencing has made significant inroads in business, education, medicine and media.

History

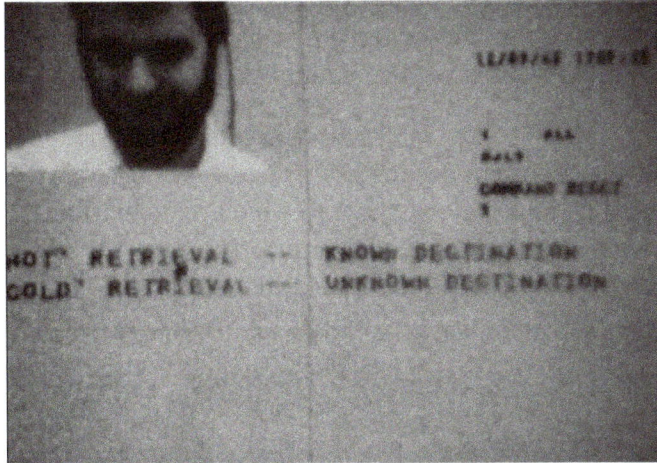

Multiple user videoconferencing first being demonstrated with Stanford R esearch Institute's NLS computer technology (1968).

Videoconferencing uses audio and video telecommunications to bring people at different sites together. This can be as simple as a conversation between people in private offices (point-to-point) or involve several (multipoint) sites in large rooms at multiple locations. Besides the audio and visual transmission of meeting activities, allied videoconferencing technologies can be used to share documents and display information on whiteboards.

Simple analog videophone communication could be established as early as the invention of the television. Such an antecedent usually consisted of two closed-circuit television systems connected via coax cable or radio. An example of that was the German Reich Postzentralamt (post office) video telephone network serving Berlin and several German cities via coaxial cables between 1936 and 1940.

During the first manned space flights, NASA used two radio-frequency (UHF or VHF) video links, one in each direction. TV channels routinely use this type of videotelephony when reporting from distant locations. The news media were to become regular users of mobile links to satellites using specially equipped trucks, and much later via special satellite videophones in a briefcase.

This technique was very expensive, though, and could not be used for applications such as telemedicine, distance education, and business meetings. Attempts at using normal telephony networks to transmit slow-scan video, such as the first systems developed by AT&T Corporation, first researched in the 1950s, failed mostly due to the poor picture quality and the lack of efficient video compression techniques. The greater 1 MHz bandwidth and 6 Mbit/s bit rate of the Picturephone in the 1970s also did not achieve commercial success, mostly due to its high cost, but also due to a lack of network effect —with only a few hundred Picturephones in the world, users had extremely few contacts they could actually call to, and interoperability with other videophone systems would not exist for decades.

It was only in the 1980s that digital telephony transmission networks became possible, such as with ISDN networks, assuring a minimum bit rate (usually 128 kilobits/s) for compressed video and audio transmission. During this time, there was also research into other forms of digital video and audio communication. Many of these technologies, such as the Media space, are not as widely used today as videoconferencing but were still an important area of research. The first dedicated systems started to appear in the market as ISDN networks were expanding throughout the world. One of the first commercial videoconferencing systems sold to companies came from PictureTel Corp., which had an Initial Public Offering in November, 1984.

In 1984 Concept Communication in the United States replaced the then-100 pound, US$100,000 computers necessary for teleconferencing, with a $12,000 circuit board that doubled the video frame rate from 15 up to 30 frames per second, and which reduced the equipment to the size of a circuit board fitting into standard personal computers. The company also secured a patent for a codec for full-motion videoconferencing, first demonstrated at AT&T Bell Labs in 1986.

Global Schoolhouse students communicating via CU-SeeMe, with a video framerate between 3-9 frames per second (1993).

Videoconferencing systems throughout the 1990s rapidly evolved from very expensive proprietary equipment, software and network requirements to a standards-based technology readily available to the general public at a reasonable cost.

Finally, in the 1990s, Internet Protocol-based videoconferencing became possible, and more efficient video compression technologies were developed, permitting desktop, or personal computer (PC)-based videoconferencing. In 1992 CU-SeeMe was developed at Cornell by Tim Dorcey et al. In 1995 the first public videoconference between North America and Africa took place, linking a technofair in San Francisco with a techno-rave and cyberdeli in Cape Town. At the Winter Olympics opening ceremony in Nagano, Japan, Seiji Ozawa conducted the Ode to Joy from Beethoven's Ninth Symphony simultaneously across five continents in near-real time.

While videoconferencing technology was initially used primarily within internal corporate communication networks, one of the first community service usages of the technology started in 1992 through a unique partnership with PictureTel and IBM Corporations which at the time were promoting a jointly developed desktop based videoconferencing product known as the PCS/1. Over the next 15 years, Project DIANE (Diversified Information and Assistance Network) grew to utilize a variety of videoconferencing platforms to create a multi-state cooperative public service and distance education network consisting of several hundred schools, neighborhood centers, libraries, science museums, zoos and parks, public assistance centers, and other community oriented organizations.

In the 2000s, videotelephony was popularized via free Internet services such as Skype and iChat, web plugins and on-line telecommunication programs that promoted low cost, albeit lower-quality, videoconferencing to virtually every location with an Internet connection.

Russian President Dmitry Medvedev attending the Singapore APEC summit, holding a videoconference with Rashid Nurgaliyev via a Tactical MXP, after an arms depot explosion in Russia (2009).

In May 2005, the first high definition video conferencing systems, produced by LifeSize Communications, were displayed at the Interop trade show in Las Vegas, Nevada, able to provide video at 30 frames per second with a 1280 by 720 display resolution. Polycom introduced its first high definition video conferencing system to the market in 2006. As of the 2010s, high definition resolution for videoconferencing became a popular feature, with most major suppliers in the video-conferencing market offering it.

Technological developments by videoconferencing developers in the 2010s have extended the capabilities of video conferencing systems beyond the boardroom for use with hand-held mobile devices that combine the use of video, audio and on-screen drawing capabilities broadcasting in real-time over secure networks, independent of location. Mobile collaboration systems now allow multiple people in previously unreachable locations, such as workers on an off-shore oil rig, the ability to view and discuss issues with colleagues thousands of miles away. Traditional videoconferencing system manufacturers have begun providing mobile applications as well, such as those that allow for live and still image streaming.

Technology

Dual display: An older Polycom VSX 7000 system and camera used for videoconferencing, with two displays for simultaneous broadcast from separate locations (2008).

Various components and the camera of a LifeSize Communications Room 220 high definition multipoint system (2010).

A video conference meeting facilitated by Google Hangouts.

The core technology used in a videoconferencing system is digital compression of audio and video streams in real time. The hardware or software that performs compression is called a codec (coder/decoder). Compression rates of up to 1:500 can be achieved. The resulting digital stream of 1s and 0s is subdivided into labeled packets, which are then transmitted through a digital network of some kind (usually ISDN or IP). The use of audio modems in the transmission line allow for the use of POTS, or the Plain Old Telephone System, in some low-speed applications, such as videotelephony, because they convert the digital pulses to/from analog waves in the audio spectrum range.

The other components required for a videoconferencing system include:

- Video input: (PTZ / 360° / Fisheye) video camera or webcam.

- Video output: computer monitor, television or projector.

- Audio input: microphones, CD/DVD player, cassette player, or any other source of PreAmp audio outlet.

- Audio output: usually loudspeakers associated with the display device or telephone.

- Data transfer: analog or digital telephone network, LAN or Internet.

- Computer: a data processing unit that ties together the other components, does the compressing and decompressing, and initiates and maintains the data linkage via the network.

There are basically two kinds of videoconferencing systems:

1. Dedicated systems have all required components packaged into a single piece of equipment, usually a console with a high quality remote controlled video camera. These cameras can be controlled at a distance to pan left and right, tilt up and down, and zoom. They became known as PTZ cameras. The console contains all electrical interfaces, the control computer, and the software or hardware-based codec. Omnidirectional microphones are connected to the console, as well as a TV monitor with loudspeakers and/or a video projector. There are several types of dedicated videoconferencing devices:

   1. Large group videoconferencing are non-portable, large, more expensive devices used for large rooms and auditoriums.

   2. Small group videoconferencing are non-portable or portable, smaller, less expensive devices used for small meeting rooms.

   3. Individual videoconferencing are usually portable devices, meant for single users, have fixed cameras, microphones and loudspeakers integrated into the console.

2. Desktop systems are add-ons (hardware boards or software codec) to normal PCs and laptops, transforming them into videoconferencing devices. A range of different cameras and microphones can be used with the codec, which contains the necessary codec and transmission interfaces. Most of the desktops systems work with the H.323 standard. Videoconferences carried out via dispersed PCs are also known as e-meetings. These can also be nonstandard, Microsoft Lync, Skype for Business, Google Hangouts, or Yahoo Messenger or standards based, Cisco Jabber.

3. WebRTC Platforms are video conferencing solutions that are not resident by using a software application but is available through the standard web browser. Solutions such as Adobe Connect and Cisco WebEX can be accessed by going to a URL sent by the meeting organizer and various degrees of security can be attached to the virtual "room". Often the user will be required to download a piece of software, called an "Add In" to enable the browser to access the local camera, microphone and establish a connection to the meeting. WebRTC technology doesn't require any software or Add On installation, instead a WebRTC compliant internet browser itself acts as a client to facilitate 1-to-1 and 1-to-many videoconferencing calls. Several enhanced flavours of WebRTC technology are being provided by Third Party vendors.

Conferencing Layers

The components within a Conferencing System can be divided up into several different layers: User Interface, Conference Control, Control or Signal Plane, and Media Plane.

Videoconferencing User Interfaces (VUI) can be either graphical or voice responsive. Many in the industry have encountered both types of interfaces, and normally graphical interfaces are encountered on a computer. User interfaces for conferencing have a number of different uses; they can be used for scheduling, setup, and making a videocall. Through the user interface the administrator is able to control the other three layers of the system.

Conference Control performs resource allocation, management and routing. This layer along with the User Interface creates meetings (scheduled or unscheduled) or adds and removes participants from a conference.

Control (Signaling) Plane contains the stacks that signal different endpoints to create a call and/or a conference. Signals can be, but aren't limited to, H.323 and Session Initiation Protocol (SIP) Protocols. These signals control incoming and outgoing connections as well as session parameters.

The Media Plane controls the audio and video mixing and streaming. This layer manages Real-Time Transport Protocols, User Datagram Packets (UDP) and Real-Time Transport Control Protocol (RTCP). The RTP and UDP normally carry information such the payload type which is the type of codec, frame rate, video size and many others. RTCP on the other hand acts as a quality control Protocol for detecting errors during streaming.

Multipoint Videoconferencing

Simultaneous videoconferencing among three or more remote points is possible by means of a Multipoint Control Unit (MCU). This is a bridge that interconnects calls from several sources (in a similar way to the audio conference call). All parties call the MCU, or the MCU can also call the parties which are going to participate, in sequence. There are MCU bridges for IP and ISDN-based videoconferencing. There are MCUs which are pure software, and others which are a combination of hardware and software. An MCU is characterised according to the number of simultaneous calls it can handle, its ability to conduct transposing of data rates and protocols, and features such as Continuous Presence, in which multiple parties can be seen on-screen at once. MCUs can be stand-alone hardware devices, or they can be embedded into dedicated videoconferencing units.

The MCU consists of two logical components:

1. A single multipoint controller (MC), and

2. Multipoint Processors (MP), sometimes referred to as the mixer.

The MC controls the conferencing while it is active on the signaling plane, which is simply where the system manages conferencing creation, endpoint signaling and in-conferencing controls. This component negotiates parameters with every endpoint in the network and controls conferencing resources. While the MC controls resources and signaling negotiations, the MP operates on the media plane and receives media from each endpoint. The MP generates output streams from each endpoint and redirects the information to other endpoints in the conference.

Some systems are capable of multipoint conferencing with no MCU, stand-alone, embedded or otherwise. These use a standards-based H.323 technique known as "decentralized multipoint", where each station in a multipoint call exchanges video and audio directly with the other stations with no central "manager" or other bottleneck. The advantages of this technique are that the video and audio will generally be of higher quality because they don't have to be relayed through a central point. Also, users can make ad-hoc multipoint calls without any concern for the availability or control of an MCU. This added convenience and quality comes at the expense of some increased network bandwidth, because every station must transmit to every other station directly.

Videoconferencing Modes

Videoconferencing systems use several common operating modes:

1. Voice-Activated Switch (VAS);
2. Continuous Presence.

In VAS mode, the MCU switches which endpoint can be seen by the other endpoints by the levels of one's voice. If there are four people in a conference, the only one that will be seen in the conference is the site which is talking; the location with the loudest voice will be seen by the other participants.

Continuous Presence mode, displays multiple participants at the same time. The MP in this mode takes the streams from the different endpoints and puts them all together into a single video image. In this mode, the MCU normally sends the same type of images to all participants. Typically these types of images are called "layouts" and can vary depending on the number of participants in a conference.

Echo Cancellation

A fundamental feature of professional videoconferencing systems is Acoustic Echo Cancellation (AEC). Echo can be defined as the reflected source wave interference with new wave created by source. AEC is an algorithm which is able to detect when sounds or utterances reenter the audio input of the videoconferencing codec, which came from the audio output of the same system, after some time delay. If unchecked, this can lead to several problems including:

1. the remote party hearing their own voice coming back at them (usually significantly delayed)
2. strong reverberation, which makes the voice channel useless, and
3. howling created by feedback.

Echo cancellation is a processor-intensive task that usually works over a narrow range of sound delays.

Cloud-based Video Conferencing

Cloud-based video conferencing can be used without the hardware generally required by other video conferencing systems, and can be designed for use by SMEs, or larger international companies like Facebook. Cloud-based systems can handle either 2D or 3D video broadcasting. Cloud-based systems can also implement mobile calls, VOIP, and other forms of video calling. They can also come with a video recording function to archive past meetings.

Technical and Other Issues

Computer security experts have shown that poorly configured or inadequately supervised videoconferencing system can permit an easy 'virtual' entry by computer hackers and criminals into company premises and corporate boardrooms, via their own videoconferencing systems. Some observers argue that three outstanding issues have prevented videoconferencing from becoming a standard form of communication, despite the ubiquity of videoconferencing-capable systems. These issues are:

1. Eye contact: Eye contact plays a large role in conversational turn-taking, perceived attention and intent, and other aspects of group communication. While traditional telephone conversations give no eye contact cues, many videoconferencing systems are arguably worse in that they provide an incorrect impression that the remote interlocutor is avoiding eye contact. Some telepresence systems have cameras located in the screens that reduce the amount of parallax observed by the users. This issue is also being addressed through research that generates a synthetic image with eye contact using stereo reconstruction. Telcordia Technologies, formerly Bell Communications Research, owns a patent for eye-to-eye videoconferencing using rear projection screens with the video camera behind it, evolved from a 1960s U.S. military system that provided videoconferencing services between the White House and various other government and military facilities. This technique eliminates the need for special cameras or image processing.

2. Appearance consciousness: A second psychological problem with videoconferencing is being on camera, with the video stream possibly even being recorded. The burden of presenting an acceptable on-screen appearance is not present in audio-only communication. Early studies by Alphonse Chapanis found that the addition of video actually impaired communication, possibly because of the consciousness of being on camera.

3. Signal latency: The information transport of digital signals in many steps need time. In a telecommunicated conversation, an increased latency (time lag) larger than about 150–300 ms becomes noticeable and is soon observed as unnatural and distracting. Therefore, next to a stable large bandwidth, a small total round-trip time is another major technical requirement for the communication channel for interactive videoconferencing.

The issue of eye-contact may be solved with advancing technology, and presumably the issue of appearance consciousness will fade as people become accustomed to videoconferencing.

Standards

The Tandberg E20 is an example of a SIP-only device. Such devices need to route calls through a Video Communication Server to be able to reach H.323 systems, a process known as "interworking" (2009).

The International Telecommunications Union (ITU) (formerly: Consultative Committee on International Telegraphy and Telephony (CCITT)) has three umbrellas of standards for videoconferencing:

- ITU H.320 is known as the standard for public switched telephone networks (PSTN) or videoconferencing over integrated services digital networks. While still prevalent in Europe, ISDN was never widely adopted in the United States and Canada.

- ITU H.264 Scalable Video Coding (SVC) is a compression standard that enables videoconferencing systems to achieve highly error resilient Internet Protocol (IP) video transmissions over the public Internet without quality-of-service enhanced lines. This standard has enabled wide scale deployment of high definition desktop videoconferencing and made possible new architectures, which reduces latency between the transmitting sources and receivers, resulting in more fluid communication without pauses. In addition, an attractive factor for IP videoconferencing is that it is easier to set up for use along with web conferencing and data collaboration. These combined technologies enable users to have a richer multimedia environment for live meetings, collaboration and presentations.

- ITU V.80: videoconferencing is generally compatibilized with H.324 standard point-to-point videotelephony over regular plain old telephone service (POTS) phone lines.

The Unified Communications Interoperability Forum (UCIF), a non-profit alliance between communications vendors, launched in May 2010. The organization's vision is to maximize the interoperability of UC based on existing standards. Founding members of UCIF include HP, Microsoft, Polycom, Logitech/LifeSize Communications and Juniper Networks.

Social and Institutional Impact

Impact on the General Public

High speed Internet connectivity has become more widely available at a reasonable cost and the cost of video capture and display technology has decreased. Consequently, personal videoconferencing systems based on a webcam, personal computer system, software compression and broadband Internet connectivity have become affordable to the general public. Also, the hardware used for this technology has continued to improve in quality, and prices have dropped dramatically. The availability of freeware (often as part of chat programs) has made software based videoconferencing accessible to many.

For over a century, futurists have envisioned a future where telephone conversations will take place as actual face-to-face encounters with video as well as audio. Sometimes it is simply not possible or practical to have face-to-face meetings with two or more people. Sometimes a telephone conversation or conference call is adequate. Other times, e-mail exchanges are adequate. However, videoconferencing adds another possible alternative, and can be considered when:

- a live conversation is needed;

- non-verbal (visual) information is an important component of the conversation;

- the parties of the conversation can't physically come to the same location; or

- the expense or time of travel is a consideration.

Deaf, hard-of-hearing and mute individuals have a particular interest in the development of affordable high-quality videoconferencing as a means of communicating with each other in sign language. Unlike Video Relay Service, which is intended to support communication between a caller using sign language and another party using spoken language, videoconferencing can be used directly between two deaf signers.

Mass adoption and use of videoconferencing is still relatively low, with the following often claimed as causes:

- Complexity of systems. Most users are not technical and want a simple interface. In hardware systems an unplugged cord or a flat battery in a remote control is seen as failure, contributing to perceived unreliability which drives users back to traditional meetings. Successful systems are backed by support teams who can pro-actively support and provide fast assistance when required.

- Perceived lack of interoperability: not all systems can readily interconnect, for example ISDN and IP systems require a gateway. Popular software solutions cannot easily connect to hardware systems. Some systems use different standards, features and qualities which can require additional configuration when connecting to dissimilar systems.

- Bandwidth and quality of service: In some countries it is difficult or expensive to get a high quality connection that is fast enough for good-quality video conferencing. Technologies such as ADSL have limited upload speeds and cannot upload and download simultaneously at full speed. As Internet speeds increase higher quality and high definition video conferencing will become more readily available.

- Expense of commercial systems: well-designed telepresence systems require specially designed rooms which can cost hundreds of thousands of dollars to fit out their rooms with codecs, integration equipment (such as Multipoint Control Units), high fidelity sound systems and furniture. Monthly charges may also be required for bridging services and high capacity broadband service.

- Self-consciousness about being on camera: especially for new users or older generations who may prefer less fidelity in their communications.

- Lack of direct eye contact, an issue being circumvented in some higher end systems.

These are some of the reasons many systems are often used for internal corporate use only, as they are less likely to result in lost sales. One alternative to companies lacking dedicated facilities is the rental of videoconferencing-equipped meeting rooms in cities around the world. Clients can book rooms and turn up for the meeting, with all technical aspects being prearranged and support being readily available if needed.

Impact on Government and Law

In the United States, videoconferencing has allowed testimony to be used for an individual who is unable or prefers not to attend the physical legal settings, or would be subjected to severe psychological stress in doing so, however there is a controversy on the use of testimony by foreign or unavailable witnesses via video transmission, regarding the violation of the Confrontation Clause of the Sixth Amendment of the U.S. Constitution.

In a military investigation in State of North Carolina, Afghan witnesses have testified via video-conferencing.

In Hall County, Georgia, videoconferencing systems are used for initial court appearances. The systems link jails with court rooms, reducing the expenses and security risks of transporting prisoners to the courtroom.

The U.S. Social Security Administration (SSA), which oversees the world's largest administrative judicial system under its Office of Disability Adjudication and Review (ODAR), has made extensive use of videoconferencing to conduct hearings at remote locations. In Fiscal Year (FY) 2009, the U.S. Social Security Administration (SSA) conducted 86,320 videoconferenced hearings, a 55% increase over FY 2008. In August 2010, the SSA opened its fifth and largest videoconferencing-only National Hearing Center (NHC), in St. Louis, Missouri. This continues the SSA's effort to use video hearings as a means to clear its substantial hearing backlog. Since 2007, the SSA has also established NHCs in Albuquerque, New Mexico, Baltimore, Maryland, Falls Church, Virginia, and Chicago, Illinois.

Impact on Education

Videoconferencing provides students with the opportunity to learn by participating in two-way communication forums. Furthermore, teachers and lecturers worldwide can be brought to remote or otherwise isolated educational facilities. Students from diverse communities and backgrounds can come together to learn about one another, although language barriers will continue to persist. Such students are able to explore, communicate, analyze and share information and ideas with one another. Through videoconferencing, students can visit other parts of the world to speak with their peers, and visit museums and educational facilities. Such virtual field trips can provide enriched learning opportunities to students, especially those in geographically isolated locations, and to the economically disadvantaged. Small schools can use these technologies to pool resources and provide courses, such as in foreign languages, which could not otherwise be offered.

A few examples of benefits that videoconferencing can provide in campus environments include:

- faculty members keeping in touch with classes while attending conferences;
- guest lecturers brought in classes from other institutions;
- researchers collaborating with colleagues at other institutions on a regular basis without loss of time due to travel;
- schools with multiple campuses collaborating and sharing professors;
- schools from two separate nations engaging in cross-cultural exchanges;
- faculty members participating in thesis defenses at other institutions;
- administrators on tight schedules collaborating on budget preparation from different parts of campus;

- faculty committee auditioning scholarship candidates;

- researchers answering questions about grant proposals from agencies or review committees;

- student interviews with an employers in other cities, and

- teleseminars.

Impact on Medicine and Health

Videoconferencing is a highly useful technology for real-time telemedicine and telenursing applications, such as diagnosis, consulting, transmission of medical images, etc... With videoconferencing, patients may contact nurses and physicians in emergency or routine situations; physicians and other paramedical professionals can discuss cases across large distances. Rural areas can use this technology for diagnostic purposes, thus saving lives and making more efficient use of health care money. For example, a rural medical center in Ohio, United States, used videoconferencing to successfully cut the number of transfers of sick infants to a hospital 70 miles (110 km) away. This had previously cost nearly $10,000 per transfer.

Special peripherals such as microscopes fitted with digital cameras, videoendoscopes, medical ultrasound imaging devices, otoscopes, etc., can be used in conjunction with videoconferencing equipment to transmit data about a patient. Recent developments in mobile collaboration on hand-held mobile devices have also extended video-conferencing capabilities to locations previously unreachable, such as a remote community, long-term care facility, or a patient's home.

Impact on Sign Language Communications

A deaf person using a video relay service at his workplace to communicate with a hearing person in London. (Courtesy: *SignVideo*)

A video relay service (VRS), also known as a 'video interpreting service' (VIS), is a service that allows deaf, hard-of-hearing and speech-impaired (D-HOH-SI) individuals to communicate by videoconferencing (or similar technologies) with hearing people in real-time, via a sign language interpreter.

A similar video interpreting service called video remote interpreting (VRI) is conducted through a different organization often called a "Video Interpreting Service Provider" (VISP). VRS is a newer form of telecommunication service to the D-HOH-SI community, which had, in the United States, started earlier in 1974 using a non-video technology called telecommunications relay service (TRS).

One of the first demonstrations of the ability for telecommunications to help sign language users communicate with each other occurred when AT&T's videophone (trademarked as the "Picturephone") was introduced to the public at the 1964 New York World's Fair –two deaf users were able to communicate freely with each other between the fair and another city. Various universities and other organizations, including British Telecom's Martlesham facility, have also conducted extensive research on signing via videotelephony. The use of sign language via videotelephony was hampered for many years due to the difficulty of its use over slow analogue copper phone lines, coupled with the high cost of better quality ISDN (data) phone lines. Those factors largely disappeared with the introduction of more efficient video codecs and the advent of lower cost high-speed ISDN data and IP (Internet) services in the 1990s.

VRS services have become well developed nationally in Sweden since 1997 and also in the United States since the first decade of the 2000s. With the exception of Sweden, VRS has been provided in Europe for only a few years since the mid-2000s, and as of 2010 has not been made available in many European Union countries, with most European countries still lacking the legislation or the financing for large-scale VRS services, and to provide the necessary telecommunication equipment to deaf users. Germany and the Nordic countries are among the other leaders in Europe, while the United States is another world leader in the provisioning of VRS services.

Impact on Business

Videoconferencing can enable individuals in distant locations to participate in meetings on short notice, with time and money savings. Technology such as VoIP can be used in conjunction with desktop videoconferencing to enable low-cost face-to-face business meetings without leaving the desk, especially for businesses with widespread offices. The technology is also used for telecommuting, in which employees work from home. One research report based on a sampling of 1,800 corporate employees showed that, as of June 2010, 54% of the respondents with access to video conferencing used it "all of the time" or "frequently".

Intel Corporation have used videoconferencing to reduce both costs and environmental impacts of its business operations.

Videoconferencing is also currently being introduced on online networking websites, in order to help businesses form profitable relationships quickly and efficiently without leaving their place of work. This has been leveraged by banks to connect busy banking professionals with customers in various locations using video banking technology.

Videoconferencing on hand-held mobile devices (mobile collaboration technology) is being used in industries such as manufacturing, energy, healthcare, insurance, government and public safety. Live, visual interaction removes traditional restrictions of distance and time, often in locations previously unreachable, such as a manufacturing plant floor a continent away.

In the increasingly globalized film industry, videoconferencing has become useful as a method by which creative talent in many different locations can collaborate closely on the complex details of film production. For example, for the 2013 award-winning animated film *Frozen*, Burbank-based Walt Disney Animation Studios hired the New York City-based husband-and-wife songwriting team of Robert Lopez and Kristen Anderson-Lopez to write the songs, which required two-hour-long transcontinental videoconferences nearly every weekday for about 14 months.

With the development of lower cost endpoints, cloud based infrastructure and technology trends such as WebRTC, Video Conferencing is moving from just a business-to-business offering, to a business-to-business and business-to-consumer offering.

Although videoconferencing has frequently proven its value, research has shown that some non-managerial employees prefer not to use it due to several factors, including anxiety. Some such anxieties can be avoided if managers use the technology as part of the normal course of business.

Remote workers can also adopt certain behaviors and best practices to stay connected with their co-workers and company.

Researchers also find that attendees of business and medical videoconferences must work harder to interpret information delivered during a conference than they would if they attended face-to-face. They recommend that those coordinating videoconferences make adjustments to their conferencing procedures and equipment.

Impact on Media Relations

The concept of press videoconferencing was developed in October 2007 by the PanAfrican Press Association (APPA), a Paris France-based non-governmental organization, to allow African journalists to participate in international press conferences on developmental and good governance issues.

Press videoconferencing permits international press conferences via videoconferencing over the Internet. Journalists can participate on an international press conference from any location, without leaving their offices or countries. They need only be seated by a computer connected to the Internet in order to ask their questions to the speaker.

In 2004, the International Monetary Fund introduced the Online Media Briefing Center, a password-protected site available only to professional journalists. The site enables the IMF to present press briefings globally and facilitates direct questions to briefers from the press. The site has been copied by other international organizations since its inception. More than 4,000 journalists worldwide are currently registered with the IMF.

Descriptive Names and Terminology

Videophone calls (also: *videocalls, video chat* as well as *Skype* and *Skyping* in verb form), differ from videoconferencing in that they expect to serve individuals, not groups. However that distinction has become increasingly blurred with technology improvements such as increased bandwidth and sophisticated software clients that can allow for multiple parties on a call. In general everyday usage the term *videoconferencing* is now frequently used instead of *videocall* for point-to-point calls between two units. Both videophone calls and videoconferencing are also now commonly referred to as a *video link*.

Webcams are popular, relatively low cost devices which can provide live video and audio streams via personal computers, and can be used with many software clients for both video calls and videoconferencing.

A *videoconference system* is generally higher cost than a videophone and deploys greater capabilities. A *videoconference* (also known as a *videoteleconference*) allows two or more locations to communicate via live, simultaneous two-way video and audio transmissions. This is often accomplished by the use of a multipoint control unit (a centralized distribution and call management system) or by a similar non-centralized multipoint capability embedded in each videoconferencing unit. Again, technology improvements have circumvented traditional definitions by allowing multiple party videoconferencing via web-based applications.

A *telepresence system* is a high-end videoconferencing system and service usually employed by enterprise-level corporate offices. Telepresence conference rooms use state-of-the art room designs, video cameras, displays, sound-systems and processors, coupled with high-to-very-high capacity bandwidth transmissions.

Typical use of the various technologies described above include calling or conferencing on a one-on-one, one-to-many or many-to-many basis for personal, business, educational, deaf Video Relay Service and tele-medical, diagnostic and rehabilitative use or services. New services utilizing videocalling and videoconferencing, such as teachers and psychologists conducting online sessions, personal videocalls to inmates incarcerated in penitentiaries, and videoconferencing to resolve airline engineering issues at maintenance facilities, are being created or evolving on an ongoing basis.

A telepresence robot (also telerobotics) is a robotically controlled and motorized videoconferencing display to help provide a better sense of remote physical presence for communication and collaboration in an office, home, school, etc... when one cannot be there in person. The robotic avatar and videoconferencing display-camera can move about and look around at the command of the remote person.

References

- Linzmayer, Owen W. (2004). Apple confidential 2.0 : the definitive history of the world's most colorful company ([Rev. 2. ed.]. ed.). San Francisco, Calif.: No Starch Press. ISBN 1-59327-010-0.

- Peterson, Kerstin Day (2000). Business telecom systems: a guide to choosing the best technologies and services. Focal Press. pp. 191–192. ISBN 1578200415. Retrieved 2011-04-02.

- Clint Smith, Curt Gervelis (2003). Wireless Network Performance Handbook. McGraw-Hill Professional. ISBN 0-07-140655-7.

- Clint Smith, Curt Gervelis (2003). Wireless Network Performance Handbook. McGraw-Hill Professional. ISBN 0-07-140655-7.

- Free radio: electronic civil disobedience by Lawrence C. Soley. Published by Westview Press, 1998. ISBN 0-8133-9064-8, ISBN 978-0-8133-9064-2

- Rebel Radio: The Full Story of British Pirate Radio by John Hind, Stephen Mosco. Published by Pluto Press, 1985. ISBN 0-7453-0055-3, ISBN 978-0-7453-0055-9

- Definition of "cable", The Macquarie Dictionary (3rd ed.). Australia: Macquarie Library. 1997. ISBN 0-949757-89-6. (n.) 4. a telegram sent abroad, especially by submarine cable. (v.) 9. to send a message by submarine cable.

- Ronalds, B.F. (2016). Sir Francis Ronalds: Father of the Electric Telegraph. London: Imperial College Press. ISBN 978-1-78326-917-4.

- Wilson, Arthur (1994). The Living Rock: The Story of Metals Since Earliest Times and Their Impact on Civilization. p. 203. Woodhead Publishing. ISBN 978-1-85573-301-5.

- "IEEE Multipath routing with adaptive playback scheduling for Voice over IP in Service Overlay Networks". Sarnoff Symposium, 2008 IEEE: 1–5. 28–30 April 2008. doi:10.1109/SARNOF.2008.4520089. ISBN 978-1-4244-1843-5.

- Firestone, Scott & Thiya Ramalingam, & Fry, Steve. Voice and Video Conferencing Fundamentals. Indianapolis, IN: Cisco Press, 2007, pg 10, ISBN 1-58705-268-7, ISBN 978-1-58705-268-2.

- Norman, Jeremy. "Francis Ronalds Builds the First Working Electric Telegraph (1816)". HistoryofInformation. com. Retrieved 1 May 2016.

- Ronalds, B.F. (Feb 2016). "The Bicentennial of Francis Ronalds's Electric Telegraph". Physics Today. doi:10.1063/PT.3.3079.

- Ronalds, B.F. (2016). "Sir Francis Ronalds and the Electric Telegraph". Int. J. for the History of Engineering & Technology. doi:10.1080/17581206.2015.1119481.

- Stuart Dredge. "How secure is your favourite messaging app? Today's Open Thread". the Guardian. Retrieved May 16, 2015.

- Alexia Tsotsis. "iMessage Has More Than 140M Users And Has Sent Over 150B Messages, With Over 1B Messages Sent Per Day". TechCrunch. AOL. Retrieved May 16, 2015.

- "WIRELESS: Carriers look to IP for back haul". www.eetimes.com. EE Times. Archived from the original on August 9, 2011. Retrieved 8 April 2015.

- "Mobile's IP challenge". www.totaltele.com. Total Telecom Online. Archived from the original on February 17, 2006. Retrieved 8 April 2015.

Essential Concepts of Telecommunications

Telecommunications have brought a transformation in communication in the lives of people. It has an equal impact on advertising. Communication channels, multiplexing, path protection and electrical length are some of the essential concepts of telecommunications. This text is a compilation of the essential concepts of telecommunications.

Channel (Communications)

In telecommunications and computer networking, a communication chanel or channel, refers either to a physical transmission medium such as a wire, or to a logical connection over a multiplexed medium such as a radio channel. A channel is used to convey an information signal, for example a digital bit stream, from one or several *senders* (or transmitters) to one or several *receivers*. A channel has a certain capacity for transmitting information, often measured by its bandwidth in Hz or its data rate in bits per second.

Old telephone wires are a challenging communications channel for modern digital communications.

Communicating data from one location to another requires some form of pathway or medium. These pathways, called communication channels, use two types of media: cable (twisted-pair wire, cable, and fiber-optic cable) and broadcast (microwave, satellite, radio, and infrared). Cable or wire line media use physical wires of cables to transmit data and information. Twisted-pair wire and coaxial cables are made of copper, and fiber-optic cable is made of glass.

In information theory, a channel refers to a theoretical *channel model* with certain error characteristics. In this more general view, a storage device is also a kind of channel, which can be sent to (written) and received from (read).

Examples

Examples of communications channels include:

1. A connection between initiating and terminating nodes of a circuit.

2. A single path provided by a transmission medium via either

    o   physical separation, such as by multipair cable or

    o   electrical separation, such as by frequency-division or time-division multiplexing.

3. A path for conveying electrical or electromagnetic signals, usually distinguished from other parallel paths.

    o   A storage which can communicate a message over time as well as space

    o   The portion of a storage medium, such as a track or band, that is accessible to a given reading or writing station or head.

    o   A buffer from which messages can be 'put' and 'got'.

4. In a communications system, the physical or logical link that connects a data source to a data sink.

5. A specific radio frequency, pair or band of frequencies, usually named with a letter, number, or codeword, and often allocated by international agreement.

6. Examples:

    o   Marine VHF radio uses some 88 channels in the VHF band for two-way FM voice communication. Channel 16, for example, is 156.800 MHz. In the US, seven additional channels, WX1 - WX7, are allocated for weather broadcasts.

    o   Television channels such as North American TV Channel 2 = 55.25 MHz, Channel 13 = 211.25 MHz. Each channel is 6 MHz wide. Besides these "physical channels", television also has "virtual channels".

    o   Wi-Fi consists of unlicensed channels 1-13 from 2412 MHz to 2484 MHz in 5 MHz steps.

    o   The radio channel between an amateur radio repeater and a ham uses two frequencies often 600 kHz (0.6 MHz) apart. For example, a repeater that transmits on 146.94 MHz typically listens for a ham transmitting on 146.34 MHz.

All of these communications channels share the property that they transfer information. The information is carried through the channel by a signal.

Channel Models

A channel can be modelled physically by trying to calculate the physical processes which modify the transmitted signal. For example, in wireless communications the channel can be modelled by calculating the reflection off every object in the environment. A sequence of random numbers might also be added in to simulate external interference and/or electronic noise in the receiver.

Statistically a communication channel is usually modelled as a triple consisting of an input alphabet, an output alphabet, and for each pair (i, o) of input and output elements a transition probability $p(i, o)$. Semantically, the transition probability is the probability that the symbol o is received given that i was transmitted over the channel.

Statistical and physical modelling can be combined. For example, in wireless communications the channel is often modelled by a random attenuation (known as fading) of the transmitted signal, followed by additive noise. The attenuation term is a simplification of the underlying physical processes and captures the change in signal power over the course of the transmission. The noise in the model captures external interference and/or electronic noise in the receiver. If the attenuation term is complex it also describes the relative time a signal takes to get through the channel. The statistics of the random attenuation are decided by previous measurements or physical simulations.

Channel models may be continuous channel models in that there is no limit to how precisely their values may be defined.

Communication channels are also studied in a discrete-alphabet setting. This corresponds to abstracting a real world communication system in which the analog $\rightarrow$ digital and digital $\rightarrow$ analog blocks are out of the control of the designer. The mathematical model consists of a transition probability that specifies an output distribution for each possible sequence of channel inputs. In information theory, it is common to start with memoryless channels in which the output probability distribution only depends on the current channel input.

A channel model may either be digital (quantified, e.g. binary) or analog.

Digital Channel Models

In a digital channel model, the transmitted message is modelled as a digital signal at a certain protocol layer. Underlying protocol layers, such as the physical layer transmission technique, is replaced by a simplified model. The model may reflect channel performance measures such as bit rate, bit errors, latency/delay, delay jitter, etc. Examples of digital channel models are:

- Binary symmetric channel (BSC), a discrete memoryless channel with a certain bit error probability

- Binary bursty bit error channel model, a channel "with memory"

- Binary erasure channel (BEC), a discrete channel with a certain bit error detection (erasure) probability

- Packet erasure channel, where packets are lost with a certain packet loss probability or packet error rate

- Arbitrarily varying channel (AVC), where the behavior and state of the channel can change randomly

Analog Channel Models

In an analog channel model, the transmitted message is modelled as an analog signal. The model can be a linear or non-linear, time-continuous or time-discrete (sampled), memoryless or dynamic (resulting in burst errors), time-invariant or time-variant (also resulting in burst errors), baseband, passband (RF signal model), real-valued or complex-valued signal model. The model may reflect the following channel impairments:

- Noise model, for example

  o Additive white Gaussian noise (AWGN) channel, a linear continuous memoryless model

  o Phase noise model

- Interference model, for example cross-talk (co-channel interference) and intersymbol interference (ISI)

- Distortion model, for example a non-linear channel model causing intermodulation distortion (IMD)

- Frequency response model, including attenuation and phase-shift

- Group delay model

- Modelling of underlying physical layer transmission techniques, for example a complex-valued equivalent baseband model of modulation and frequency response

- Radio frequency propagation model, for example

  o Log-distance path loss model

  o Fading model, for example Rayleigh fading, Ricean fading, log-normal shadow fading and frequency selective (dispersive) fading

  o Doppler shift model, which combined with fading results in a time-variant system

  o Ray tracing models, which attempt to model the signal propagation and distortions for specified transmitter-receiver geometries, terrain types, and antennas

  o Mobility models, which also causes a time-variant system

Types

- Digital (discrete) or analog (continuous) channel

- Transmission medium, for example a fibre channel

- Multiplexed channel

- Computer network virtual channel

- Simplex communication, duplex communication or half duplex communication channel

- Return channel

- Uplink or downlink (upstream or downstream channel)

- Broadcast channel, unicast channel or multicast channel

Channel Performance Measures

These are examples of commonly used channel capacity and performance measures:

- Spectral bandwidth in Hertz

- Symbol rate in baud, pulses/s or symbols/s

- Digital bandwidth bit/s measures: gross bit rate (signalling rate), net bit rate (information rate), channel capacity, and maximum throughput

- Channel utilization

- Link spectral efficiency

- Signal-to-noise ratio measures: signal-to-interference ratio, Eb/No, carrier-to-interference ratio in decibel

- Bit-error rate (BER), packet-error rate (PER)

- Latency in seconds: propagation time, transmission time

- Delay jitter

Multi-terminal Channels, with Application to Cellular Systems

In networks, as opposed to point-to-point communication, the communication media is shared between multiple nodes (terminals). Depending on the type of communication, different terminals can cooperate or interfere on each other. In general, any complex multi-terminal network can be considered as a combination of simplified multi-terminal channels. The following channels are the principal multi-terminal channels which was first introduced in the field of information theory:

- A point-to-multipoint channel, also known as broadcasting medium: In this channel, a single sender transmits multiple messages to different destination nodes. All wireless channels except radio links can be considered as broadcasting media, but may not always provide broadcasting service. The downlink of a cellular system can be considered as a point-to-multipoint channel, if only one cell is considered and inter-cell co-channel interference is neglected. However, the communication service of a phone call is unicasting.

- Multiple access channel: In this channel, multiple senders transmit multiple possible different messages over a shared physical medium to one or several destination nodes. This requires a channel access scheme, including a media access control (MAC) protocol combined with a multiplexing scheme. This channel model has applications in the uplink of the cellular networks.

- Relay channel: In this channel, one or several intermediate nodes (called relay, repeater or gap filler nodes) cooperate with a sender to send the message to an ultimate destination node. Relay nodes are considered as a possible add-on in the upcoming cellular standards like 3GPP Long Term Evolution (LTE).

- Interference channel: In this channel, two different senders transmit their data to different destination nodes. Hence, the different senders can have a possible cross-talk or co-channel interference on the signal of each other. The inter-cell interference in the cellular wireless communications is an example of the interference channel. In spread spectrum systems like 3G, interference also occur inside the cell if non-orthogonal codes are used.

- A unicasting channel is a channel that provides a unicasting service, i.e. that sends data addressed to one specific user. An established phone call is an example.

- A broadcasting channel is a channel that provides a broadcasting service, i.e. that sends data addressed to all users in the network. Cellular network examples are the paging service as well as the Multimedia Broadcast Multicast Service.

- A multicasting channel is a channel where data is addressed to a group of subscribing users. LTE examples are the Physical Multicast Channel (PMCH) and MBSFN (Multicast Broadcast Single Frequency Network).

From the above 4 basic multi-terminal channels, multiple access channel is the only one whose capacity region is known. Even for the special case of the Gaussian scenario, the capacity region of the other 3 channels except the broadcast channel is unknown in general.

Multiplexing

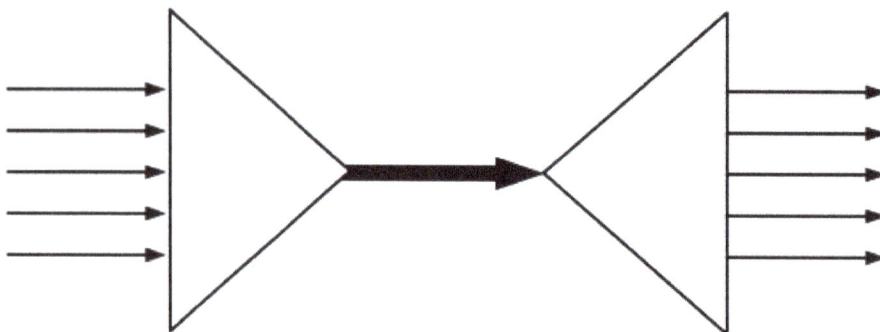

Multiple low data rate signals are multiplexed over a single high data rate link, then demultiplexed at the other end

In telecommunications and computer networks, multiplexing (sometimes contracted to muxing) is a method by which multiple analog or digital signals are combined into one signal over a shared medium. The aim is to share an expensive resource. For example, in telecommunications, several telephone calls may be carried using one wire. Multiplexing originated in telegraphy in the 1870s, and is now widely applied in communications. In telephony, George Owen Squier is credited with the development of telephone carrier multiplexing in 1910.

The multiplexed signal is transmitted over a communication channel such as a cable. The multiplexing divides the capacity of the communication channel into several logical channels, one for each message signal or data stream to be transferred. A reverse process, known as demultiplexing, extracts the original channels on the receiver end.

A device that performs the multiplexing is called a multiplexer (MUX), and a device that performs the reverse process is called a demultiplexer (DEMUX or DMX).

Inverse multiplexing (IMUX) has the opposite aim as multiplexing, namely to break one data stream into several streams, transfer them simultaneously over several communication channels, and recreate the original data stream.

Types

Multiple variable bit rate digital bit streams may be transferred efficiently over a single fixed bandwidth channel by means of statistical multiplexing. This is an asynchronous mode time-domain multiplexing which is a form of time-division multiplexing.

Digital bit streams can be transferred over an analog channel by means of code-division multiplexing techniques such as frequency-hopping spread spectrum (FHSS) and direct-sequence spread spectrum (DSSS).

In wireless communications, multiplexing can also be accomplished through alternating polarization (horizontal/vertical or clockwise/counterclockwise) on each adjacent channel and satellite, or through phased multi-antenna array combined with a multiple-input multiple-output communications (MIMO) scheme.

Space-division Multiplexing

In wired communication, space-division multiplexing, also known as Space-division multiple access is the use of separate point-to-point electrical conductors for each transmitted channel. Examples include an analogue stereo audio cable, with one pair of wires for the left channel and another for the right channel, and a multi-pair telephone cable, a switched star network such as a telephone access network, a switched Ethernet network, and a mesh network.

In wireless communication, space-division multiplexing is achieved with multiple antenna elements forming a phased array antenna. Examples are multiple-input and multiple-output (MIMO), single-input and multiple-output (SIMO) and multiple-input and single-output (MISO) multiplexing. An IEEE 802.11n wireless router with k antennas makes it in principle possible to communicate with k multiplexed channels, each with a peak bit rate of 54 Mbit/s, thus increasing the total peak bit rate by the factor k. Different antennas would give different multi-path propagation (echo) signatures, making it possible for digital signal processing techniques to separate different signals from each other. These techniques may also be utilized for space diversity (improved robustness to fading) or beamforming (improved selectivity) rather than multiplexing

Frequency-division Multiplexing

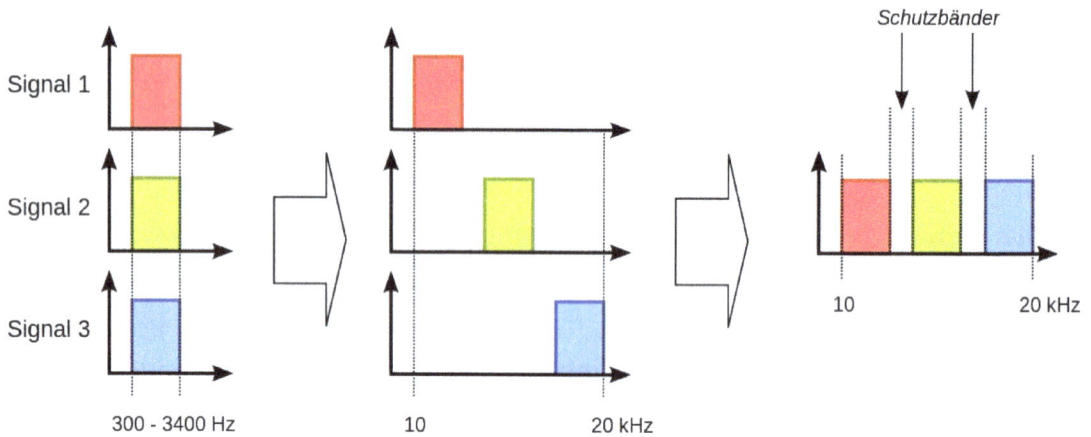

Frequency-division multiplexing (FDM): The spectrum of each input signal is shifted to a distinct frequency range.

Frequency-division multiplexing (FDM) is inherently an analog technology. FDM achieves the combining of several signals into one medium by sending signals in several distinct frequency ranges over a single medium. In FDM the signals are electrical signals. One of the most common applications for FDM is traditional radio and television broadcasting from terrestrial, mobile or satellite stations, or cable television. Only one cable reaches a customer's residential area, but the service provider can send multiple television channels or signals simultaneously over that cable to all subscribers without interference. Receivers must tune to the appropriate frequency (channel) to access the desired signal.

A variant technology, called wavelength-division multiplexing (WDM) is used in optical communications.

Time-division Multiplexing

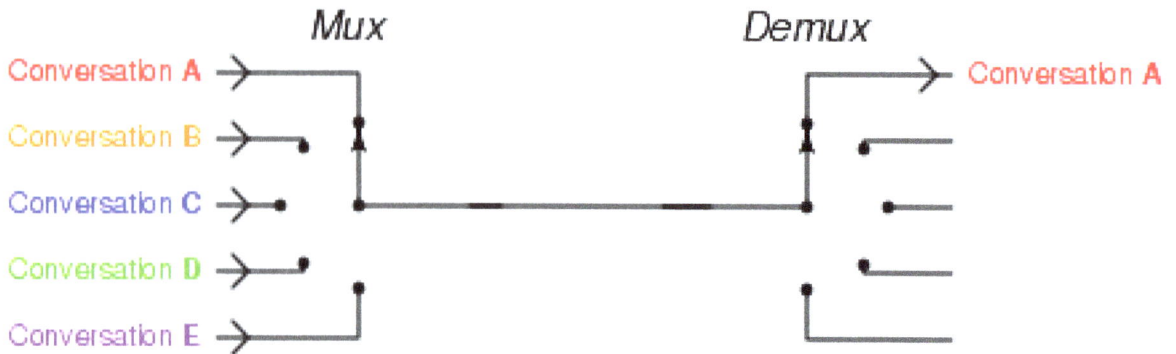

Time-division multiplexing (TDM).

Time-division multiplexing (TDM) is a digital (or in rare cases, analog) technology which uses time, instead of space or frequency, to separate the different data streams. TDM involves sequencing groups of a few bits or bytes from each individual input stream, one after the other, and in such a way that they can be associated with the appropriate receiver. If done sufficiently quickly, the receiving devices will not detect that some of the circuit time was used to serve another logical communication path.

Consider an application requiring four terminals at an airport to reach a central computer. Each terminal communicated at 2400 baud, so rather than acquire four individual circuits to carry such a low-speed transmission, the airline has installed a pair of multiplexers. A pair of 9600 baud modems and one dedicated analog communications circuit from the airport ticket desk back to the airline data center are also installed. Some web proxy servers (e.g. polipo) use TDM in HTTP pipelining of multiple HTTP transactions onto the same TCP/IP connection.

Carrier sense multiple access and multidrop communication methods are similar to time-division multiplexing in that multiple data streams are separated by time on the same medium, but because the signals have separate origins instead of being combined into a single signal, are best viewed as channel access methods, rather than a form of multiplexing.

Polarization-division Multiplexing

Polarization-division multiplexing uses the polarization of electromagnetic radiation to separate orthogonal channels. It is in practical use in both radio and optical communications, particularly in 100 Gbit/s per channel fiber optic transmission systems.

Orbital Angular Momentum Multiplexing

Orbital angular momentum multiplexing is a relatively new and experimental technique for multiplexing multiple channels of signals carried using electromagnetic radiation over a single path. It can potentially be used in addition to other physical multiplexing methods to greatly expand the transmission capacity of such systems. As of 2012 it is still in its early research phase, with small-scale laboratory demonstrations of bandwidths of up to 2.5 Tbit/s over a single light path.

Code-division Multiplexing

Code division multiplexing (CDM), Code division multiple access (CDMA) or spread spectrum is a class of techniques where several channels simultaneously share the same frequency spectrum, and this spectral bandwidth is much higher than the bit rate or symbol rate. One form is frequency hopping, another is direct sequence spread spectrum. In the latter case, each channel transmits its bits as a coded channel-specific sequence of pulses called chips. Number of chips per bit, or chips per symbol, is the spreading factor. This coded transmission typically is accomplished by transmitting a unique time-dependent series of short pulses, which are placed within chip times within the larger bit time. All channels, each with a different code, can be transmitted on the same fiber or radio channel or other medium, and asynchronously demultiplexed. Advantages over conventional techniques are that variable bandwidth is possible (just as in statistical multiplexing), that the wide bandwidth allows poor signal-to-noise ratio according to Shannon-Hartley theorem, and that multi-path propagation in wireless communication can be combated by rake receivers.

A significant application of CDMA is the Global Positioning System (GPS).

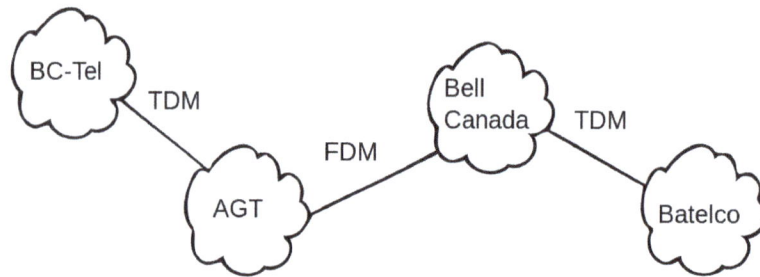

Telecommunication multiplexing

Multiple Access Method

A multiplexing technique may be further extended into a multiple access method or channel access method, for example, TDM into time division multiple access (TDMA) and statistical multiplexing into carrier sense multiple access (CSMA). A multiple access method makes it possible for several transmitters connected to the same physical medium to share its capacity.

Multiplexing is provided by the Physical Layer of the OSI model, while multiple access also involves a media access control protocol, which is part of the Data Link Layer.

The Transport layer in the OSI model as well as TCP/IP model provides statistical multiplexing of several application layer data flows to/from the same computer.

Code Division Multiplexing (CDM) is a technique in which each channel transmits its bits as a coded channel-specific sequence of pulses. This coded transmission typically is accomplished by transmitting a unique time-dependent series of short pulses, which are placed within chip times within the larger bit time. All channels, each with a different code, can be transmitted on the same fiber and asynchronously demultiplexed. Other widely used multiple access techniques are Time Division Multiple Access (TDMA) and Frequency Division Multiple Access (FDMA). Code Division Multiplex techniques are used as an access technology, namely Code Division Multiple Access (CDMA), in Universal Mobile Telecommunications System (UMTS) standard for the third generation (3G) mobile communication identified by the ITU.

Application Areas

Telegraphy

The earliest communication technology using electrical wires, and therefore sharing an interest in the economies afforded by multiplexing, was the electric telegraph. Early experiments allowed two separate messages to travel in opposite directions simultaneously, first using an electric battery at both ends, then at only one end.

- Émile Baudot developed a time-multiplexing system of multiple Hughes machines in the 1870s.

- In 1874, the quadruplex telegraph developed by Thomas Edison transmitted two messages in each direction simultaneously, for a total of four messages transiting the same wire at the same time.

- Several workers were investigating acoustic telegraphy, a frequency-division multiplexing technique, which led to the invention of the telephone.

Telephony

In telephony, a customer's telephone line now typically ends at the remote concentrator box, where it is multiplexed along with other telephone lines for that neighborhood or other similar area. The multiplexed signal is then carried to the central switching office on significantly fewer wires and for much further distances than a customer's line can practically go. This is likewise also true for digital subscriber lines (DSL).

Fiber in the loop (FITL) is a common method of multiplexing, which uses optical fiber as the backbone. It not only connects POTS phone lines with the rest of the PSTN, but also replaces DSL by connecting directly to Ethernet wired into the home. Asynchronous Transfer Mode is often the communications protocol used.

Because all the phone (and data) lines have been clumped together, none of them can be accessed except through a demultiplexer. Where such demultiplexers are uncommon, this provides for more-secure communications, though the connections are not typically encrypted.

Cable TV has long carried multiplexed television channels, and late in the 20th century began offering the same services as telephone companies. IPTV also depends on multiplexing.

Video Processing

In video editing and processing systems, multiplexing refers to the process of interleaving audio and video into one coherent data stream.

In digital video, such a transport stream is normally a feature of a container format which may include metadata and other information, such as subtitles. The audio and video streams may have variable bit rate. Software that produces such a transport stream and/or container is commonly called a statistical multiplexer or muxer. A demuxer is software that extracts or otherwise makes available for separate processing the components of such a stream or container.

Digital Broadcasting

In digital television and digital radio systems, several variable bit-rate data streams are multiplexed together to a fixed bitrate transport stream by means of statistical multiplexing. This makes it possible to transfer several video and audio channels simultaneously over the same frequency channel, together with various services.

In the digital television systems, this may involve several standard definition television (SDTV) programmes (particularly on DVB-T, DVB-S2, ISDB and ATSC-C), or one HDTV, possibly with a single SDTV companion channel over one 6 to 8 MHz-wide TV channel. The device that accomplishes this is called a statistical multiplexer. In several of these systems, the multiplexing results in an MPEG transport stream. The newer DVB standards DVB-S2 and DVB-T2 has the capacity to carry several HDTV channels in one multiplex. Even the original DVB standards can carry more HDTV channels in a multiplex if the most advanced MPEG-4 compressions hardware is used.

On communications satellites which carry broadcast television networks and radio networks, this is known as multiple channel per carrier or MCPC. Where multiplexing is not practical (such as where there are different sources using a single transponder), single channel per carrier mode is used.

Signal multiplexing of satellite TV and radio channels is typically carried out in a central signal playout and uplink centre, such as SES Platform Services in Germany, which provides playout, digital archiving, encryption, and satellite uplinks, as well as multiplexing, for hundreds of digital TV and radio channels.

In digital radio, both the Digital Audio Broadcasting (DAB) Eureka 147 system of digital audio broadcasting and the in-band on-channel HD Radio, FMeXtra, and Digital Radio Mondiale systems can multiplex channels. This is essentially required with DAB-type transmissions (where a multiplex is called a DAB ensemble), but is entirely optional with IBOC systems.

Analog Broadcasting

In FM broadcasting and other analog radio media, multiplexing is a term commonly given to the process of adding subcarriers to the audio signal before it enters the transmitter, where modulation occurs. (In fact, the stereo multiplex signal can be generated using time-division multiplexing, by switching between the two (left channel and right channel) input signals at an ultrasonic rate (the subcarrier), and then filtering out the higher harmonics.) Multiplexing in this sense is sometimes known as MPX, which in turn is also an old term for stereophonic FM, seen on stereo systems since the 1960s.

Other Meanings

In spectroscopy the term is used to indicate that the experiment is performed with a mixture of frequencies at once and their respective response unravelled afterwards using the Fourier transform principle.

In computer programming, it may refer to using a single in-memory resource (such as a file handle) to handle multiple external resources (such as on-disk files).

Some electrical multiplexing techniques do not require a physical "multiplexer" device, they refer to a "keyboard matrix" or "Charlieplexing" design style:

- Multiplexing may refer to the design of a multiplexed display (non-multiplexed displays are immune to break up).

- Multiplexing may refer to the design of a "switch matrix" (non-multiplexed buttons are immune to "phantom keys" and also immune to "phantom key blocking").

Types of Multiplexing

Frequency-division Multiplexing

In telecommunications, frequency-division multiplexing (FDM) is a technique by which the total bandwidth available in a communication medium is divided into a series of non-overlapping frequency sub-bands, each of which is used to carry a separate signal. This allows a single transmission medium such as the radio spectrum, a cable or optical fiber to be shared by multiple independent signals. Another use is to carry separate serial bits or segments of a higher rate signal in parallel.

The most natural example of frequency-division multiplexing is radio and television broadcasting, in which multiple radio signals at different frequencies pass through the air at the same time. Another example is cable television, in which many television channels are carried simultaneously on a single cable. FDM is also used by telephone systems to transmit multiple telephone calls through high capacity trunklines, communications satellites to transmit multiple channels of data on uplink and downlink radio beams, and broadband DSL modems to transmit large amounts of computer data through twisted pair telephone lines, among many other uses.

An analogous technique called wavelength division multiplexing is used in fiber-optic communication, in which multiple channels of data are transmitted over a single optical fiber using different wavelengths (frequencies) of light.

How It Works

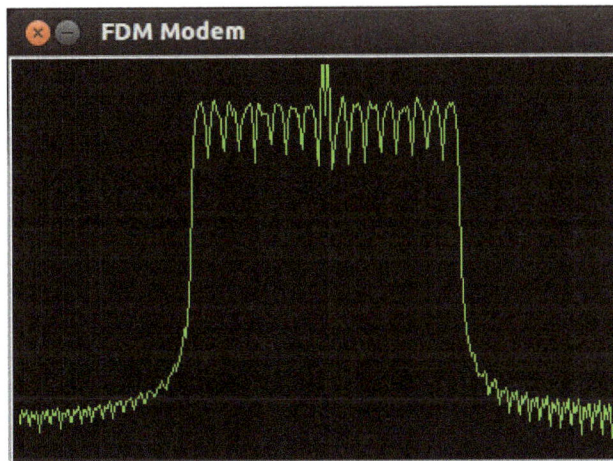

The passband of an FDM channel carrying digital data, modulated by QPSK quadrature phase-shift keying.

The multiple separate information (modulation) signals that are sent over an FDM system, such as the video signals of the television channels that are sent over a cable TV system, are called baseband signals. At the source end, for each frequency channel, an electronic oscillator generates a *carrier* signal, a steady oscillating waveform at a single frequency that serves to "carry" information. The carrier is much higher in frequency than the baseband signal. The carrier signal and the baseband signal are combined in a modulator circuit. The modulator alters some aspect of the carrier signal, such as its amplitude, frequency, or phase, with the baseband signal, "piggybacking" the data onto the carrier.

The result of modulating (mixing) the carrier with the baseband signal is to generate sub-frequencies near the carrier frequency, at the sum $(f_C + f_B)$ and difference $(f_C - f_B)$ of the frequencies. The information from the modulated signal is carried in sidebands on each side of the carrier frequency. Therefore, all the information carried by the channel is in a narrow band of frequencies clustered around the carrier frequency, this is called the passband of the channel.

Similarly, additional baseband signals are used to modulate carriers at other frequencies, creating other channels of information. The carriers are spaced far enough apart in frequency that the band of frequencies occupied by each channel, the passbands of the separate channels, do not overlap. All the channels are sent through the transmission medium, such as a coaxial cable, optical fiber, or through the air using a radio transmitter. As long as the channel frequencies are spaced far enough apart that none of the passbands overlap, the separate channels will not interfere with each another. Thus the available bandwidth is divided into "slots" or channels, each of which can carry a separate modulated signal.

For example, the coaxial cable used by cable television systems has a bandwidth of about 1000 MHz, but the passband of each television channel is only 6 MHz wide, so there is room for many channels on the cable (in modern digital cable systems each channel in turn is subdivided into subchannels and can carry up to 10 digital television channels).

At the destination end of the cable or fiber, or the radio receiver, for each channel a local oscillator produces a signal at the carrier frequency of that channel, that is mixed with the incoming modulated signal. The frequencies subtract, producing the baseband signal for that channel again. This is called demodulation. The resulting baseband signal is filtered out of the other frequencies and output to the user.

Telephone

For long distance telephone connections, 20th century telephone companies used L-carrier and similar co-axial cable systems carrying thousands of voice circuits multiplexed in multiple stages by channel banks.

For shorter distances, cheaper balanced pair cables were used for various systems including Bell System K- and N-Carrier. Those cables didn't allow such large bandwidths, so only 12 voice chan-nels (double sideband) and later 24 (single sideband) were multiplexed into four wires, one pair for each direction with repeaters every several miles, approximately 10 km. By the end of the 20th Century, FDM voice circuits had become rare. Modern telephone systems employ digital transmission, in which time-division multiplexing (TDM) is used instead of FDM.

Since the late 20th century digital subscriber lines (DSL) have used a Discrete multitone (DMT) system to divide their spectrum into frequency channels.

The concept corresponding to frequency-division multiplexing in the optical domain is known as wavelength-division multiplexing.

Group and Supergroup

A once commonplace FDM system, used for example in L-carrier, uses crystal filters which operate at the 8 MHz range to form a Channel Group of 12 channels, 48 kHz bandwidth in the range 8140 to 8188 kHz by selecting carriers in the range 8140 to 8184 kHz selecting upper sideband this group can then be translated to the standard range 60 to 108 kHz by a carrier of 8248 kHz. Such systems are used in DTL (Direct To Line) and DFSG (Directly formed super group).

132 voice channels (2SG + 1G) can be formed using DTL plane the modulation and frequency plan are given in FIG1 and FIG2 use of DTL technique allows the formation of a maximum of 132 voice channels that can be placed direct to line. DTL eliminates group and super group equipment.

DFSG can take similar steps where a direct formation of a number of super groups can be obtained in the 8 kHz the DFSG also eliminates group equipment and can offer:

- Reduction in cost 7% to 13%

- Less equipment to install and maintain

- Increased reliability due to less equipment

Both DTL and DFSG can fit the requirement of low density system (using DTL) and higher density system (using DFSG). The DFSG terminal is similar to DTL terminal except instead of two super groups many super groups are combined. A Mastergroup of 600 channels (10 super-groups) is an example based on DFSG.

Other Examples

FDM can also be used to combine signals before final modulation onto a carrier wave. In this case the carrier signals are referred to as subcarriers: an example is stereo FM transmission, where a 38 kHz subcarrier is used to separate the left-right difference signal from the central left-right sum channel, prior to the frequency modulation of the composite signal. An analog NTSC television channel is divided into subcarrier frequencies for video, color, and audio. DSL uses different frequencies for voice and for upstream and downstream data transmission on the same conductors, which is also an example of frequency duplex.

Where frequency-division multiplexing is used as to allow multiple users to share a physical communications channel, it is called frequency-division multiple access (FDMA).

FDMA is the traditional way of separating radio signals from different transmitters.

In the 1860s and 70s, several inventors attempted FDM under the names of acoustic telegraphy and harmonic telegraphy. Practical FDM was only achieved in the electronic age. Meanwhile, their efforts led to an elementary understanding of electroacoustic technology, resulting in the invention of the telephone.

Wavelength-division Multiplexing

In fiber-optic communications, wavelength-division multiplexing (WDM) is a technology which multiplexes a number of optical carrier signals onto a single optical fiber by using different wavelengths (i.e., colors) of laser light. This technique enables bidirectional communications over one strand of fiber, as well as multiplication of capacity.

The term *wavelength-division multiplexing* is commonly applied to an optical carrier (which is typically described by its wavelength), whereas frequency-division multiplexing typically applies to a radio carrier (which is more often described by frequency). Since *wavelength* and *frequency* are tied together through a simple directly inverse relationship, in which the product of frequency and wavelength equals **c** (the propagation speed of light), the two terms actually describe the same concept.

WDM Systems

wavelength-division multiplexing (WDM)

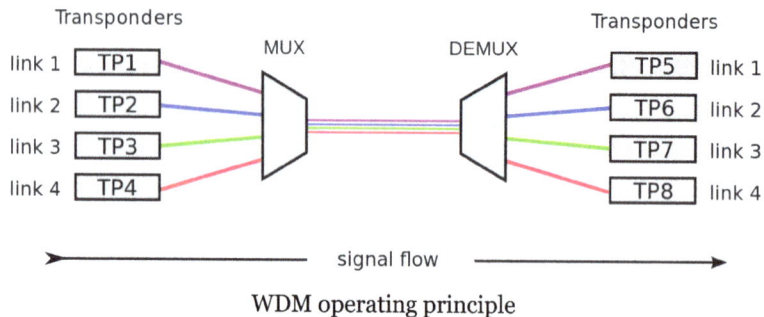

WDM operating principle

Nortel's WDM System

A WDM system uses a multiplexer at the transmitter to join the several signals together, and a demultiplexer at the receiver to split them apart. With the right type of fiber it is possible to have a device that does both simultaneously, and can function as an optical add-drop multiplexer. The optical filtering devices used have conventionally been etalons (stable solid-state single-frequency Fabry–Pérot interferometers in the form of thin-film-coated optical glass). As there are three different WDM types, whereof one is called "WDM", we would normally use the notation "xWDM" when discussing the technology as such.

The concept was first published in 1978, and by 1980 WDM systems were being realized in the laboratory. The first WDM systems combined only two signals. Modern systems can handle 160 signals and can thus expand a basic 100 Gbit/s system over a single fiber pair to over 16 Tbit/s. A system of 320 channels in also present.

WDM systems are popular with telecommunications companies because they allow them to expand the capacity of the network without laying more fiber. By using WDM and optical amplifiers, they can accommodate several generations of technology development in their optical infrastructure without having to overhaul the backbone network. Capacity of a given link can be expanded simply by upgrading the multiplexers and demultiplexers at each end.

This is often done by use of optical-to-electrical-to-optical (O/E/O) translation at the very edge of the transport network, thus permitting interoperation with existing equipment with optical interfaces.

Most WDM systems operate on single-mode fiber optical cables, which have a core diameter of 9 μm. Certain forms of WDM can also be used in multi-mode fiber cables (also known as premises cables) which have core diameters of 50 or 62.5 μm.

Early WDM systems were expensive and complicated to run. However, recent standardization and better understanding of the dynamics of WDM systems have made WDM less expensive to deploy.

Optical receivers, in contrast to laser sources, tend to be wideband devices. Therefore, the demultiplexer must provide the wavelength selectivity of the receiver in the WDM system.

WDM systems are divided into three different wavelength patterns, normal (WDM), coarse (CWDM) and dense (DWDM). Normal WDM (sometimes called BWDM) uses the two normal wavelengths 1310 and 1550 on one fiber. Coarse WDM provides up to 16 channels across multiple transmission windows of silica fibers. *Dense wavelength division multiplexing* (DWDM) uses the C-Band(1530 nm-1560 nm) transmission window but with denser channel spacing. Channel plans vary, but a typical DWDM system would use 40 channels at 100 GHz spacing or 80 channels with 50 GHz spacing. Some technologies are capable of 12.5 GHz spacing (sometimes called ultra dense WDM). New amplification options (Raman amplification) enable the extension of the usable wave-lengths to the L-band, more or less doubling these numbers.

Coarse wavelength division multiplexing (CWDM) in contrast to DWDM uses increased channel spacing to allow less sophisticated and thus cheaper transceiver designs. To provide 16 channels on a single fiber CWDM uses the entire frequency band spanning the second and third transmission window (1310/1550 nm respectively) including both windows (minimum dispersion window and minimum attenuation window) but also the critical area where OH scattering may occur, recommending the use of OH-free silica fibers in case the wavelengths between second and third transmission windows are to be used. Avoiding this region, the channels 47, 49, 51, 53, 55, 57, 59, 61 remain and these are the most commonly used. With OS2 fibers the water peak problem is overcome, and all possible 18 channels can be used.

WDM, DWDM and CWDM are based on the same concept of using multiple wavelengths of light on a single fiber, but differ in the spacing of the wavelengths, number of channels, and the ability to amplify the multiplexed signals in the optical space. EDFA provide an efficient wideband amplification for the C-band, Raman amplification adds a mechanism for amplification in the L-band. For CWDM, wideband optical amplification is not available, limiting the optical spans to several tens of kilometres.

Coarse WDM

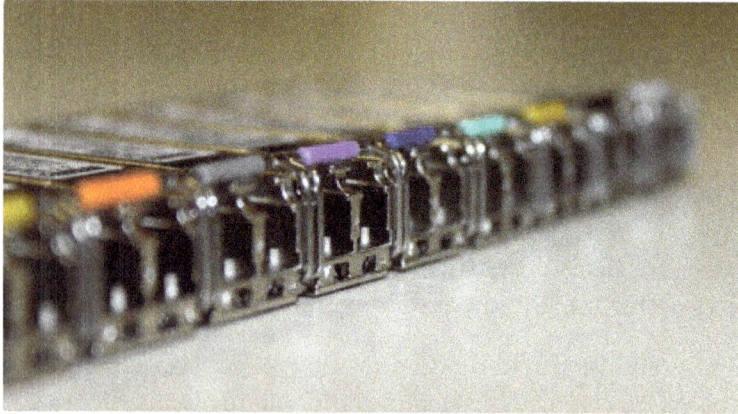

Series of SFP+ transceivers for 10 Gbit/s WDM communications

Originally, the term "coarse wavelength division multiplexing" was fairly generic, and meant a number of different things. In general, these things shared the fact that the choice of channel spacings and frequency stability was such that erbium doped fiber amplifiers (EDFAs) could not be utilized. Prior to the relatively recent ITU standardization of the term, one common meaning for coarse WDM meant two (or possibly more) signals multiplexed onto a single fiber, where one signal was in the 1550 nm band, and the other in the 1310 nm band.

In 2002 the ITU standardized a channel spacing grid for use with CWDM (ITU-T G.694.2), using the wavelengths from 1270 nm through 1610 nm with a channel spacing of 20 nm. (G.694.2 was revised in 2003 to shift the actual channel centers by 1 nm, so that strictly speaking the center wavelengths are 1271 to 1611 nm). Many CWDM wavelengths below 1470 nm are considered "unusable" on older G.652 specification fibers, due to the increased attenuation in the 1270–1470 nm bands. Newer fibers which conform to the G.652.C and G.652.D standards, such as Corning SMF-28e and Samsung Widepass nearly eliminate the "water peak" attenuation peak and allow for full operation of all 18 ITU CWDM channels in metropolitan networks.

The 10GBASE-LX4 10 Gbit/s physical layer standard is an example of a CWDM system in which four wavelengths near 1310 nm, each carrying a 3.125 gigabit-per-second (Gbit/s) data stream, are used to carry 10 Gbit/s of aggregate data.

The main characteristic of the recent ITU CWDM standard is that the signals are not spaced appropriately for amplification by EDFAs. This therefore limits the total CWDM optical span to somewhere near 60 km for a 2.5 Gbit/s signal, which is suitable for use in metropolitan applications. The relaxed optical frequency stabilization requirements allow the associated costs of CWDM to approach those of non-WDM optical components.

CWDM is also being used in cable television networks, where different wavelengths are used for the *downstream* and *upstream* signals. In these systems, the wavelengths used are often widely separated, for example the downstream signal might be at 1310 nm while the upstream signal is at 1550 nm.

An interesting and relatively recent development relating coarse WDM is the creation of GBIC and small form factor pluggable (SFP) transceivers utilizing standardized CWDM wavelengths. GBIC and SFP optics allow for something very close to a seamless upgrade in even legacy systems that support SFP interfaces. Thus, a legacy switch system can be easily "converted" to allow wavelength multiplexed transport over a fiber simply by judicious choice of transceiver wavelengths, combined with an inexpensive passive optical multiplexing device.

Passive CWDM is an implementation of CWDM that uses no electrical power. It separates the wavelengths using passive optical components such as bandpass filters and prisms. Many manufacturers are promoting passive CWDM to deploy fiber to the home.

Dense WDM

Dense wavelength division multiplexing (DWDM) refers originally to optical signals multiplexed within the 1550 nm band so as to leverage the capabilities (and cost) of erbium doped fiber amplifiers (EDFAs), which are effective for wavelengths between approximately 1525–1565 nm (C band), or 1570–1610 nm (L band). EDFAs were originally developed to replace SONET/SDH optical-electrical-optical (OEO) regenerators, which they have made practically obsolete. EDFAs can amplify any optical signal in their operating range, regardless of the modulated bit rate. In terms of multi-wavelength signals, so long as the EDFA has enough pump energy available to it, it can amplify as many optical signals as can be multiplexed into its amplification band (though signal densities are limited by choice of modulation format). EDFAs therefore allow a single-channel optical link to be upgraded in bit rate by replacing only equipment at the ends of the link, while retaining the existing EDFA or series of EDFAs through a long haul route. Furthermore, single-wavelength links using EDFAs can similarly be upgraded to WDM links at reasonable cost. The EDFA's cost is thus leveraged across as many channels as can be multiplexed into the 1550 nm band.

DWDM Systems

At this stage, a basic DWDM system contains several main components:

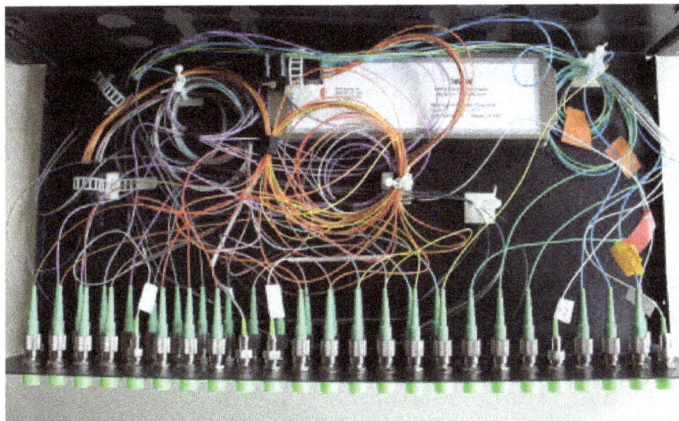

WDM multiplexer for DWDM communications

1. A DWDM terminal multiplexer. The terminal multiplexer contains a wavelength-converting transponder for each data signal, an optical multiplexer and where necessary an optical amplifier (EDFA). Each wavelength-converting transponder receives an optical data signal from the client-layer, such as Synchronous optical networking [SONET /SDH] or another type of data signal, converts this signal into the electrical domain and re-transmits the

signal at a specific wavelength using a 1,550 nm band laser. These data signals are then combined together into a multi-wavelength optical signal using an optical multiplexer, for transmission over a single fiber (e.g., SMF-28 fiber). The terminal multiplexer may or may not also include a local transmit EDFA for power amplification of the multi-wavelength optical signal. In the mid-1990s DWDM systems contained 4 or 8 wavelength-converting transponders; by 2000 or so, commercial systems capable of carrying 128 signals were available.

2. An intermediate line repeater is placed approximately every 80–100 km to compensate for the loss of optical power as the signal travels along the fiber. The 'multi-wavelength optical signal' is amplified by an EDFA, which usually consists of several amplifier stages.

3. An intermediate optical terminal, or optical add-drop multiplexer. This is a remote amplification site that amplifies the multi-wavelength signal that may have traversed up to 140 km or more before reaching the remote site. Optical diagnostics and telemetry are often extracted or inserted at such a site, to allow for localization of any fiber breaks or signal impairments. In more sophisticated systems (which are no longer point-to-point), several signals out of the multi-wavelength optical signal may be removed and dropped locally.

4. A DWDM terminal demultiplexer. At the remote site, the terminal de-multiplexer consisting of an optical de-multiplexer and one or more wavelength-converting transponders separates the multi-wavelength optical signal back into individual data signals and outputs them on separate fibers for client-layer systems (such as SONET/SDH). Originally, this de-multiplexing was performed entirely passively, except for some telemetry, as most SONET systems can receive 1,550 nm signals. However, in order to allow for transmission to remote client-layer systems (and to allow for digital domain signal integrity determination) such de-multiplexed signals are usually sent to O/E/O output transponders prior to being relayed to their client-layer systems. Often, the functionality of output transponder has been integrated into that of input transponder, so that most commercial systems have transponders that support bi-directional interfaces on both their 1,550 nm (i.e., internal) side, and external (i.e., client-facing) side. Transponders in some systems supporting 40 GHz nominal operation may also perform forward error correction (FEC) via digital wrapper technology, as described in the ITU-T G.709 standard.

5. Optical Supervisory Channel (OSC). This is data channel which uses an additional wavelength usually outside the EDFA amplification band (at 1,510 nm, 1,620 nm, 1,310 nm or another proprietary wavelength). The OSC carries information about the multi-wavelength optical signal as well as remote conditions at the optical terminal or EDFA site. It is also normally used for remote software upgrades and user (i.e., network operator) Network Management information. It is the multi-wavelength analogue to SONET's DCC (or supervisory channel). ITU standards suggest that the OSC should utilize an OC-3 signal structure, though some vendors have opted to use 100 megabit Ethernet or another signal format. Unlike the 1550 nm multi-wavelength signal containing client data, the OSC is always terminated at intermediate amplifier sites, where it receives local information before re-transmission.

The introduction of the ITU-T G.694.1 frequency grid in 2002 has made it easier to integrate WDM with older but more standard SONET/SDH systems. WDM wavelengths are positioned in a grid having exactly 100 GHz (about 0.8 nm) spacing in optical frequency, with a reference frequency fixed at 193.10 THz (1,552.52 nm). The main grid is placed inside the optical fiber amplifier bandwidth, but can be extended to wider bandwidths. Today's DWDM systems use 50 GHz or even 25 GHz channel spacing for up to 160 channel operation.

DWDM systems have to maintain more stable wavelength or frequency than those needed for CWDM because of the closer spacing of the wavelengths. Precision temperature control of laser transmitter is required in DWDM systems to prevent "drift" off a very narrow frequency window of the order of a few GHz. In addition, since DWDM provides greater maximum capacity it tends to be used at a higher level in the communications hierarchy than CWDM, for example on the Internet backbone and is therefore associated with higher modulation rates, thus creating a smaller market for DWDM devices with very high performance. These factors of smaller volume and higher performance result in DWDM systems typically being more expensive than CWDM.

Recent innovations in DWDM transport systems include pluggable and software-tunable transceiver modules capable of operating on 40 or 80 channels. This dramatically reduces the need for discrete spare pluggable modules, when a handful of pluggable devices can handle the full range of wavelengths.

Wavelength-converting Transponders

At this stage, some details concerning wavelength-converting transponders should be discussed, as this will clarify the role played by current DWDM technology as an additional optical transport layer. It will also serve to outline the evolution of such systems over the last 10 or so years.

As stated above, wavelength-converting transponders served originally to translate the transmit wavelength of a client-layer signal into one of the DWDM system's internal wavelengths in the 1,550 nm band (note that even external wavelengths in the 1,550 nm will most likely need to be translated, as they will almost certainly not have the required frequency stability tolerances nor will it have the optical power necessary for the system's EDFA).

In the mid-1990s, however, wavelength converting transponders rapidly took on the additional function of signal regeneration. Signal regeneration in transponders quickly evolved through 1R to 2R to 3R and into overhead-monitoring multi-bitrate 3R regenerators. These differences are outlined below:

1R

Retransmission. Basically, early transponders were "garbage in garbage out" in that their output was nearly an analogue "copy" of the received optical signal, with little signal cleanup occurring. This limited the reach of early DWDM systems because the signal had to be handed off to a client-layer receiver (likely from a different vendor) before the signal deteriorated too far. Signal monitoring was basically confined to optical domain parameters such as received power.

2R

Re-time and re-transmit. Transponders of this type were not very common and utilized a quasi-digital Schmitt-triggering method for signal clean-up. Some rudimentary signal-quality monitoring was done by such transmitters that basically looked at analogue parameters.

3R

Re-time, re-transmit, re-shape. 3R Transponders were fully digital and normally able to view SONET/SDH section layer overhead bytes such as A1 and A2 to determine signal quality health. Many systems will offer 2.5 Gbit/s transponders, which will normally mean the transponder is able to perform 3R regeneration on OC-3/12/48 signals, and possibly gigabit Ethernet, and reporting on signal health by monitoring SONET/SDH section layer overhead bytes. Many transponders will be able to perform full multi-rate 3R in both directions. Some vendors offer 10 Gbit/s transponders, which will perform Section layer overhead monitoring to all rates up to and including OC-192.

Muxponder

The muxponder (from multiplexed transponder) has different names depending on vendor. It essentially performs some relatively simple time-division multiplexing of lower-rate signals into a higher-rate carrier within the system (a common example is the ability to accept 4 OC-48s and then output a single OC-192 in the 1,550 nm band). More recent muxponder designs have absorbed more and more TDM functionality, in some cases obviating the need for traditional SONET/SDH transport equipment.

Reconfigurable Optical Add-drop Multiplexer (ROADM)

As mentioned above, intermediate optical amplification sites in DWDM systems may allow for the dropping and adding of certain wavelength channels. In most systems deployed as of August 2006 this is done infrequently, because adding or dropping wavelengths requires manually inserting or replacing wavelength-selective cards. This is costly, and in some systems requires that all active traffic be removed from the DWDM system, because inserting or removing the wavelength-specific cards interrupts the multi-wavelength optical signal.

With a ROADM, network operators can remotely reconfigure the multiplexer by sending soft commands. The architecture of the ROADM is such that dropping or adding wavelengths does not interrupt the "pass-through" channels. Numerous technological approaches are utilized for various commercial ROADMs, the tradeoff being between cost, optical power, and flexibility.

Optical Cross Connects (OXCs)

When the network topology is a mesh, where nodes are interconnected by fibers to form an arbitrary graph, an additional fiber interconnection device is needed to route the signals from an input port to the desired output port. These devices are called optical crossconnectors (OXCs). Various categories of OXCs include electronic ("opaque"), optical ("transparent"), and wavelength selective devices.

Enhanced WDM

Cisco's Enhanced WDM system combines 1 Gb Coarse Wave Division Multiplexing (CWDM) connections using SFPs and GBICs with 10 Gb Dense Wave Division Multiplexing (DWDM) connections using XENPAK, X2 or XFP DWDM modules. These DWDM connections can either be passive or boosted to allow a longer range for the connection. In addition to this, CFP modules deliver 100 Gbit/s Ethernet suitable for high speed Internet backbone connections.

Transceivers Versus Transponders

- *Transceivers* – Since communication over a single wavelength is one-way (simplex communication), and most practical communication systems require two-way (duplex communication) communication, two wavelengths will be required (which might or might not be on the same fiber, but typically they will be each on a separate fiber in a so-called fiber pair). As a result, at each end both a transmitter (to send a signal over a first wavelength) and a receiver (to receive a signal over a second wavelength) will be required. A combination of a transmitter and a receiver is called a transceiver; it converts an electrical signal to and from an optical signal. There are usually transreceiver types based on WDM technology.

  - Coarse WDM (CWDM) Transceivers: Wavelength 1270 nm, 1290 nm, 1310 nm, 1330 nm, 1350 nm, 1370 nm, 1390 nm, 1410 nm, 1430 nm, 1450 nm, 1470 nm, 1490 nm, 1510 nm, 1530 nm, 1550 nm, 1570 nm, 1590 nm, 1610 nm.

  - Dense WDM (DWDM) Transceivers: Channel 17 to Channel 61 according to ITU-T.

- *Transponder* – In practice, the signal inputs and outputs will not be electrical but optical instead (typically at 1550 nm). This means that in effect we need wavelength converters instead, which is exactly what a transponder is. A transponder can be made up of two transceivers placed after each other: the first transceiver converting the 1550 nm optical signal to/from an electrical signal, and the second transceiver converting the electrical signal to/from an optical signal at the required wavelength. Transponders that don't use an intermediate electrical signal (all-optical transponders) are in development.

Implementations

There are several simulation tools that can be used to design WDM systems.

Time-division Multiplexing

Time-division multiplexing (TDM) is a method of transmitting and receiving independent signals over a common signal path by means of synchronized switches at each end of the transmission line so that each signal appears on the line only a fraction of time in an alternating pattern. It is used when the data rate of the transmission medium exceeds that of signal to be transmitted. This form of signal multiplexing was developed in telecommunications for telegraphy systems in the late 19th century, but found its most common application in digital telephony in the second half of the 20th century.

History

Time-division multiplexing was first developed for applications in telegraphy to route multiple transmissions simultaneously over a single transmission line. In the 1870s, Émile Baudot developed a time-multiplexing system of multiple Hughes telegraph machines.

In 1953 a 24-channel TDM was placed in commercial operation by RCA Communications to send audio information between RCA's facility on Broad Street, New York, their transmitting station at Rocky Point and the receiving station at Riverhead, Long Island, New York. The communication was by a microwave system throughout Long Island. The experimental TDM system was developed by RCA Laboratories between 1950 and 1953.

In 1962, engineers from Bell Labs developed the first D1 channel banks, which combined 24 digitized voice calls over a four-wire copper trunk between Bell central office analogue switches. A channel bank sliced a 1.544 Mbit/s digital signal into 8,000 separate frames, each composed of 24 contiguous bytes. Each byte represented a single telephone call encoded into a constant bit rate signal of 64 kbit/s. Channel banks used the fixed position (temporal alignment) of one byte in the frame to identify the call it belonged to.

Technology

Time-division multiplexing is used primarily for digital signals, but may be applied in analog multiplexing in which two or more signals or bit streams are transferred appearing simultaneously as sub-channels in one communication channel, but are physically taking turns on the channel. The time domain is divided into several recurrent *time slots* of fixed length, one for each sub-channel. A sample byte or data block of sub-channel 1 is transmitted during time slot 1, sub-channel 2 during time slot 2, etc. One TDM frame consists of one time slot per sub-channel plus a synchronization channel and sometimes error correction channel before the synchronization. After the last sub-channel, error correction, and synchronization, the cycle starts all over again with a new frame, starting with the second sample, byte or data block from sub-channel 1, etc.

Application Examples

- The plesiochronous digital hierarchy (PDH) system, also known as the PCM system, for digital transmission of several telephone calls over the same four-wire copper cable (T-carrier or E-carrier) or fiber cable in the circuit switched digital telephone network

- The synchronous digital hierarchy (SDH)/synchronous optical networking (SONET) network transmission standards that have replaced PDH.

- The Basic Rate Interface and Primary Rate Interface for the Integrated Services Digital Network (ISDN).

- The RIFF (WAV) audio standard interleaves left and right stereo signals on a per-sample basis

TDM can be further extended into the time division multiple access (TDMA) scheme, where several stations connected to the same physical medium, for example sharing the same frequency channel, can communicate. Application examples include:

- The GSM telephone system

- The Tactical Data Links Link 16 and Link 22

Multiplexed Digital Transmission

In circuit-switched networks, such as the public switched telephone network (PSTN), it is desirable to transmit multiple subscriber calls over the same transmission medium to effectively utilize the bandwidth of the medium. TDM allows transmitting and receiving telephone switches to create channels (*tributaries*) within a transmission stream. A standard DS0 voice signal has a data bit rate of 64 kbit/s. A TDM circuit runs at a much higher signal bandwidth, permitting the bandwidth to be divided into time frames (time slots) for each voice signal which is multiplexed onto the line by the transmitter. If the TDM frame consists of n voice frames, the line bandwidth is $n*64$ kbit/s.

Each voice time slot in the TDM frame is called a channel. In European systems, standard TDM frames contain 30 digital voice channels (E1), and in American systems (T1), they contain 24 channels. Both standards also contain extra bits (or bit time slots) for signaling and synchronization bits.

Multiplexing more than 24 or 30 digital voice channels is called *higher order multiplexing*. Higher order multiplexing is accomplished by multiplexing the standard TDM frames. For example, a European 120 channel TDM frame is formed by multiplexing four standard 30 channel TDM frames. At each higher order multiplex, four TDM frames from the immediate lower order are combined, creating multiplexes with a bandwidth of $n*64$ kbit/s, where n = 120, 480, 1920, etc.

Telecommunications Systems

There are three types of synchronous TDM: T1, SONET/SDH, and ISDN.

Plesiochronous digital hierarchy (PDH) was developed as a standard for multiplexing higher order frames. PDH created larger numbers of channels by multiplexing the standard Europeans 30 channel TDM frames. This solution worked for a while; however PDH suffered from several inherent drawbacks which ultimately resulted in the development of the Synchronous Digital Hierarchy (SDH). The requirements which drove the development of SDH were these:

- Be synchronous – All clocks in the system must align with a reference clock.

- Be service-oriented – SDH must route traffic from End Exchange to End Exchange without worrying about exchanges in between, where the bandwidth can be reserved at a fixed level for a fixed period of time.

- Allow frames of any size to be removed or inserted into an SDH frame of any size.

- Easily manageable with the capability of transferring management data across links.

- Provide high levels of recovery from faults.

- Provide high data rates by multiplexing any size frame, limited only by technology.

- Give reduced bit rate errors.

SDH has become the primary transmission protocol in most PSTN networks. It was developed to allow streams 1.544 Mbit/s and above to be multiplexed, in order to create larger SDH frames known as Synchronous Transport Modules (STM). The STM-1 frame consists of smaller streams that are multiplexed to create a 155.52 Mbit/s frame. SDH can also multiplex packet based frames e.g. Ethernet, PPP and ATM.

While SDH is considered to be a transmission protocol (Layer 1 in the OSI Reference Model), it also performs some switching functions, as stated in the third bullet point requirement listed above. The most common SDH Networking functions are these:

- *SDH Crossconnect* – The SDH Crossconnect is the SDH version of a Time-Space-Time crosspoint switch. It connects any channel on any of its inputs to any channel on any of its outputs. The SDH Crossconnect is used in Transit Exchanges, where all inputs and outputs are connected to other exchanges.

- *SDH Add-Drop Multiplexer* – The SDH Add-Drop Multiplexer (ADM) can add or remove any multiplexed frame down to 1.544Mb. Below this level, standard TDM can be performed. SDH ADMs can also perform the task of an SDH Crossconnect and are used in End Exchanges where the channels from subscribers are connected to the core PSTN network.

SDH network functions are connected using high-speed optic fibre. Optic fibre uses light pulses to transmit data and is therefore extremely fast. Modern optic fibre transmission makes use of wavelength-division multiplexing (WDM) where signals transmitted across the fibre are transmitted at different wavelengths, creating additional channels for transmission. This increases the speed and capacity of the link, which in turn reduces both unit and total costs.

Statistical Time-division Multiplexing

Statistical time division multiplexing (STDM) is an advanced version of TDM in which both the address of the terminal and the data itself are transmitted together for better routing. Using STDM allows bandwidth to be split over one line. Many college and corporate campuses use this type of TDM to distribute bandwidth.

On a 10-Mbit line entering a network, STDM can be used to provide 178 terminals with a dedicated 56k connection (178 * 56k = 9.96Mb). A more common use however is to only grant the bandwidth when that much is needed. STDM does not reserve a time slot for each terminal, rather it assigns a slot when the terminal is requiring data to be sent or received.

In its primary form, TDM is used for circuit mode communication with a fixed number of channels and constant bandwidth per channel. Bandwidth reservation distinguishes time-division multiplexing from statistical multiplexing such as statistical time division multiplexing. In pure TDM, the time slots are recurrent in a fixed order and pre-allocated to the channels, rather than scheduled on a packet-by-packet basis.

In dynamic TDMA, a scheduling algorithm dynamically reserves a variable number of time slots in each frame to variable bit-rate data streams, based on the traffic demand of each data stream. Dynamic TDMA is used in:

- HIPERLAN/2

- Dynamic synchronous transfer mode

- IEEE 802.16a

Asynchronous time-division multiplexing (ATDM), is an alternative nomenclature in which STDM designates synchronous time-division multiplexing, the older method that uses fixed time slots.

Polarization-division Multiplexing

Polarization-division multiplexing (PDM) is a physical layer method for multiplexing signals carried on electromagnetic waves using the polarization of the electromagnetic waves to distinguish between the different orthogonal signals.

Radio

Polarization techniques have long been used in radio transmission to reduce interference between channels, particularly at VHF frequencies and beyond.

Photonics

Polarization-division multiplexing is typically used together with phase modulation or optical QAM, allowing transmission speeds of 100 Gbit/s or more over a single wavelength. Sets of PDM wavelength signals can then be carried over wavelength-division multiplexing infrastructure, potentially substantially expanding its capacity. Multiple polarization signals can be combined to form new states of polarization, which is known as parallel polarization state generation.

The major problem with the practical use of PDM over fiber-optic transmission systems are the drifts in polarization state that occur continuously over time due to physical changes in the fibre environment. Over a long-distance system, these drifts accumulate progressively without limit, resulting in rapid and erratic rotation of the polarized light's Jones vector over the entire Poincaré sphere. Polarization mode dispersion, polarization-dependent loss. and cross-polarization modulation are other phenomena that can cause problems in PDM systems.

For this reason, PDM is generally used in conjunction with advanced channel coding techniques, allowing the use of digital signal processing to decode the signal in a way that is resilient to polarization-related signal artifacts. Modulations used include PM-QPSK and PM-DQPSK.

Companies working on commercial PDM technology include Alcatel-Lucent, Ciena, Cisco Systems, Huawei and Infinera.

Orbital Angular Momentum Multiplexing

Orbital angular momentum (OAM) multiplexing is a physical layer method for multiplexing signals carried on electromagnetic waves using the orbital angular momentum of the electromagnetic waves to distinguish between the different orthogonal signals.

Orbital angular momentum is one of two forms of angular momentum of light. OAM is distinct from, and should not be confused with, light spin angular momentum. The spin angular momentum of light offers only two orthogonal quantum states corresponding to the two states of circular polarization, and can be demonstrated to be equivalent to a combination of polarization multiplexing and phase shifting. OAM multiplexing can (at least in theory) access a potentially unbounded set of OAM quantum states, and thus offer a much larger number of channels, subject only to the constraints of real-world optics.

As of 2013, although OAM multiplexing promises very significant improvements in bandwidth when used in concert with other existing modulation and multiplexing schemes, it is still an experimental technique, and has so far only been demonstrated in the laboratory.

History

OAM multiplexing was demonstrated using light beams in free space as early as 2004. Since then, research into OAM has proceeded in two areas: radio frequency and optical transmission.

Radio Frequency

An experiment in 2011 demonstrated OAM multiplexing of two incoherent radio signals over a distance of 442 m. It has been claimed that OAM does not improve on what can achieved with conventional linear-momentum based RF systems which already use MIMO, since theoretical work suggests that, at radio frequencies, conventional MIMO techniques can be shown to duplicate many of the linear-momentum properties of OAM-carrying radio beam, leaving little or no extra performance gain.

In November 2012, there were reports of disagreement about the basic theoretical concept of OAM multiplexing at radio frequencies between the research groups of Tamburini and Thide, and many different camps of communications engineers and physicists, with some declaring their belief that OAM multiplexing was just an implementation of MIMO, and others holding to their assertion that OAM multiplexing is a distinct, experimentally confirmed phenomenon.

In 2014, a group of researchers described an implementation of a communication link over eight millimetre wave channels multiplexed using a combination of OAM and polarization mode multiplexing to achieve an aggregate bandwidth of 32 Gbit/s over a distance of 2.5 metres. These results agree well with predictions about severely limited distances made by Edfors et al.

The hype seems to have cooled down lately. Even the original promoters of OAM based communication at radio frequencies have realized that there is no real gain beyond traditional MIMO.

Optical

OAM multiplexing is used in the optical domain. In 2012, researchers demonstrated OAM-multiplexed optical transmission speeds of up to 2.5 Tbits/s using eight distinct OAM channels in a single beam of light, but only over a very short free-space path of roughly one metre. Work is ongoing on applying OAM techniques to long-range practical free-space optical communication links.

OAM multiplexing can not be implemented in the existing long-haul optical fiber systems, since these systems are based on single-mode fibers, which inherently do not support OAM states of light. Instead, few-mode or multi-mode fibers need to be used. Additional problem for OAM multiplexing implementation is caused by the mode coupling that is present in conventional fibers, which cause changes in the spin angular momentum of modes under normal conditions and changes in orbital angular momentum when fibers are bent or stressed. Because of this mode-instability, direct-detection OAM multiplexing has not yet been realized in long-haul communications. In 2012, transmission of OAM states with 97% purity after 20 meters over specialty fibers was demonstrated by researchers at Boston University. Later experiments have shown stable propagation of these modes over distances of 50 meters, and further improvements of this distance are the subject of ongoing work. Other ongoing research on making OAM multiplexing work over future fibre optic transmission systems includes the possibility of using similar techniques to those used to compensate mode rotation in optical polarization multiplexing.

Alternative to direct-detection OAM multiplexing is a computationally complex coherent-detection with (MIMO) digital signal processing (DSP) approach, that can be used to achieve long-haul communication, where strong mode coupling is suggested to be beneficial for coherent-detection based systems.

Practical Demonstration in Optical Fiber System

A paper by Bozinovic. et al. published in *Science* in 2013 claims the successful demonstration of an OAM multiplexed fiber optic transmission system over a 1.1 km test path. The test system was capable of using up to four different OAM channels simultaneously, using a fiber with a "vortex" refractive index profile. They also demonstrated combined OAM and WDM using the same apparatus, but using only two OAM modes.

Practical Demonstration in Conventional Optical Fiber Systems

In 2014, papers by G. Milione et al. and H. Huang et al. claimed the first and successful demonstration of an OAM multiplexed fiber optic transmission system over a 5km of conventional optical fiber , i.e., an optical fiber having a circular core and a graded index profile. In contrast to the work of Bozinovic et al. which used a custom optical fiber that had a"vortex" refractive index profile, the work by G. Milione et al. and H. Huang et al. showed that OAM multiplexing could be used in commercially available optical fibers.

Path Protection

Path protection in Telecommunications is an end-to-end protection scheme used in connection oriented circuits in different network architectures to protect against inevitable failures on service providers' network that might affect the services offered to end customers. Any failure occurred at any point along the path of a circuit will cause the end nodes to move/pick the traffic to/from a new route.

Other techniques to protect telecommunications networks against failures are: Channel Protection, Link Protection, Segment-Protection, and P-cycle Protection

Path Protection in Ring Based Networks

In Ring Based networks topology where the setup is to form a closed loop among the Add Drop Multiplexers, there is basically one path related ring protection scheme available in Unidirectional Path-Switched Ring architecture. In SDH networks, the equivalent of UPSR is Sub-Network Connection Protection (SNCP). Note that SNCP does not assume a ring topology, and can also be used in mesh topologies.

In UPSR, the data is transmitted in both directions, clock and counter clock wise, at the source ADM. At the destination then, both signals are compared and the best one of the two is selected. If a failure occurs then the destination just needs to switch to the unaffected path.

Path Protection in Optical Mesh Network

Circuits in Optical Mesh Networks can be unprotected, protected to a single failure, and protected to multiple failures. The end optical switches in protected circuits are in charge of detecting the failure, in some cases requesting digital cross connects or optical cross-connects in intermediate devices, and switching the traffic to/from the backup path. When the primary and backup paths are calculated, it is important that they are at least link diverse so that a single link failure does not affect both of them at the same time. They can also be node diverse, which offers more protection in case a node failure occurs; depending on the network sometimes the primary and backup path cannot be provisioned to be node diverse at the edges, ingress and egress, node.

There are two types of path protection in Optical Mesh Networks: Dedicated Backup Path Protection and Shared Backup Path Protection

Dedicated Backup Path Protection or DBPP (1+1)

In DBPP, both the primary and backup path carry the traffic end to end, then it is up to the receiver to decide which of the two incoming traffic it is going to pick; this is exactly the same concept as in Ring Based Path Protection. Since the optics along both paths are already active, DBPP is the fastest protection scheme available, usually in the order of a few tens of milliseconds, because there is no signaling involved in between ingress and egress nodes thus only needing the egress node to detect the failure and switch the traffic over to the unaffected path. Being the fastest protection scheme also makes it the most expensive; normally using more than double of the provisioned capacity for the primary because the backup path is usually longer due to the link and/or node diversity rule of thumb.

Shared Backup Path-Protected or SBPP

The concept behind this protection scheme is to share a backup channel among different, link/node diverse, primary paths. In other words, one backup channel can be used to protect various primary paths as shown on the figure below where the link between S and T is used to protect both AB and CD primaries. Under normal operations, assuming no failure on the network, the traffic is carried on the primary paths only; the shared backup path is only used when there is a failure in one of those primary paths.

There are two approaches to provision or reserve backups channels. First, there is the failure dependent assignment or approach also known as restoration in which the backup path is calculated in real time after the failure occurs. This technique is found in early versions of Mesh networks. However, in today's Optical Mesh Network it can be used as a re-provisioning technique to help recover a second failure when the backup resources are already in use. The down side to restoration as a protection technique is that the recovery time is not fast enough.

The second approach is to have a predefined backup path computed before the failure. This approach is said to be failure independent and it takes less processing time to recover as compared to the failure dependent approach. Here the backup path is calculated together with the primary at provisioning time. Even though the backup path is calculated, it is not assigned to a specific circuit before a failure occurs; cross connect requests are initiated after the fact on a first-come, first-served basis. Since this approach can only protect from a single failure at a time, if a second primary path fails and at least a portion of its backup path is already in used, this path won't be able to recover unless restoration technique is in place for such cases.

There is a general down side to both of the above approaches and is that assuming there is a link failure with several paths running through it, each path in that link is going to be recovered individually. This implies that the total time the last path on that link is going to take to be back in service through the secondary path will be the sum of all other previous recovery times plus its own. This could affect the committed SLA (Service Level Agreement) to the customer.

Path Protection in MPLS Networks

Multi-Protocol Label Switching (MPLS) architecture is described in the RFC-3031. It is a packet-based network technology that provides a framework for recovery through the creation of point to point paths called Label Switched Paths (LSP). These LSPs creation are between a head-end and a tail-end Label Switch Router (LSR). In the former case, the head-end router is the input or ingress router. In the latter case the tail-end represents the output or egress router in the path. There are a few protection techniques for MPLS very similar in the general concept to those for Optical Mesh Networks, such as link protection (e.g., MPLS local protection) and path protection. The path protection schemes for MPLS are as follow:

Packet Protection Scheme (1+1)

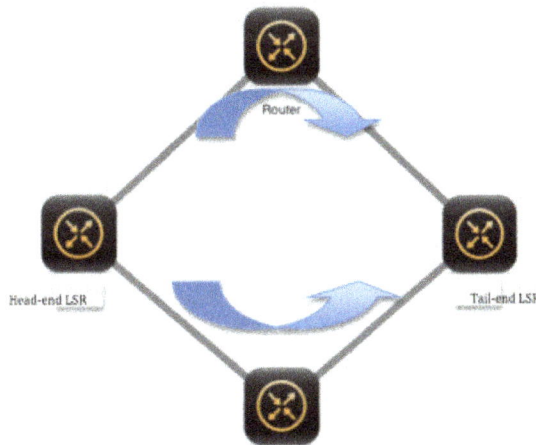

Packet Protection Scheme (1+1)

This protection scheme is similar in a sense to Ring-based path protection and Dedicated Backup Path Protection (DBPP) schemes described before. Here, same traffic is transmitted over two, link and/or node disjoint, LSPs; primary and backup. The transmission is done by the head-end LSR. The tail-end LSR then receives and compares both traffics; when a failure occurs, the tail-end detects it and switches the traffic to the secondary LSP. As with DBPP in Optical Mesh Network, there is no signaling involved in this protection scheme. This technique is the simplest and fastest of all, but as it reserves and transmits packets on both LSP, it takes away bandwidth that could be shared and used by other LSPs.

Global Path Protection (1:1)

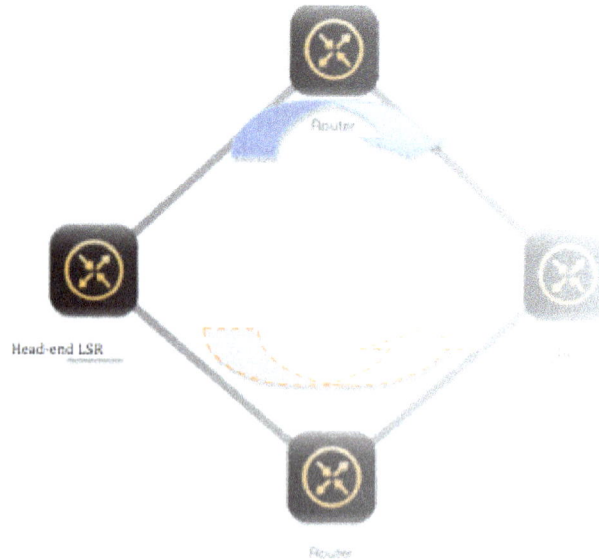

Global Path Protection (1:1)

In this protection scheme, a primary and a backup LSP are computed and setup at the provisioning time prior to failures. The backup LSP does not necessarily need to have the same constrain in terms of bandwidth as the primary; it is possible to reserve less bandwidth on the backup LSP and not incur in packet loss when in use. This is because the bandwidth of the link is shared among the different LSPs and the reason why the previous explained protection scheme is not preferred. It is also true that the Backup LSP does not necessarily carry traffic unless the primary LSP fails. When this occurs, a fault indication signal (FIS) is sent back to the head-end LSR that will immediately switch the traffic to the backup LSP. The drawback in this protection scheme is that the longer the LSPs, the longer the recovery time will be because of the travel time of the FIS notification.

Electrical Length

In telecommunications and electrical engineering, electrical length (or phase length) refers to the length of an electrical conductor in terms of the phase shift introduced by transmission over that conductor at some frequency.

Usage of the Term

Depending on the specific usage, the term "electrical length" is used rather than simple physical length to incorporate one or more of the following three concepts:

- When one is concerned with the number of wavelengths, or phase, involved in a wave's transit over a segment of transmission line especially, one may simply specify that electrical length, while specification of a physical length, frequency, or velocity factor is omitted. The electrical length is then typically expressed as N wavelengths or as the phase φ expressed in degrees or radians. Thus in a microstrip design one might specify a shorted stub of 60°

phase length, which will correspond to different physical lengths when applied to different frequencies. Or one might consider a 2-meter section of coax which has an electrical length of one quarter wavelength (90°) at 25 MHz and ask what its electrical length becomes when the circuit is operated at a different frequency.

- Due to the velocity factor of a particular transmission line, for instance, the transit time of a signal in a certain length of cable is equal to the transit time over a *longer* distance when travelling at the speed of light. So a pulse sent down a 2-meter section of coax (whose velocity factor is 2/3) would arrive at the end of the coax at the same time that the same pulse arrives at the end of a bare wire of length 3 meters (over which it propagates at the speed of light), and one might refer to the 2 meter section of coax as having an electrical length of 3 meters, or an electrical length of 1/2 wavelength at 50 MHz (since a 50 MHz radio wave has a wavelength of 6 meters).

- Since resonant antennas are usually specified in terms of the electrical length of their conductors (such as the *half wave* dipole), the attainment of such an electrical length is loosely equated with electrical resonance, that is, a purely resistive impedance at the antenna's input, as is usually desired. An antenna that has been made slightly too long, for instance, will present an inductive reactance, which can be corrected by physically shortening the antenna. Based on this understanding, a common jargon in the antenna trade refers to the achievement of resonance (cancellation of reactance) at the antenna terminals as *electrically shortening* that too-long antenna (or *electrically lengthening* a too-short antenna) when an electrical matching network (or antenna tuner) has performed that task without *physically* altering the antenna's length. Although a very inexact use of terminology, this usage is widespread, especially as applied to the use of a loading coil at the bottom of a short monopole (a vertical, or whip antenna) to "electrically lengthen" it and achieve electrical resonance as seen through the loading coil.

Phase Length

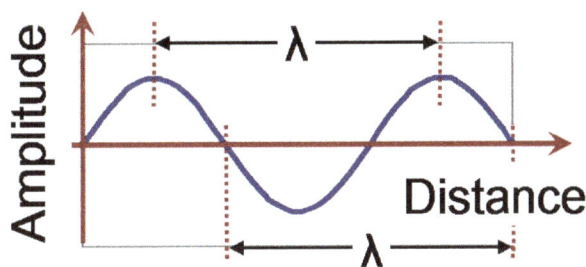

In the figure, the wave shown is seen to be N=1.5 wavelengths long. A wave crest at the beginning of the graph, moving towards the right, will arrive at the end after a time $1.5T$. The *electrical length* of that segment is said to be "1.5 wavelengths" or, expressed as a phase angle, "540°" (or 3π radians) where N wavelengths corresponds to $\varphi = 360°{\cdot}N$ (or $\varphi = 2\pi{\cdot}N$ radians). In radio frequency applications, when a delay is introduced due to a transmission line, it is often the phase shift φ that is of importance, so specifying a design in terms of the phase or electrical length allows one to adapt that design to an arbitrary frequency by employing the wavelength λ applying to that frequency.

The first usage of the term "electrical length" assumes a sine wave of some frequency, or at least a narrowband waveform centered around some frequency f. The sine wave will repeat with a period of $T = 1/f$. The frequency f will correspond to a particular wavelength λ along a particular conductor. For conductors (such as bare wire or air-filled coax) which transmit signals at the speed of light c, the wavelength is given by $\lambda=c/f$. A distance L along that conductor corresponds to N wavelengths where $N= L / \lambda$.

Velocity Factor

In a transmission line, a signal travels at a rate controlled by the effective capacitance and inductance per unit of length of the transmission line. Some transmission lines consist only of bare conductors, in which case their signals propagate at the speed of light, c. More often the signal travels at a reduced velocity κc, where κ is the *velocity factor*, a number less than 1 representing the ratio of that velocity to the speed of light.

Most transmission lines contain a dielectric material (insulator) filling some or all of the space in between the conductors. The relative permittivity or *dielectric constant* of that material increases the distributed capacitance in the cable, which reduces the velocity factor below unity. It is also possible for κ to be reduced due to a relative permeability of that material which increases the distributed inductance, but this is almost never the case. Now, if one fills a space with a dielectric of relative permittivity ϵ_r, then the velocity of an electromagnetic plane wave is reduced by the velocity factor:

$$\kappa = \frac{v_p}{c} = \frac{1}{\sqrt{\epsilon_r}}.$$

This reduced velocity factor would also apply to propagation of signals along wires immersed in a large space filled with that dielectric. However, with only part of the space around the conductors filled with that dielectric, there is less reduction of the wave velocity. Part of the electromagnetic wave surrounding each conductor "feels" the effect of the dielectric, and part is in free space. Then it is possible to define an *effective relative permittivity* ϵ_{eff} which then predicts the velocity factor according to

$$\kappa = \frac{1}{\sqrt{\epsilon_{eff}}} \quad \epsilon_{eff}$$

is computed as a weighted average of the relative permittivity of free space (1) and that of the dielectric:

$$\epsilon_{eff} = (1-F)+F\epsilon_r$$

where the *fill factor* F expresses the effective proportion of space so affected by the dielectric.

In the case of coaxial cable, where all of the volume in between the inner conductor and the shield is filled with a dielectric, the fill factor is unity, since the electromagnetic wave is confined to that region. In other types of cable, such as twin lead, the fill factor can be much smaller. Regardless, any cable intended for radio frequencies will have its velocity factor (as well as its characteristic impedance) specified by the manufacturer. In the case of coaxial cable, where F=1, the velocity factor is solely determined by the sort of dielectric used as specified here.

For example, a typical velocity factor for coaxial cable is .66, corresponding to a dielectric constant of 2.25. Suppose we wish to send a 30 MHz signal down a short section of such a cable, and delay it by a quarter wave (90°). In free space, this frequency corresponds to a wavelength of λ_o=10m, so a delay of .25λ would require an *electrical length* of 2.5 m. Applying the velocity factor of .66, this results in a *physical* length of cable 1.67 m long.

The velocity factor likewise applies to antennas in cases where the antenna conductors are (partly) surrounded by a dieletric. This particularly applies to microstrip antennas such as the patch antenna. Waves on microstrip are affected by the dielectric of the circuit board beneath them, but not the air above them. Their velocity factors thus depend not directly on the permittivity of the circuit board material but on the *effective* permittivity $\grave{o}_{eff}$ which is often specified for a circuit board material (or can be calculated). Note that the fill factor and therefore ϵ_{eff} are somewhat dependent on the width of the trace compared to the thickness of the board.

Antennas

While there are certain wideband antenna designs, many antennas are classified as resonant and perform according to design around a particular frequency. This applies especially to broadcasting stations and communication systems which are confined to one frequency or narrow frequency band. This includes the dipole and monopole antennas and all of the designs based on them (Yagi, dipole or monopole arrays, folded dipole, etc.). In addition to the directive gain in beam antennas suffering away from the design frequency, the antenna feedpoint impedance is very sensitive to frequency offsets. Especially for transmitting, the antenna is often intended to operate at the resonant frequency. At the resonant frequency, by definition, that impedance is a pure resistance which matches the characteristic impedance of the transmission line and the output (or input) impedance of the transmitter (or receiver). At frequencies away from the resonant frequency, the impedance includes some reactance (capacitance or inductance). It is possible for an antenna tuner to be used to cancel that reactance (and to change the resistance to match the transmission line), however that is often avoided as an extra complication (and needs to be controlled at the antenna side of the transmission line).

The condition for resonance in a monopole antenna is for the element to be an odd multiple of a quarter-wavelength, $\lambda/4$. In a dipole antenna both driven conductors must be that long, for a total dipole length of *(2N+1)λ/2*.

The electrical length of an antenna element is, in general, different from its physical length For example, increasing the diameter of the conductor, or the presence of nearby metal objects, will decrease the velocity of the waves in the element, increasing the electrical length.

An antenna which is shorter than its resonant length is described as *"electrically short"*, and exhibits capacitive reactance. Similarly, an antenna which is longer than its resonant length is described as *"electrically long"* and exhibits inductive reactance.

Changing Electrical Length by Loading

An antenna's effective electrical length can be changed without changing its physical length by adding reactance, (inductance or capacitance) in series with it. This is called *lumped-impedance matching* or *loading*.

Loading coil in a cellphone antenna mounted on the roof of a car. The coil allows the antenna to be shorter than a quarter wavelength and still be resonant.

For example, a monopole antenna such as a metal rod fed at one end, will be resonant when its electrical length is equal to a quarter wavelength, $\lambda/4$, of the frequency used. If the antenna is shorter than a quarter wavelength, the feedpoint impedance will include capacitive reactance; this causes reflections on the feedline and a mismatch at the transmitter or receiver, even if the resistive component of the impedance is correct. To cancel the capacitive reactance, an inductance, called a loading coil, is inserted in between the feedline and the antenna terminal. Selecting an inductance with the same reactance as the (negative) capacitive reactance seen at the antenna terminal, cancels that capacitance, and the *antenna system* (antenna and coil) will again be resonant. The feedline sees a purely resistive impedance. Since an antenna which had been too short now appears as if it were resonant, the addition of the loading coil is sometimes referred to as "electrically lengthening" the antenna.

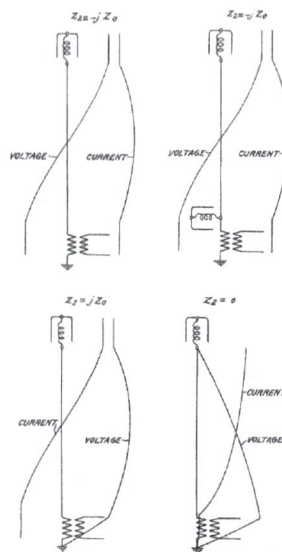

Vertical antenna which may be of any desired height : less than about one-half wavelength of the frequency at which the antenna operates. These antennas may operate either as transmitting or receiving antennas

Similarly, the feedpoint impedance of a monopole antenna longer than $\lambda/4$ (or a dipole with arms longer than $\lambda/4$) will include inductive reactance. A capacitor in series with the antenna can cancel this reactance to make it resonant, which can be referred to as "electrically shortening" the antenna.

Inductive loading is widely used to reduce the length of whip antennas on portable radios such as walkie-talkies and short wave antennas on cars, to meet physical requirements.

Advantages

The electrical lengthening allows the construction of shorter aerials. It is applied in particular for aerials for VLF, longwave and medium-wave transmitters. Because those radio waves are several hundred meters to many kilometers long, mast radiators of the necessary height cannot be realised economically. It is also used widely for whip antennas on portable devices such as walkie-talkies to allow antennas much shorter than the standard quarter-wavelength to be used. The most widely used example is the rubber ducky antenna.

Disadvantages

The electrical lengthening reduces the bandwidth of the antenna if other phase control measures are not undertaken. An electrically extended aerial is less efficient than a non-extended antenna.

Technical Realization

There are two possibilities for the realisation of the electric lengthening.

1. switching in inductive coils in series with the aerial

2. switching in metal surfaces, known as roof capacitance, at the aerial ends which form capacitors to earth.

Often both measures are combined. The coils switched in series must sometimes be placed in the middle of the aerial construction. The cabin installed at a height of 150-metres on the Blosenbergturm in Beromünster is such a construction, in which a lengthening coil is installed for the supply of the upper tower part (the Blosenbergturm has in addition a ring-shaped roof capacitor on its top)

Application

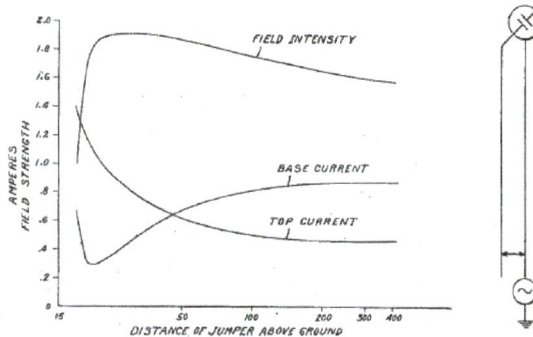

On the left, characteristics plotted from experimentally obtained data on coordinates with logarithmic abscissa. On the right, an antenna with increased effective inductance between the two points in accordance with the well known operation of shunt tuned circuits adjusted somewhat off resonance.

Transmission aerials of transmitters working at frequencies below the longwave broadcasting band always apply electric lengthening. Broadcasting aerials of longwave broadcasting stations apply it often. However, for transmission aerials of NDBs electrical lengthening is extensively applied, because these use antennas which are considerably less tall than a quarter of the radiated wavelength.

Reference

- Siva Ram Murthy C.; Guruswamy M., "WDM Optical Networks, Concepts, Design, and Algorithms", Prentice Hall India, ISBN 81-203-2129-4.

- White, Curt (2007). Data Communications and Computer Networks. Boston, MA: Thomson Course Technology. pp. 143–152. ISBN 1-4188-3610-9.

- Guowang Miao; Jens Zander; Ki Won Sung; Ben Slimane (2016). Fundamentals of Mobile Data Networks. Cambridge University Press. ISBN 1107143217.

- Jean Philippe Vasseur, Mario Pickavet & Piet Demeester (2004). Network Recovery, Protection and Restoration of Optical, SONET-SDH, IP, and MPLS. Morgan Kaufmann Publishers. ISBN 0-12-715051-X.

- Eric Bouillet; Georgios Ellinas; Jean-Francois Labourdette & Ramu Ramamurthy (2007). Path Routing in Mesh Optical Networks. John Wiley & Sons, Ltd. ISBN 978-0-470-01565-0.

- Eric Bouillet; Georgios Ellinas; Jean-Francois Labourdette & Ramu Ramamurthy (2007). Path Routing in Mesh Optical Networks. John Wiley & Sons, Ltd. pp. 31, 43, 84. ISBN 978-0-470-01565-0.

- Eric Bouillet; Georgios Ellinas; Jean-Francois Labourdette & Ramu Ramamurthy (2007). Path Routing in Mesh Optical Networks. John Wiley & Sons, Ltd. pp. 32, 44, 86. ISBN 978-0-470-01565-0.

- Bruce S. Davie & Adrian Farrel (2008). MPLS: Next Steps. Morgan Kaufmann Publishers. ISBN 978-0-12-374400-5.

- Weik, Martin (1997). Fiber Optics Standard Dictionary. Springer Science & Business Media. p. 270. ISBN 0412122413.

- Helfrick, Albert D. (2012). Electrical Spectrum & Network Analyzers: A Practical Approach. Academic Press. p. 192. ISBN 0080918670.

- She, Alan; Capasso, Federico (17 May 2016). "Parallel Polarization State Generation". Scientific Reports. Nature. doi:10.1038/srep26019. Retrieved 27 June 2016.

- Gregg, Patrick; Poul Kristensen; Siddharth Ramachandran (January 2015). "Conservation of orbital angular momentum in air-core optical fibers". Optica. 2 (3): 267–270. doi:10.1364/optica.2.000267.

- Lewis, Geoff (2013). Newnes Communications Technology Handbook. Elsevier. p. 46. ISBN 1483101029.

Telecommunications Techniques

Techniques are an important part of any field of study. The major techniques related to telecommunications are communications satellite, arbitrary slice ordering, diversity combining and phantom circuit. Lesser known techniques of telecommunications include through-the-earth mine communications and two-way communications. The aspects elucidated are of vital importance and provide a better understanding of telecommunications.

Communications Satellite

A communications satellite is an artificial satellite that relays and amplifies radio telecommunications signals via a transponder; it creates a communication channel between a source transmitter and a receiver at different locations on Earth. Communications satellites are used for television, telephone, radio, internet, and military applications. There are over 2,000 communications satellites in Earth's orbit, used by both private and government organizations.

An Advanced Extremely High Frequency communications satellite relays secure communications for the United States and other allied countries.

Wireless communication uses electromagnetic waves to carry signals. These waves require line-of-sight, and are thus obstructed by the curvature of the Earth. The purpose of communications satellites is to relay the signal around the curve of the Earth allowing communication between widely separated points. Communications satellites use a wide range of radio and microwave frequencies. To avoid signal interference, international organizations have regulations for which frequency ranges or "bands" certain organizations are allowed to use. This allocation of bands minimizes the risk of signal interference.

History

The concept of the geostationary communications satellite was first proposed by Arthur C. Clarke, building on work by Konstantin Tsiolkovsky and on the 1929 work by Herman Potočnik (writing as Herman Noordung) *Das Problem der Befahrung des Weltraums — der Raketen-motor*. In October 1945 Clarke published an article titled "Extraterrestrial Relays" in the British magazine *Wireless World*. The article described the fundamentals behind the deployment of artificial satellites in geostationary orbits for the purpose of relaying radio signals. Thus, Arthur C. Clarke is often quoted as being the inventor of the communications satellite and the term 'Clarke Belt' employed as a description of the orbit.

Decades later a project named Communication Moon Relay was a telecommunication project carried out by the United States Navy. Its objective was to develop a secure and reliable method of wireless communication by using the Moon as a passive reflector and natural communications satellite.

The first artificial Earth satellite was Sputnik 1. Put into orbit by the Soviet Union on October 4, 1957, it was equipped with an on-board radio-transmitter that worked on two frequencies: 20.005 and 40.002 MHz. Sputnik 1 was launched as a step in the exploration of space and rocket development. While incredibly important it was not placed in orbit for the purpose of sending data from one point on earth to another. And it was the first artificial satellite in the steps leading to today's satellite communications.

The first artificial satellite used solely to further advances in global communications was a balloon named Echo 1. Echo 1 was the world's first artificial communications satellite capable of relaying signals to other points on Earth. It soared 1,600 kilometres (1,000 mi) above the planet after its Aug. 12, 1960 launch, yet relied on humanity's oldest flight technology — ballooning. Launched by NASA, Echo 1 was a 30-metre (100 ft) aluminised PET film balloon that served as a passive reflector for radio communications. The world's first inflatable satellite — or "satelloon", as they were informally known — helped lay the foundation of today's satellite communications. The idea behind a communications satellite is simple: Send data up into space and beam it back down to another spot on the globe. Echo 1 accomplished this by essentially serving as an enormous mirror, 10 stories tall, that could be used to reflect communications signals.

The first American satellite to relay communications was Project SCORE in 1958, which used a tape recorder to store and forward voice messages. It was used to send a Christmas greeting to the world from U.S. President Dwight D. Eisenhower.; Courier 1B, built by Philco, launched in 1960, was the world's first active repeater satellite.

There are two major classes of communications satellites, *passive* and *active*. Passive satellites only reflect the signal coming from the source, toward the direction of the receiver. With passive satellites, the reflected signal is not amplified at the satellite, and only a very small amount of the transmitted energy actually reaches the receiver. Since the satellite is so far above Earth, the radio signal is attenuated due to free-space path loss, so the signal received on Earth is very weak. Active satellites, on the other hand, amplify the received signal before re-transmitting it to the receiver on the ground. Passive satellites were the first communications satellites, but are little used now. Telstar was the second active, direct relay communications satellite. Belonging to AT&T as part of a multi-national agreement between AT&T, Bell Telephone Laboratories, NASA, the British General Post Office, and the French National PTT (Post Office) to develop satellite communications, it was launched by NASA from Cape Canaveral on July 10, 1962, the first privately sponsored space launch. Relay 1 was launched on December 13, 1962, and became the first satellite to broadcast across the Pacific on November 22, 1963.

An immediate antecedent of the geostationary satellites was Hughes' Syncom 2, launched on July 26, 1963. Syncom 2 was the first communications satellite in a geosynchronous orbit. It revolved around the earth once per day at constant speed, but because it still had north-south motion, special equipment was needed to track it. Its successor, Syncom 3 was the first geostationary communications satellite. Syncom 3 obtained a geosynchronous orbit, without a north-south motion, making it appear from the ground as a stationary object in the sky.

Beginning with the Mars Exploration Rovers, probes on the surface of Mars have used orbiting spacecraft as communications satellites for relaying their data to Earth. The orbiters were designed for this relay purpose to allow the landers to conserve power. The Orbiters with their solar power arrays, large antennas and more powerful transmitters enable them to transmit data to Earth with a much stronger, and as a result, clearer signal than a lander could manage on its own from the surface.

Satellite Orbits

Communications satellites usually have one of three primary types of orbit, while other orbital classifications are used to further specify orbital details.

- Geostationary satellites have a *geostationary orbit* (GEO), which is 35,786 kilometres (22,236 mi) from Earth's surface. This orbit has the special characteristic that the apparent position of the satellite in the sky when viewed by a ground observer does not change, the satellite appears to "stand still" in the sky. This is because the satellite's orbital period is the same as the rotation rate of the Earth. The advantage of this orbit is that ground antennas do not have to track the satellite across the sky, they can be fixed to point at the location in the sky the satellite appears.

- *Medium Earth orbit* (MEO) satellites are closer to Earth. Orbital altitudes range from 2,000 to 35,786 kilometres (1,243 to 22,236 mi) above Earth.

- The region below medium orbits is referred to as *low Earth orbit* (LEO), and is about 160 to 2,000 kilometres (99 to 1,243 mi) above Earth.

As satellites in MEO and LEO orbit the Earth faster, they do not remain visible in the sky to a fixed point on Earth continually like a geostationary satellite, but appear to a ground observer to cross the sky and "set" when they go behind the Earth. Therefore, to provide continuous communications capability with these lower orbits requires a larger number of satellites, so one will always be in the sky for transmission of communication signals. However, due to their relatively small distance to the Earth their signals are stronger.

Low Earth Orbiting (LEO) Satellites

A low Earth orbit (LEO) typically is a circular orbit about 160 to 2,000 kilometres (99 to 1,243 mi) above the earth's surface and, correspondingly, a period (time to revolve around the earth) of about 90 minutes.

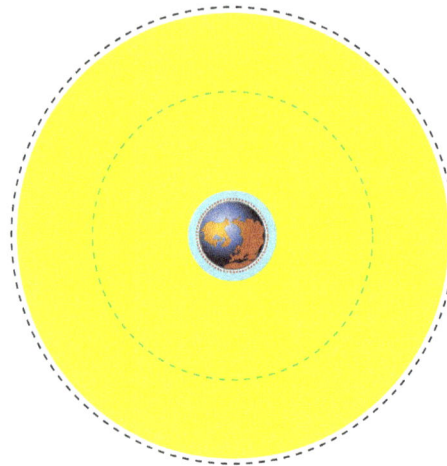

Low Earth orbit in Cyan

Because of their low altitude, these satellites are only visible from within a radius of roughly 1,000 kilometres (620 mi) from the sub-satellite point. In addition, satellites in low earth orbit change their position relative to the ground position quickly. So even for local applications, a large number of satellites are needed if the mission requires uninterrupted connectivity.

Low-Earth-orbiting satellites are less expensive to launch into orbit than geostationary satellites and, due to proximity to the ground, do not require as high signal strength (*Recall that signal strength falls off as the square of the distance from the source, so the effect is dramatic*). Thus there is a trade off between the number of satellites and their cost.

In addition, there are important differences in the onboard and ground equipment needed to support the two types of missions.

Satellite Constellation

A group of satellites working in concert is known as a satellite constellation. Two such constellations, intended to provide satellite phone services, primarily to remote areas, are the Iridium and Globalstar systems. The Iridium system has 66 satellites.

It is also possible to offer discontinuous coverage using a low-Earth-orbit satellite capable of storing data received while passing over one part of Earth and transmitting it later while passing over another part. This will be the case with the CASCADE system of Canada's CASSIOPE communications satellite. Another system using this store and forward method is Orbcomm.

Medium Earth Orbit (MEO)

A MEO is a satellite in orbit somewhere between 2,000 and 35,786 kilometres (1,243 and 22,236 mi) above the earth's surface. MEO satellites are similar to LEO satellites in functionality. MEO satellites are visible for much longer periods of time than LEO satellites, usually between 2 and 8 hours. MEO satellites have a larger coverage area than LEO satellites. A MEO satellite's longer duration of visibility and wider footprint means fewer satellites are needed in a MEO network than a LEO network. One disadvantage is that a MEO satellite's distance gives it a longer time delay and weaker signal than a LEO satellite, although these limitations are not as severe as those of a GEO satellite.

Like LEOs, these satellites don't maintain a stationary distance from the earth. This is in contrast to the geostationary orbit, where satellites are always approximately 35,786 kilometres (22,236 mi) from the earth.

Typically the orbit of a medium earth orbit satellite is about 16,000 kilometres (10,000 mi) above earth. In various patterns, these satellites make the trip around earth in anywhere from 2–12 hours, which provides better coverage to wider areas than that provided by LEOs.

Example

In 1962, the first communications satellite, Telstar, was launched. It was a medium earth orbit satellite designed to help facilitate high-speed telephone signals. Although it was the first practical way to transmit signals over the horizon, its major drawback was soon realized. Because its orbital period of about 2.5 hours did not match the Earth's rotational period of 24 hours, continuous coverage was impossible. It was apparent that multiple MEOs needed to be used in order to provide continuous coverage.

Geostationary Orbits (GEO)

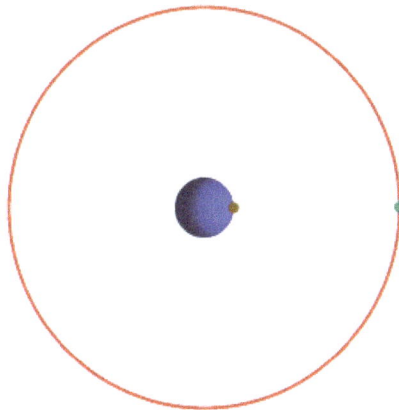

Geostationary orbit

To an observer on the earth, a satellite in a geostationary orbit appears motionless, in a fixed position in the sky. This is because it revolves around the earth at the earth's own angular velocity (360 degrees every 24 hours, in an equatorial orbit).

A geostationary orbit is useful for communications because ground antennas can be aimed at the satellite without their having to track the satellite's motion. This is relatively inexpensive.

In applications that require a large number of ground antennas, such as DirecTV distribution, the savings in ground equipment can more than outweigh the cost and complexity of placing a satellite into orbit.

Examples

- The first geostationary satellite was Syncom 3, launched on August 19, 1964, and used for communication across the Pacific starting with television coverage of the 1964 Summer Olympics. Shortly after Syncom 3, Intelsat I, aka *Early Bird*, was launched on April 6, 1965 and placed in orbit at 28° west longitude. It was the first geostationary satellite for telecommunications over the Atlantic Ocean.

- On November 9, 1972, Canada's first geostationary satellite serving the continent, Anik A1, was launched by Telesat Canada, with the United States following suit with the launch of Westar 1 by Western Union on April 13, 1974.

- On May 30, 1974, the first geostationary communications satellite in the world to be three-axis stabilized was launched: the experimental satellite ATS-6 built for NASA.

- After the launches of the Telstar through Westar 1 satellites, RCA Americom (later GE Americom, now SES) launched Satcom 1 in 1975. It was Satcom 1 that was instrumental in helping early cable TV channels such as WTBS (now TBS Superstation), HBO, CBN (now ABC Family) and The Weather Channel become successful, because these channels distributed their programming to all of the local cable TV headends using the satellite. Additionally, it was the first satellite used by broadcast television networks in the United States, like ABC, NBC, and CBS, to distribute programming to their local affiliate stations. Satcom 1 was widely used because it had twice the communications capacity of the competing Westar 1 in America (24 transponders as opposed to the 12 of Westar 1), resulting in lower transponder-usage costs. Satellites in later decades tended to have even higher transponder numbers.

By 2000, Hughes Space and Communications (now Boeing Satellite Development Center) had built nearly 40 percent of the more than one hundred satellites in service worldwide. Other major satellite manufacturers include Space Systems/Loral, Orbital Sciences Corporation with the STAR Bus series, Indian Space Research Organisation, Lockheed Martin (owns the former RCA Astro Electronics/GE Astro Space business), Northrop Grumman, Alcatel Space, now Thales Alenia Space, with the Spacebus series, and Astrium.

Molniya Satellites

Geostationary satellites must operate above the equator and therefore appear lower on the horizon as the receiver gets the farther from the equator. This will cause problems for extreme northerly latitudes, affecting connectivity and causing multipath interference (caused by signals reflecting off the ground and into the ground antenna).

Thus, for areas close to the North (and South) Pole, a geostationary satellite may appear below the horizon. Therefore, Molniya orbit satellites have been launched, mainly in Russia, to alleviate this problem.

Molniya orbits can be an appealing alternative in such cases. The Molniya orbit is highly inclined, guaranteeing good elevation over selected positions during the northern portion of the orbit. (Elevation is the extent of the satellite's position above the horizon. Thus, a satellite at the horizon has zero elevation and a satellite directly overhead has elevation of 90 degrees).

The Molniya orbit is designed so that the satellite spends the great majority of its time over the far northern latitudes, during which its ground footprint moves only slightly. Its period is one half day, so that the satellite is available for operation over the targeted region for six to nine hours every second revolution. In this way a constellation of three Molniya satellites (plus in-orbit spares) can provide uninterrupted coverage.

The first satellite of the Molniya series was launched on April 23, 1965 and was used for experimental transmission of TV signals from a Moscow uplink station to downlink stations located in Siberia and the Russian Far East, in Norilsk, Khabarovsk, Magadan and Vladivostok. In November 1967 Soviet engineers created a unique system of national TV network of satellite television, called Orbita, that was based on Molniya satellites.

Polar Orbit

In the United States, the National Polar-orbiting Operational Environmental Satellite System (NPOESS) was established in 1994 to consolidate the polar satellite operations of NASA (National Aeronautics and Space Administration) NOAA (National Oceanic and Atmospheric Administration). NPOESS manages a number of satellites for various purposes; for example, METSAT for meteorological satellite, EUMETSAT for the European branch of the program, and METOP for meteorological operations.

These orbits are sun synchronous, meaning that they cross the equator at the same local time each day. For example, the satellites in the NPOESS (civilian) orbit will cross the equator, going from south to north, at times 1:30 P.M., 5:30 P.M., and 9:30 P.M.

Structure

Communications Satellites are usually composed of the following subsystems:

- Communication Payload, normally composed of transponders, antennas, and switching systems

- Engines used to bring the satellite to its desired orbit

- Station Keeping Tracking and stabilization subsystem used to keep the satellite in the right orbit, with its antennas pointed in the right direction, and its power system pointed towards the sun

- Power subsystem, used to power the Satellite systems, normally composed of solar cells, and batteries that maintain power during solar eclipse

- Command and Control subsystem, which maintains communications with ground control stations. The ground control earth stations monitor the satellite performance and control its functionality during various phases of its life-cycle.

The bandwidth available from a satellite depends upon the number of transponders provided by the satellite. Each service (TV, Voice, Internet, radio) requires a different amount of bandwidth for transmission. This is typically known as link budgeting and a network simulator can be used to arrive at the exact value.

Frequency Allocation for Satellite Systems

Allocating frequencies to satellite services is a complicated process which requires international coordination and planning. This is carried out under the auspices of the International Telecommunication Union (ITU). To facilitate frequency planning, the world is divided into three regions:

Region 1: Europe, Africa, what was formerly the Soviet Union, and Mongolia Region 2: North and South America and Greenland Region 3: Asia (excluding region 1 areas), Australia, and the south-west Pacific

Within these regions, frequency bands are allocated to various satellite services, although a given service may be allocated different frequency bands in different regions. Some of the services provided by satellites are:

- Fixed satellite service (FSS)

- Broadcasting satellite service (BSS)

- Mobile satellite service

- Radionavigation-satellite service

- Meteorological-satellite service

- Amateur-satellite service

Applications

Telephone

An Iridium satellite

The first and historically most important application for communication satellites was in intercontinental long distance telephony. The fixed Public Switched Telephone Network relays telephone calls from land line telephones to an earth station, where they are then transmitted to a geostationary satellite. The downlink follows an analogous path. Improvements in submarine communications cables through the use of fiber-optics caused some decline in the use of satellites for fixed telephony in the late 20th century.

Satellite communications are still used in many applications today. Remote islands such as Ascension Island, Saint Helena, Diego Garcia, and Easter Island, where no submarine cables are in service, need satellite telephones. There are also regions of some continents and countries where landline telecommunications are rare to nonexistent, for example large regions of South America, Africa, Canada, China, Russia, and Australia. Satellite communications also provide connection to the edges of Antarctica and Greenland. Other land use for satellite phones are rigs at sea, a back up for hospitals, military, and recreation. Ships at sea, as well as planes, often use satellite phones.

Satellite phone systems can be accomplished by a number of means. On a large scale, often there will be a local telephone system in an isolated area with a link to the telephone system in a main land area. There are also services that will patch a radio signal to a telephone system. In this example, almost any type of satellite can be used. Satellite phones connect directly to a constellation of either geostationary or low-earth-orbit satellites. Calls are then forwarded to a satellite teleport connected to the Public Switched Telephone Network .

Television

As television became the main market, its demand for simultaneous delivery of relatively few signals of large bandwidth to many receivers being a more precise match for the capabilities of geosynchronous comsats. Two satellite types are used for North American television and radio: Direct broadcast satellite (DBS), and Fixed Service Satellite (FSS).

The definitions of FSS and DBS satellites outside of North America, especially in Europe, are a bit more ambiguous. Most satellites used for direct-to-home television in Europe have the same high power output as DBS-class satellites in North America, but use the same linear polarization as FSS-class satellites. Examples of these are the Astra, Eutelsat, and Hotbird spacecraft in orbit over the European continent. Because of this, the terms FSS and DBS are more so used throughout the North American continent, and are uncommon in Europe.

Fixed Service Satellites use the C band, and the lower portions of the K_u band. They are normally used for broadcast feeds to and from television networks and local affiliate stations (such as program feeds for network and syndicated programming, live shots, and backhauls), as well as being used for distance learning by schools and universities, business television (BTV), Videoconferencing, and general commercial telecommunications. FSS satellites are also used to distribute national cable channels to cable television headends.

Free-to-air satellite TV channels are also usually distributed on FSS satellites in the K_u band. The Intelsat Americas 5, Galaxy 10R and AMC 3 satellites over North America provide a quite large amount of FTA channels on their K_u band transponders.

The American Dish Network DBS service has also recently utilized FSS technology as well for their programming packages requiring their SuperDish antenna, due to Dish Network needing more capacity to carry local television stations per the FCC's "must-carry" regulations, and for more bandwidth to carry HDTV channels.

A direct broadcast satellite is a communications satellite that transmits to small DBS satellite dishes (usually 18 to 24 inches or 45 to 60 cm in diameter). Direct broadcast satellites generally operate in the upper portion of the microwave K_u band. DBS technology is used for DTH-oriented (Direct-To-Home) satellite TV services, such as DirecTV and DISH Network in the United States, Bell TV and Shaw Direct in Canada, Freesat and Sky in the UK, Ireland, and New Zealand and DSTV in South Africa.

Operating at lower frequency and lower power than DBS, FSS satellites require a much larger dish for reception (3 to 8 feet (1 to 2.5 m) in diameter for K_u band, and 12 feet (3.6 m) or larger for C band). They use linear polarization for each of the transponders' RF input and output (as opposed to circular polarization used by DBS satellites), but this is a minor technical difference that users do not notice. FSS satellite technology was also originally used for DTH satellite TV from the late 1970s to the early 1990s in the United States in the form of TVRO (TeleVision Receive Only) receivers and dishes. It was also used in its K_u band form for the now-defunct Primestar satellite TV service.

Some satellites have been launched that have transponders in the K_a band, such as DirecTV's SPACEWAY-1 satellite, and Anik F2. NASA and ISRO have also launched experimental satellites carrying K_a band beacons recently.

Some manufacturers have also introduced special antennas for mobile reception of DBS television. Using Global Positioning System (GPS) technology as a reference, these antennas automatically re-aim to the satellite no matter where or how the vehicle (on which the antenna is mounted) is situated. These mobile satellite antennas are popular with some recreational vehicle owners. Such mobile DBS antennas are also used by JetBlue Airways for DirecTV (supplied by LiveTV, a subsidiary of JetBlue), which passengers can view on-board on LCD screens mounted in the seats.

Radio Broadcasting

Satellite radio offers audio broadcast services in some countries, notably the United States. Mobile services allow listeners to roam a continent, listening to the same audio programming anywhere.

A satellite radio or subscription radio (SR) is a digital radio signal that is broadcast by a communications satellite, which covers a much wider geographical range than terrestrial radio signals.

Satellite radio offers a meaningful alternative to ground-based radio services in some countries, notably the United States. Mobile services, such as SiriusXM, and Worldspace, allow listeners to roam across an entire continent, listening to the same audio programming anywhere they go. Other services, such as Music Choice or Muzak's satellite-delivered content, require a fixed-location receiver and a dish antenna. In all cases, the antenna must have a clear view to the satellites. In areas where tall buildings, bridges, or even parking garages obscure the signal, repeaters can be placed to make the signal available to listeners.

Initially available for broadcast to stationary TV receivers, by 2004 popular mobile direct broadcast applications made their appearance with the arrival of two satellite radio systems in the United States: Sirius and XM Satellite Radio Holdings. Later they merged to become the conglomerate SiriusXM.

Radio services are usually provided by commercial ventures and are subscription-based. The various services are proprietary signals, requiring specialized hardware for decoding and playback. Providers usually carry a variety of news, weather, sports, and music channels, with the music channels generally being commercial-free.

In areas with a relatively high population density, it is easier and less expensive to reach the bulk of the population with terrestrial broadcasts. Thus in the UK and some other countries, the contemporary evolution of radio services is focused on Digital Audio Broadcasting (DAB) services or HD Radio, rather than satellite radio.

Amateur Radio

Amateur radio operators have access to amateur satellites, which have been designed specifically to carry amateur radio traffic. Most such satellites operate as spaceborne repeaters, and are generally accessed by amateurs equipped with UHF or VHF radio equipment and highly directional antennas such as Yagis or dish antennas. Due to launch costs, most current amateur satellites are launched into fairly low Earth orbits, and are designed to deal with only a limited number of brief contacts at any given time. Some satellites also provide data-forwarding services using the X.25 or similar protocols.

Internet Access

After the 1990s, satellite communication technology has been used as a means to connect to the Internet via broadband data connections. This can be very useful for users who are located in remote areas, and cannot access a broadband connection, or require high availability of services.

Military

Communications satellites are used for military communications applications, such as Global Command and Control Systems. Examples of military systems that use communication satellites are the MILSTAR, the DSCS, and the FLTSATCOM of the United States, NATO satellites, United Kingdom satellites (for instance Skynet), and satellites of the former Soviet Union. India has launched its first Military Communication satellite GSAT-7, its transponders operate in UHF, F, C and K_u band bands. Typically military satellites operate in the UHF, SHF (also known as X-band) or EHF (also known as K_a band) frequency bands.

Channel Access Method

In telecommunications and computer networks, a channel access method or multiple access method allows several terminals connected to the same multi-point transmission medium to transmit over it and to share its capacity. Examples of shared physical media are wireless networks, bus networks, ring networks and point-to-point links operating in half-duplex mode.

A channel-access scheme is based on a multiplexing method, that allows several data streams or signals to share the same communication channel or physical medium. In this context. multiplexing is provided by the physical layer.

A channel-access scheme is also based on a multiple access protocol and control mechanism, also known as media access control (MAC). Media access control deals with issues such as addressing, assigning multiplex channels to different users, and avoiding collisions. Media access control is a sub-layer in Layer 2 (data link layer) of the OSI model and a component of the link layer of the TCP/IP model.

Fundamental Types of Channel Access Schemes

These numerous channel access schemes which generally fall into the following categories:

Frequency-division Multiple Access (FDMA)

The frequency-division multiple access (FDMA) channel-access scheme is based on the frequency-division multiplexing (FDM) scheme, which provides different frequency bands to different data-streams. In the FDMA case, the data streams are allocated to different nodes or devices. An example of FDMA systems were the first-generation (1G) cell-phone systems, where each phone call was assigned to a specific uplink frequency channel, and another downlink frequency channel. Each message signal (each phone call) is modulated on a specific carrier frequency.

A related technique is wavelength division multiple access (WDMA), based on wavelength-division multiplexing (WDM), where different datastreams get different colors in fiber-optical communications. In the WDMA case, different network nodes in a bus or hub network get a different color.

An advanced form of FDMA is the orthogonal frequency-division multiple access (OFDMA) scheme, for example used in 4G cellular communication systems. In OFDMA, each node may use several sub-carriers, making it possible to provide different quality of service (different data rates) to different users. The assignment of sub-carriers to users may be changed dynamically, based on the current radio channel conditions and traffic load.

Time Division Multiple Access (TDMA)

The time division multiple access (TDMA) channel access scheme is based on the time-division multiplexing (TDM) scheme, which provides different time-slots to different data-streams (in the TDMA case to different transmitters) in a cyclically repetitive frame structure. For example, node 1 may use time slot 1, node 2 time slot 2, etc. until the last transmitter. Then it starts all over again, in a repetitive pattern, until a connection is ended and that slot becomes free or assigned to another node. An advanced form is Dynamic TDMA (DTDMA), where a scheduling may give different time sometimes but some times node 1 may use time slot 1 in first frame and use another time slot in next frame.

As an example, 2G cellular systems are based on a combination of TDMA and FDMA. Each frequency channel is divided into eight timeslots, of which seven are used for seven phone calls, and one for signalling data.

Statistical time division multiplexing multiple-access is typically also based on time-domain multiplexing, but not in a cyclically repetitive frame structure. Due to its random character it can be categorised as statistical multiplexing methods, making it possible to provide dynamic bandwidth allocation. This requires a media access control (MAC) protocol, i.e. a principle for the nodes to take turns on the channel and to avoid collisions. Common examples are CSMA/CD, used in Ethernet bus networks and hub networks, and CSMA/CA, used in wireless networks such as IEEE 802.11.

Code Division Multiple Access (CDMA)/Spread Spectrum Multiple Access (SSMA)

The code division multiple access (CDMA) scheme is based on spread spectrum, meaning that a wider radio spectrum in Hertz is used than the data rate of each of the transferred bit streams, and several message signals are transferred simultaneously over the same carrier frequency, utilizing different spreading codes. The wide bandwidth makes it possible to send with a very poor signal-to-noise ratio of much less than 1 (less than 0 dB) according to the Shannon-Heartly formula, meaning that the transmission power can be reduced to a level below the level of the noise and co-channel interference (cross talk) from other message signals sharing the same frequency.

One form is direct sequence spread spectrum (DS-CDMA), used for example in 3G cell phone systems. Each information bit (or each symbol) is represented by a long code sequence of several pulses, called chips. The sequence is the spreading code, and each message signal (for example each phone call) uses a different spreading code.

Another form is frequency-hopping (FH-CDMA), where the channel frequency is changing very rapidly according to a sequence that constitutes the spreading code. As an example, the Bluetooth communication system is based on a combination of frequency-hopping and either CSMA/CA statistical time division multiplexing communication (for data communication applications) or TDMA (for audio transmission). All nodes belonging to the same user (to the same virtual private area network or piconet) use the same frequency hopping sequency synchronously, meaning that they send on the same frequency channel, but CDMA/CA or TDMA is used to avoid collisions within the VPAN. Frequency-hopping is used to reduce the cross-talk and collision probability between nodes in different VPANs.

Subdivisions of FH-CDMA are "fast hopping" where the frequency of hopping is much higher than the message frequency content and "slow hopping" where the hopping frequency is comparable to message frequency content. The subdivision is necessary as they are considerably different.

Space Division Multiple Access (SDMA)

Space-division multiple access (SDMA) transmits different information in different physical areas. Examples include simple cellular radio systems and more advanced cellular systems which use directional antennas and power modulation to refine spatial transmission patterns.

Power Division Multiple Access (PDMA)

Power-division multiple access (PDMA) scheme is based on using variable transmission power between users in order to share the available power on the channel. Examples include multiple SCPC modems on a satellite transponder, where users get on demand a larger share of the power budget to transmit at higher data rates.

List of Channel Access Methods

Circuit Mode and Channelization Methods

The following are common circuit mode and channelization channel access methods:

- *Frequency-division multiple access (FDMA)*, based on frequency-division multiplexing (FDM)
    - o Wavelength division multiple access (WDMA)
    - o Orthogonal frequency-division multiple access (OFDMA), based on Orthogonal frequency-division multiplexing (OFDM)

- Single-carrier FDMA (SC-FDMA), a.k.a. linearly-precoded OFDMA (LP-OFDMA), based on single-carrier frequency-domain-equalization (SC-FDE).

- *Time-division multiple access (TDMA)*, based on time-division multiplexing (TDM)

  - Multi-Frequency Time Division Multiple Access (MF-TDMA)

- *Code division multiple access (CDMA)*, a.k.a. Spread spectrum multiple access (SSMA)

  - Direct-sequence CDMA (DS-CDMA), based on Direct-sequence spread spectrum (DSSS)

  - Frequency-hopping CDMA (FH-CDMA), based on Frequency-hopping spread spectrum (FHSS)

  - Orthogonal frequency-hopping multiple access (OFHMA)

  - Multi-carrier code division multiple access (MC-CDMA)

- *Space-division multiple access (SDMA)*

- Power-division multiple access (PDMA)

Packet Mode Methods

The following are examples of packet mode channel access methods:

- *Contention based random multiple access methods*

  - Aloha

  - Slotted Aloha

  - Multiple Access with Collision Avoidance (MACA)

  - Multiple Access with Collision Avoidance for Wireless (MACAW)

  - Carrier sense multiple access (CSMA)

  - Carrier sense multiple access with collision detection (CSMA/CD) - suitable for wired networks

  - Carrier sense multiple access with collision avoidance (CSMA/CA) - suitable for wireless networks

    - Distributed Coordination Function (DCF)

  - Carrier sense multiple access with collision avoidance and Resolution using Priorities (CSMA/CARP)

  - Carrier Sense Multiple Access/Bitwise Arbitration (CSMA/BA) Based on constructive interference (CAN-bus)

- *Token passing*:
    - Token ring
    - Token bus
- *Polling*
- *Resource reservation (scheduled) packet-mode protocols*
    - Dynamic Time Division Multiple Access (Dynamic TDMA)
    - Packet reservation multiple access (PRMA)
    - Reservation ALOHA (R-ALOHA)

Duplexing Methods

Where these methods are used for dividing forward and reverse communication channels, they are known as duplexing methods, such as:

- Time division duplex (TDD)
- Frequency division duplex (FDD)

Hybrid Channel Access Scheme Application Examples

Note that hybrids of these techniques can be - and frequently are - used. Some examples:

- The GSM cellular system combines the use of frequency division duplex (FDD) to prevent interference between outward and return signals, with FDMA and TDMA to allow multiple handsets to work in a single cell.

- GSM with the GPRS packet switched service combines FDD and FDMA with slotted Aloha for reservation inquiries, and a Dynamic TDMA scheme for transferring the actual data.

- Bluetooth packet mode communication combines frequency hopping (for shared channel access among several private area networks in the same room) with CSMA/CA (for shared channel access inside a medium).

- IEEE 802.11b wireless local area networks (WLANs) are based on FDMA and DS-CDMA for avoiding interference among adjacent WLAN cells or access points. This is combined with CSMA/CA for multiple access within the cell.

- HIPERLAN/2 wireless networks combine FDMA with dynamic TDMA, meaning that resource reservation is achieved by packet scheduling.

- G.hn, an ITU-T standard for high-speed networking over home wiring (power lines, phone lines and coaxial cables) employs a combination of TDMA, Token passing and CSMA/CARP to allow multiple devices to share the medium.

Definition Within Certain Application Areas

Local and Metropolitan Area Networks

In local area networks (LANs) and metropolitan area networks (MANs), multiple access methods enable bus networks, ring networks, hubbed networks, wireless networks and half duplex point-to-point communication, but are not required in full duplex point-to-point serial lines between network switches and routers, or in switched networks (logical star topology). The most common multiple access method is CSMA/CD, which is used in Ethernet. Although today's Ethernet installations typically are switched, CSMA/CD is utilized anyway to achieve compatibility with hubs.

Satellite Communications

In satellite communications, multiple access is the capability of a communications satellite to function as a portion of a communications link between more than one pair of satellite terminals concurrently. Three types of multiple access presently used with communications satellites are code-division, frequency-division, and time-division multiple access.

Switching Centers

In telecommunication switching centers, multiple access is the connection of a user to two or more switching centers by separate access lines using a single message routing indicator or telephone number.

Classifications in the Literature

Several ways of categorizing multiple-access schemes and protocols have been used in the literature. For example, Daniel Minoli (2009) identifies five principal types of multiple-access schemes: FDMA, TDMA, CDMA, SDMA, and Random access. R. Rom and M. Sidi (1990) categorize the protocols into *Conflict-free access protocols*, *Aloha protocols*, and *Carrier Sensing protocols*.

The Telecommunications Handbook (Terplan and Morreale, 2000) identifies the following MAC categories:

- Fixed assigned: TDMA, FDMA+WDMA, CDMA, SDMA

- Demand assigned (DA)

    o Reservation: DA/TDMA, DA/FDMA+DA/WDMA, DA/CDMA, DA/SDMA

    o Polling: Generalized polling, Distributed polling, Token Passing, Implicit polling, Slotted access

- Random access (RA): Pure RA (ALOHA, GRA), Adaptive RA (TRA), CSMA, CSMA/CD, CSMA/CA

Code Division Multiple Access

Code division multiple access (CDMA) is a channel access method used by various radio communication technologies.

CDMA is an example of multiple access, where several transmitters can send information simultaneously over a single communication channel. This allows several users to share a band of frequencies. To permit this without undue interference between the users, CDMA employs spread-spectrum technology and a special coding scheme (where each transmitter is assigned a code).

CDMA is used as the access method in many mobile phone standards. IS-95, also called "cdma-One", and its 3G evolution CDMA2000, are often simply referred to as "CDMA"', but UMTS, the 3G standard used by GSM carriers, also uses "wideband CDMA", or W-CDMA, as well as TD-CDMA and TD-SCDMA, as its radio technologies.

History

The technology of code division multiple access channels has long been known. In the Soviet Union (USSR), the first work devoted to this subject was published in 1935 by professor Dmitriy V. Ageev. It was shown that through the use of linear methods, there are three types of signal separation: frequency, time and compensatory. The technology of CDMA was used in 1957, when the young military radio engineer Leonid Kupriyanovich in Moscow, made an experimental model of a wearable automatic mobile phone, called LK-1 by him, with a base station. LK-1 has a weight of 3 kg, 20–30 km operating distance, and 20–30 hours of battery life. The base station, as described by the author, could serve several customers. In 1958, Kupriyanovich made the new experimental "pocket" model of mobile phone. This phone weighed 0.5 kg. To serve more customers, Kupriyanovich proposed the device, named by him as correllator. In 1958, the USSR also started the development of the "Altai" national civil mobile phone service for cars, based on the Soviet MRT-1327 standard. The phone system weighed 11 kg (24 lb). It was placed in the trunk of the vehicles of high-ranking officials and used a standard handset in the passenger compartment. The main developers of the Altai system were VNIIS (Voronezh Science Research Institute of Communications) and GSPI (State Specialized Project Institute). In 1963 this service started in Moscow and in 1970 Altai service was used in 30 USSR cities.

Uses

A CDMA2000 mobile phone

- One of the early applications for code division multiplexing is in the Global Positioning System (GPS). This predates and is distinct from its use in mobile phones.

- The Qualcomm standard IS-95, marketed as cdmaOne.

- The Qualcomm standard IS-2000, known as CDMA2000, is used by several mobile phone companies, including the Globalstar satellite phone network.

- The UMTS 3G mobile phone standard, which uses W-CDMA.

- CDMA has been used in the OmniTRACS satellite system for transportation logistics.

Steps in CDMA Modulation

CDMA is a spread-spectrum multiple access technique. A spread spectrum technique spreads the bandwidth of the data uniformly for the same transmitted power. A spreading code is a pseudo-random code that has a narrow ambiguity function, unlike other narrow pulse codes. In CDMA a locally generated code runs at a much higher rate than the data to be transmitted. Data for transmission is combined via bitwise XOR (exclusive OR) with the faster code. The figure shows how a spread spectrum signal is generated. The data signal with pulse duration of (symbol period) is XOR'ed with the code signal with pulse duration of T_b (chip period). (Note: bandwidth is proportional to $1/T$, where T = bit time.) Therefore, the bandwidth of the data signal is $1/T_b$ and the bandwidth of the spread spectrum signal is $1/T_c$. Since T_c is much smaller than T_b, the bandwidth of the spread spectrum signal is much larger than the bandwidth of the original signal. The ratio T_b/T_c is called the spreading factor or processing gain and determines to a certain extent the upper limit of the total number of users supported simultaneously by a base station.

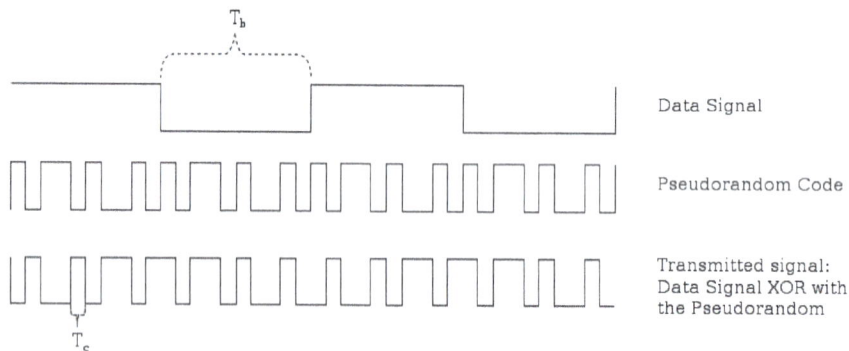

Generation of a CDMA signal

Each user in a CDMA system uses a different code to modulate their signal. Choosing the codes used to modulate the signal is very important in the performance of CDMA systems. The best performance will occur when there is good separation between the signal of a desired user and the signals of other users. The separation of the signals is made by correlating the received signal with the locally generated code of the desired user. If the signal matches the desired user's code then the correlation function will be high and the system can extract that signal. If the desired user's code has nothing in common with the signal the correlation should be as close to zero as possible (thus eliminating the signal); this is referred to as cross-correlation. If the code is correlated with the signal at any time offset other than zero, the correlation should be as close to zero as possible. This is referred to as auto-correlation and is used to reject multi-path interference.

An analogy to the problem of multiple access is a room (channel) in which people wish to talk to each other simultaneously. To avoid confusion, people could take turns speaking (time division), speak at different pitches (frequency division), or speak in different languages (code division). CDMA is analogous to the last example where people speaking the same language can understand each other, but other languages are perceived as noise and rejected. Similarly, in radio CDMA, each group of users is given a shared code. Many codes occupy the same channel, but only users associated with a particular code can communicate.

In general, CDMA belongs to two basic categories: synchronous (orthogonal codes) and asynchronous (pseudorandom codes).

Code Division Multiplexing (Synchronous CDMA)

The digital modulation method is analogous to those used in simple radio transceivers. In the analog case, a low frequency data signal is time multiplied with a high frequency pure sine wave carrier, and transmitted. This is effectively a frequency convolution (Wiener–Khinchin theorem) of the two signals, resulting in a carrier with narrow sidebands. In the digital case, the sinusoidal carrier is replaced by Walsh functions. These are binary square waves that form a complete orthonormal set. The data signal is also binary and the time multiplication is achieved with a simple XOR function. This is usually a Gilbert cell mixer in the circuitry.

Synchronous CDMA exploits mathematical properties of orthogonality between vectors representing the data strings. For example, binary string *1011* is represented by the vector (1, 0, 1, 1). Vectors can be multiplied by taking their dot product, by summing the products of their respective components (for example, if u = (a, b) and v = (c, d), then their dot product u·v = ac + bd). If the dot product is zero, the two vectors are said to be *orthogonal* to each other. Some properties of the dot product aid understanding of how W-CDMA works. If vectors a and b are orthogonal, then $\mathbf{a} \cdot \mathbf{b} = 0$ and:

$$\mathbf{a} \cdot (\mathbf{a} + \mathbf{b}) = \|\mathbf{a}\|^2 \quad \text{since} \quad \mathbf{a} \cdot \mathbf{a} + \mathbf{a} \cdot \mathbf{b} = \|\mathbf{a}\|^2 + 0$$

$$\mathbf{a} \cdot (-\mathbf{a} + \mathbf{b}) = -\|\mathbf{a}\|^2 \quad \text{since} \quad -\mathbf{a} \cdot \mathbf{a} + \mathbf{a} \cdot \mathbf{b} = -\|\mathbf{a}\|^2 + 0$$

$$\mathbf{b} \cdot (\mathbf{a} + \mathbf{b}) = \|\mathbf{b}\|^2 \quad \text{since} \quad \mathbf{b} \cdot \mathbf{a} + \mathbf{b} \cdot \mathbf{b} = 0 + \|\mathbf{b}\|^2$$

$$\mathbf{b} \cdot (\mathbf{a} - \mathbf{b}) = -\|\mathbf{b}\|^2 \quad \text{since} \quad \mathbf{b} \cdot \mathbf{a} - \mathbf{b} \cdot \mathbf{b} = 0 - \|\mathbf{b}\|^2$$

Each user in synchronous CDMA uses a code orthogonal to the others' codes to modulate their signal. An example of four mutually orthogonal digital signals is shown in the figure. Orthogonal codes have a cross-correlation equal to zero; in other words, they do not interfere with each other. In the case of IS-95 64 bit Walsh codes are used to encode the signal to separate different users. Since each of the 64 Walsh codes are orthogonal to one another, the signals are channelized into 64 orthogonal signals. The following example demonstrates how each user's signal can be encoded and decoded.

Example

Start with a set of vectors that are mutually orthogonal. (Although mutual orthogonality is the only condition, these vectors are usually constructed for ease of decoding, for example columns or rows from Walsh matrices.) An example of orthogonal functions is shown in the picture on the right. These vectors will be assigned to individual users and are called the *code*, *chip code*, or *chipping code*. In the interest of brevity, the rest of this example uses codes, **v**, with only two bits.

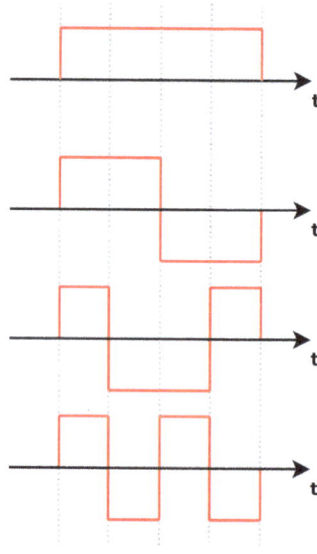

An example of four mutually orthogonal digital signals.

Each user is associated with a different code, say **v**. A 1 bit is represented by transmitting a positive code, **v**, and a 0 bit is represented by a negative code, $-\mathbf{v}$. For example, if $\mathbf{v} = (v_0, v_1) = (1, -1)$ and the data that the user wishes to transmit is $(1, 0, 1, 1)$, then the transmitted symbols would be $(\mathbf{v}, -\mathbf{v}, \mathbf{v}, \mathbf{v}) = (v_0, v_1, -v_0, -v_1, v_0, v_1, v_0, v_1) = (1, -1, -1, 1, 1, -1, 1, -1)$. For the purposes of this article, we call this constructed vector the *transmitted vector*.

Each sender has a different, unique vector **v** chosen from that set, but the construction method of the transmitted vector is identical.

Now, due to physical properties of interference, if two signals at a point are in phase, they add to give twice the amplitude of each signal, but if they are out of phase, they subtract and give a signal that is the difference of the amplitudes. Digitally, this behaviour can be modelled by the addition of the transmission vectors, component by component.

If sender0 has code $(1, -1)$ and data $(1, 0, 1, 1)$, and sender1 has code $(1, 1)$ and data $(0, 0, 1, 1)$, and both senders transmit simultaneously, then this table describes the coding steps:

Step	Encode sender0	Encode sender1
0	code0 = (1, −1), data0 = (1, 0, 1, 1)	code1 = (1, 1), data1 = (0, 0, 1, 1)
1	encode0 = 2(1, 0, 1, 1) − (1, 1, 1, 1) = (1, −1, 1, 1)	encode1 = 2(0, 0, 1, 1) − (1, 1, 1, 1) = (−1, −1, 1, 1)
2	signal0 = encode0 ⊗ code0 = (1, −1, 1, 1) ⊗ (1, −1) = (1, −1, −1, 1, 1, −1, 1, −1)	signal1 = encode1 ⊗ code1 = (−1, −1, 1, 1) ⊗ (1, 1) = (−1, −1, −1, −1, 1, 1, 1, 1)

Because signal0 and signal1 are transmitted at the same time into the air, they add to produce the raw signal:

$$(1, -1, -1, 1, 1, -1, 1, -1) + (-1, -1, -1, -1, 1, 1, 1, 1) = (0, -2, -2, 0, 2, 0, 2, 0)$$

This raw signal is called an interference pattern. The receiver then extracts an intelligible signal for any known sender by combining the sender's code with the interference pattern. The following table explains how this works, and shows that the signals do not interfere with one another:

Step	Decode sender0	Decode sender1
0	code0 = (1, −1), signal = (0, −2, −2, 0, 2, 0, 2, 0)	code1 = (1, 1), signal = (0, −2, −2, 0, 2, 0, 2, 0)
1	decode0 = pattern.vector0	decode1 = pattern.vector1
2	decode0 = ((0, −2), (−2, 0), (2, 0), (2, 0)).(1, −1)	decode1 = ((0, −2), (−2, 0), (2, 0), (2, 0)).(1, 1)
3	decode0 = ((0 + 2), (−2 + 0), (2 + 0), (2 + 0))	decode1 = ((0 − 2), (−2 + 0), (2 + 0), (2 + 0))
4	data0=(2, −2, 2, 2), meaning (1, 0, 1, 1)	data1=(−2, −2, 2, 2), meaning (0, 0, 1, 1)

Further, after decoding, all values greater than 0 are interpreted as 1 while all values less than zero are interpreted as 0. For example, after decoding, data0 is (2, −2, 2, 2), but the receiver interprets this as (1, 0, 1, 1). Values of exactly 0 means that the sender did not transmit any data, as in the following example:

Assume signal0 = (1, −1, −1, 1, 1, −1, 1, −1) is transmitted alone. The following table shows the decode at the receiver:

Step	Decode sender0	Decode sender1
0	code0 = (1, −1), signal = (1, −1, −1, 1, 1, −1, 1, −1)	code1 = (1, 1), signal = (1, −1, −1, 1, 1, −1, 1, −1)
1	decode0 = pattern.vector0	decode1 = pattern.vector1
2	decode0 = ((1, −1), (−1, 1), (1, −1), (1, −1)).(1, −1)	decode1 = ((1, −1), (−1, 1), (1, −1), (1, −1)).(1, 1)
3	decode0 = ((1 + 1), (−1 − 1),(1 + 1), (1 + 1))	decode1 = ((1 − 1), (−1 + 1),(1 − 1), (1 − 1))
4	data0 = (2, −2, 2, 2), meaning (1, 0, 1, 1)	data1 = (0, 0, 0, 0), meaning no data

When the receiver attempts to decode the signal using sender1's code, the data is all zeros, therefore the cross correlation is equal to zero and it is clear that sender1 did not transmit any data.

Asynchronous CDMA

When mobile-to-base links cannot be precisely coordinated, particularly due to the mobility of the handsets, a different approach is required. Since it is not mathematically possible to create signature sequences that are both orthogonal for arbitrarily random starting points and which make full use of the code space, unique "pseudo-random" or "pseudo-noise" (PN) sequences are used in *asynchronous* CDMA systems. A PN code is a binary sequence that appears random but can be reproduced in a deterministic manner by intended receivers. These PN codes are used to encode and decode a user's signal in Asynchronous CDMA in the same manner as the orthogonal codes in synchronous CDMA. These PN sequences are statistically uncorrelated, and the sum of a large number of PN sequences results in *multiple access interference* (MAI) that is approximated by a Gaussian noise process (following the central limit theorem in statistics). Gold codes are an example of a PN suitable for this purpose, as there is low correlation between the codes. If all of the users are received with the same power level, then the variance (e.g., the noise power) of the MAI increases in direct proportion to the number of users. In other words, unlike synchronous CDMA, the signals of other users will appear as noise to the signal of interest and interfere slightly with the desired signal in proportion to number of users.

All forms of CDMA use spread spectrum process gain to allow receivers to partially discriminate against unwanted signals. Signals encoded with the specified PN sequence (code) are received, while signals with different codes (or the same code but a different timing offset) appear as wideband noise reduced by the process gain.

Since each user generates MAI, controlling the signal strength is an important issue with CDMA transmitters. A CDM (synchronous CDMA), TDMA, or FDMA receiver can in theory completely reject arbitrarily strong signals using different codes, time slots or frequency channels due to the orthogonality of these systems. This is not true for Asynchronous CDMA; rejection of unwanted signals is only partial. If any or all of the unwanted signals are much stronger than the desired signal, they will overwhelm it. This leads to a general requirement in any asynchronous CDMA system to approximately match the various signal power levels as seen at the receiver. In CDMA cellular, the base station uses a fast closed-loop power control scheme to tightly control each mobile's transmit power.

Advantages of Asynchronous CDMA Over Other Techniques

Efficient Practical Utilization of the Fixed Frequency Spectrum

In theory CDMA, TDMA and FDMA have exactly the same spectral efficiency but practically, each has its own challenges – power control in the case of CDMA, timing in the case of TDMA, and frequency generation/filtering in the case of FDMA.

TDMA systems must carefully synchronize the transmission times of all the users to ensure that they are received in the correct time slot and do not cause interference. Since this cannot be perfectly controlled in a mobile environment, each time slot must have a guard-time, which reduces the probability that users will interfere, but decreases the spectral efficiency. Similarly, FDMA systems must use a guard-band between adjacent channels, due to the unpredictable doppler shift of the signal spectrum because of user mobility. The guard-bands will reduce the probability that adjacent channels will interfere, but decrease the utilization of the spectrum.

Flexible Allocation of Resources

Asynchronous CDMA offers a key advantage in the flexible allocation of resources i.e. allocation of a PN codes to active users. In the case of CDM (synchronous CDMA), TDMA, and FDMA the number of simultaneous orthogonal codes, time slots and frequency slots respectively are fixed hence the capacity in terms of number of simultaneous users is limited. There are a fixed number of orthogonal codes, time slots or frequency bands that can be allocated for CDM, TDMA, and FDMA systems, which remain underutilized due to the bursty nature of telephony and packetized data transmissions. There is no strict limit to the number of users that can be supported in an asynchronous CDMA system, only a practical limit governed by the desired bit error probability, since the SIR (Signal to Interference Ratio) varies inversely with the number of users. In a bursty traffic environment like mobile telephony, the advantage afforded by asynchronous CDMA is that the performance (bit error rate) is allowed to fluctuate randomly, with an average value determined by the number of users times the percentage of utilization. Suppose there are 2N users that only talk half of the time, then 2N users can be accommodated with the same *average* bit error probability as N users that talk all of the time. The key difference here is that the bit error probability for N users talking all of the time is constant, whereas it is a *random* quantity (with the same mean) for 2N users talking half of the time.

In other words, asynchronous CDMA is ideally suited to a mobile network where large numbers of transmitters each generate a relatively small amount of traffic at irregular intervals. CDM (synchronous CDMA), TDMA, and FDMA systems cannot recover the underutilized resources inherent to bursty traffic due to the fixed number of orthogonal codes, time slots or frequency channels that can be assigned to individual transmitters. For instance, if there are N time slots in a TDMA system and 2N users that talk half of the time, then half of the time there will be more than N users needing to use more than N time slots. Furthermore, it would require significant overhead to continually allocate and deallocate the orthogonal code, time slot or frequency channel resources. By comparison, asynchronous CDMA transmitters simply send when they have something to say, and go off the air when they don't, keeping the same PN signature sequence as long as they are connected to the system.

Spread-spectrum Characteristics of CDMA

Most modulation schemes try to minimize the bandwidth of this signal since bandwidth is a limited resource. However, spread spectrum techniques use a transmission bandwidth that is several orders of magnitude greater than the minimum required signal bandwidth. One of the initial reasons for doing this was military applications including guidance and communication systems. These systems were designed using spread spectrum because of its security and resistance to jamming. Asynchronous CDMA has some level of privacy built in because the signal is spread using a pseudo-random code; this code makes the spread spectrum signals appear random or have noise-like properties. A receiver cannot demodulate this transmission without knowledge of the pseudo-random sequence used to encode the data. CDMA is also resistant to jamming. A jamming signal only has a finite amount of power available to jam the signal. The jammer can either spread its energy over the entire bandwidth of the signal or jam only part of the entire signal.

CDMA can also effectively reject narrow band interference. Since narrow band interference affects only a small portion of the spread spectrum signal, it can easily be removed through notch filtering without much loss of information. Convolution encoding and interleaving can be used to assist in recovering this lost data. CDMA signals are also resistant to multipath fading. Since the spread spectrum signal occupies a large bandwidth only a small portion of this will undergo fading due to multipath at any given time. Like the narrow band interference this will result in only a small loss of data and can be overcome.

Another reason CDMA is resistant to multipath interference is because the delayed versions of the transmitted pseudo-random codes will have poor correlation with the original pseudo-random code, and will thus appear as another user, which is ignored at the receiver. In other words, as long as the multipath channel induces at least one chip of delay, the multipath signals will arrive at the receiver such that they are shifted in time by at least one chip from the intended signal. The correlation properties of the pseudo-random codes are such that this slight delay causes the multipath to appear uncorrelated with the intended signal, and it is thus ignored.

Some CDMA devices use a rake receiver, which exploits multipath delay components to improve the performance of the system. A rake receiver combines the information from several correlators, each one tuned to a different path delay, producing a stronger version of the signal than a simple receiver with a single correlation tuned to the path delay of the strongest signal.

Frequency reuse is the ability to reuse the same radio channel frequency at other cell sites within a cellular system. In the FDMA and TDMA systems frequency planning is an important consideration. The frequencies used in different cells must be planned carefully to ensure signals from different cells do not interfere with each other. In a CDMA system, the same frequency can be used in every cell, because channelization is done using the pseudo-random codes. Reusing the same frequency in every cell eliminates the need for frequency planning in a CDMA system; however, planning of the different pseudo-random sequences must be done to ensure that the received signal from one cell does not correlate with the signal from a nearby cell.

Since adjacent cells use the same frequencies, CDMA systems have the ability to perform soft hand offs. Soft hand offs allow the mobile telephone to communicate simultaneously with two or more cells. The best signal quality is selected until the hand off is complete. This is different from hard hand offs utilized in other cellular systems. In a hard hand off situation, as the mobile telephone approaches a hand off, signal strength may vary abruptly. In contrast, CDMA systems use the soft hand off, which is undetectable and provides a more reliable and higher quality signal.

Collaborative CDMA

In a recent study, a novel collaborative multi-user transmission and detection scheme called Collaborative CDMA has been investigated for the uplink that exploits the differences between users' fading channel signatures to increase the user capacity well beyond the spreading length in multiple access interference (MAI) limited environment. The authors show that it is possible to achieve this increase at a low complexity and high bit error rate performance in flat fading channels, which is a major research challenge for overloaded CDMA systems. In this approach, instead of using one sequence per user as in conventional CDMA, the authors group a small number of users to share the same spreading sequence and enable group spreading and despreading operations. The new collaborative multi-user receiver consists of two stages: group multi-user detection (MUD) stage to suppress the MAI between the groups and a low complexity maximum-likelihood detection stage to recover jointly the co-spread users' data using minimum Euclidean distance measure and users' channel gain coefficients. In CDMA, signal security is high.

Space-division Multiple Access

Space-division multiple access (SDMA) is a channel access method based on creating parallel spatial pipes next to higher capacity pipes through spatial multiplexing and/or diversity, by which it is able to offer superior performance in radio multiple access communication systems.In traditional mobile cellular network systems, the base station has no information on the position of the mobile units within the cell and radiates the signal in all directions within the cell in order to provide radio coverage. This results in wasting power on transmissions when there are no mobile units to reach, in addition to causing interference for adjacent cells using the same frequency, so called co-channel cells. Likewise, in reception, the antenna receives signals coming from all directions including noise and interference signals. By using smart antenna technology and differing spatial locations of mobile units within the cell, space-division multiple access techniques offer attractive performance enhancements. The radiation pattern of the base station, both in transmission and reception, is adapted to each user to obtain highest gain in the direction of that user. This is often done using phased array techniques.

In GSM cellular networks, the base station is aware of the distance (but not direction) of a mobile phone by use of a technique called "timing advance" (TA). The base transceiver station (BTS) can determine how distant the mobile station (MS) is by interpreting the reported TA. This information, along with other parameters, can then be used to power down the BTS or MS, if a power control feature is implemented in the network. The power control in either BTS or MS is implemented in most modern networks, especially on the MS, as this ensures a better battery life for the MS. This is also why having a BTS close to the user results in less exposure to electromagnetic radiation.

This is why one may actually be safer to have a BTS close to them as their MS will be powered down as much as possible. For example, there is more power being transmitted from the MS than what one would receive from the BTS even if they were 6 meters away from a BTS mast. However, this estimation might not consider all the Mobile stations that a particular BTS is supporting with EM radiation at any given time.

In the same manner, 5th generation mobile networks will be focused in utilizing the given position of the MS in relation to BTS in order to focus all MS Radio frequency power to the BTS direction and vice versa, thus enabling power savings for the Mobile Operator, reducing MS SAR index, reducing the EM field around base stations since beam forming will concentrate rf power when it will be actually used rather than spread uniformly around the BTS, reducing health and safety concerns, enhancing spectral efficiency, and decreased MS battery consumption.

Arbitrary Slice Ordering

Arbitrary slice ordering (ASO) is an algorithm for loss prevention. It is used for restructuring the ordering of the representation of the fundamental regions (macroblocks) in pictures. This type of algorithm avoids the need to wait for a full set of scenes to get all sources. Typically considered as an error/loss robustness feature.

This type of algorithm is included as tool in baseline profile the H.264/MPEG-4 AVC encoder with I *Slices*, P *Slices*, Context Adaptive Variable Length Coding (CAVLC), grouping of *slices* (Slice Group), arbitrary slice order (ASO) and Redundancy *slices*.

Applications

Primarily for lower-cost applications with limited computing resources, this profile is used widely in videoconferencing, mobile applications and security applications also.

Arbitrary Slice Ordering (ASO) relaxes the constraint that all macroblocks must be sequenced in decoding order, and thus enhances flexibility for low-delay performance important in teleconferencing applications and interactive Internet applications.

Problems

If ASO across pictures is supported in AVC, serious issues arise: *slices* from different pictures are interleaved. One possible way to solve these issues is to limit ASO within a picture, i.e. slices from different pictures are not interleaved.

However, even if we limit ASO within a picture, the decoder complexity is significantly increased. Because Flexible Macroblock Order FMO extend the concept of slices by allowing non-consecutive macroblocks to belong to the same *slice*, this section also addresses the decoder complexity introduced by (FMO).

Types of Decoding ASO

Association of Macroblocks to Slice

- *Impact of ASO on AVC decoders complexity*

An example of how macroblocks can be associated to different *slices* is shown in Figure. When ASO is supported, the four slices of this example can be received by the decoder in a random order. Figure shown the following receiving order: *slice #4*, *slice #3*, *slice #1*, and *slice #2*. The same figure presents the AVC decoder blocks required to support ASO decoding.

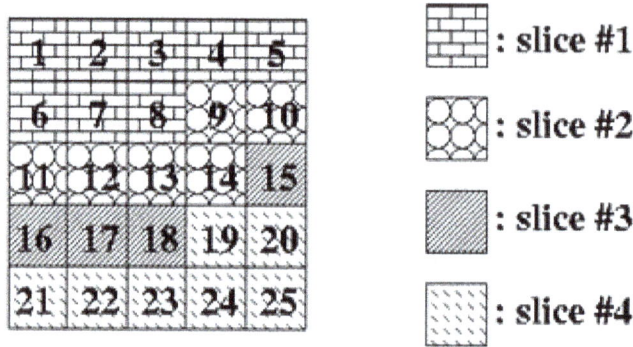

Figure: An example of macroblock assignment to four slices. Each slice is represented by a different texture.

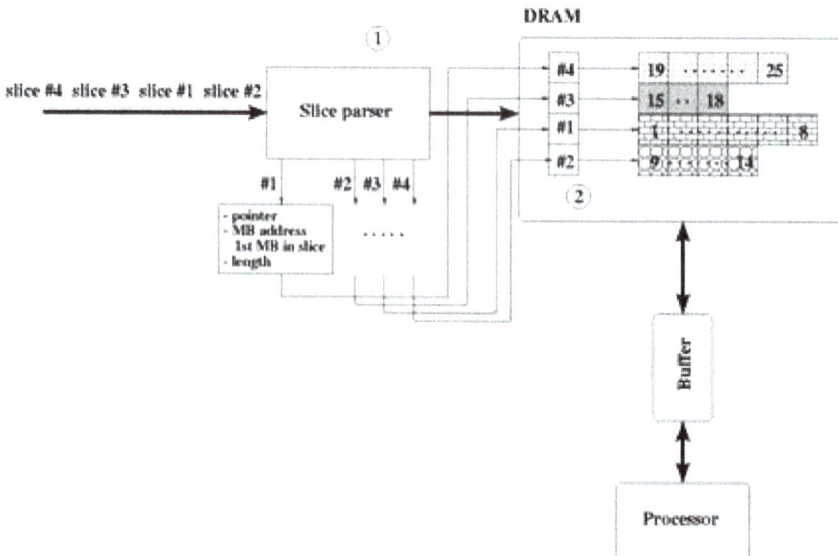

Figure: The AVC decoder blocks need to support ASO decoding.

For each *slice*, the *slice* length and the macroblock address (i.e. index with respect to the raster scan order) of the first macroblock (MB) of the *slice* are extracted by the *slice* parser (Figure). This information, together with the *slice* itself, is stored in memory In addition, a list of pointers (Figure

2, a pointer for each slice, and each pointing to the memory location where a *slice* is stored), should be generated. The list of pointers, together with the address of the first macroblock of the *slice*, will be used to navigate through the out of order *slices*. The *slice* length will be used to transfer the *slice* data from the DRAM to the decoder's internal memory.

Faced with the necessity to decode out of order *slices*, a decoder may:

- 1) wait for all the *slices* of each picture to arrive before start decoding and de-blocking the picture.

- 2) decode the *slices* in the order in which they come to the decoder.

The first method increases latency, but allows performing decoding and de-blocking in parallel. However, managing a large number of pointers (in the worst case, one pointer for each MB) and increasing the intelligence of the DRAM access unit increase the decoder complexity.

The second method hurts significantly the decoder performance. In addition, by performing the de-blocking in a second pass, the DRAM to processor's memory bandwidth is increased.

Decoding *slices* in the order they are received can result in additional memory consumption or impose higher throughput requirements on the decoder and local memory to run at higher clock speed. Consider an application in which the display operation reads the pictures to be displayed right from the section of memory where the decoder stored the pictures.

Association of Macroblocks to Slice and Slices to Group of Slices

- *Impact of ASO and FMO on AVC decoders complexity*

An example of how *slices* can be associated to different *slice* group is shown in Figure 3. When ASO and FMO are supported, the four *slices* of this example can be received by the decoder in a random order. Figure 2 shown the following order: *slice #4*, *slice #2*, *slice #1*, and *slice #3*. The same figure presents the AVC decoder blocks required to support ASO and FMO decoding.

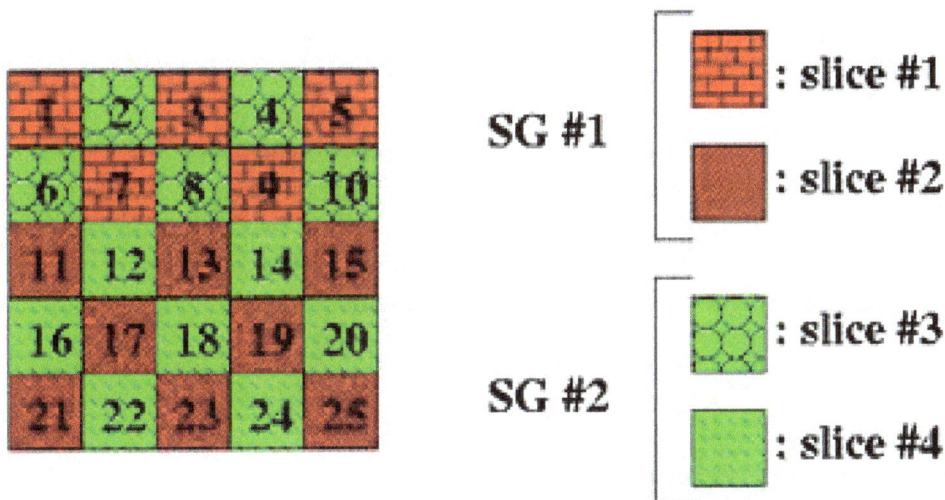

Figure: An example of macroblock assignment to four *slices* and to two "Slice" Group (SG in the figure). Each slice is represented by a different texture, and each *Slice* Group is represented a different color.

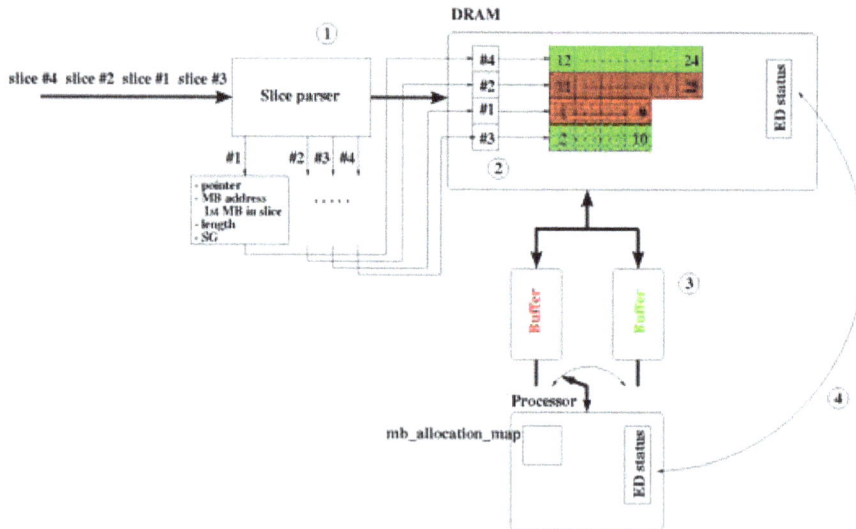

Figure: The AVC decoder blocks need to support ASO and FMO decoding.

In addition to the *slice* length and the macroblock address of the 1st macroblock (MB) of the *slice*, the *slice* parser (Figure) need to extract the *Slice* Group (SG) of each *slice*. These informations, together with the *slice* itself, are stored in DRAM. As in the ASO case, the list of pointers (Figure) should be generated.

The list of pointers, together with the address of the 1st MB of the *slice*, the SG, and the mb_allocation_map (stored in the processor's local memory), will be used to navigate through the *slices*. The *slice* length will be used to transfer the *slice* data from the DRAM to the processor local memory.

Similarly to the ASO case, in the combined ASO and FMO case the decoder may:

- 1) wait for all the *slices* of each picture to arrive before start decoding and de-blocking the picture.

- 2) decode the *slices* in the order in which they come to the decoder.

The first approach is still the preferred one. Because of FMO, decoding macroblocks in raster scan order may require to switch between different *slices* and/or *slice* groups. To speed up the DRAM access, one buffer for each *Slice* Group must be used (Figure 4). This additional intelligence of the DRAM access unit further increase the decoder complexity. Moreover, switching between different *slices* and/or *slice* groups requires swapping the Entropy Decoder (ED) status information. In the worst case, swapping occurs after decoding each macroblock. If the entire Entropy Decoder status information is too large to be stored in the processor local memory, each ED status need to be loaded from and stored into DRAM, thus further increasing the DRAM to processor's memory bandwidth (Figure 4).

Diversity Combining

Diversity combining is the technique applied to combine the multiple received signals of a diversity reception device into a single improved signal.

Various Techniques

Various diversity combining techniques can be distinguished:

- Equal-gain combining: All the received signals are summed coherently.

- Maximal-ratio combining is often used in large phased-array systems: The received signals are weighted with respect to their SNR and then summed. The resulting SNR yields $\sum_{k=1}^{N} SNR_k$ where SNR_k is SNR of the received signal k.

- Switched combining: The receiver switches to another signal when the currently selected signal drops below a predefined threshold. This is also often called "Scanning Combining".

- Selection combining: Of the N received signals, the strongest signal is selected. When the N signals are independent and Rayleigh distributed, the expected diversity gain has been shown to be $\sum_{k=1}^{N} \frac{1}{k}$, expressed as a power ratio. Therefore, any additional gain diminishes rapidly with the increasing number of channels. This is a more efficient technique than switched combining.

Sometimes more than one combining technique is used – for example, lucky imaging uses selection combining to choose (typically) the best 10% images, followed by equal-gain combining of the selected images.

Other signal combination techniques have been designed for noise reduction and have found applications in single molecule biophysics, chemometrics among other disciplines.

Timing Combining

When the focus is on the wireless transmission of longer signal sequences, such as for example Ethernet packets, specific performance characteristics regarding diversity gain on a parallel redundant wireless transmission system can be observed. For "timing combining", data packets are redundantly and simultaneously sent over parallel paths. On the receiving side, out of the branches the first arriving packet is selected and immediately processed towards the end application. Further copies of the packet -if arriving- are discarded. This type of postdetection combiner is called a "timing combiner", since a significant performance improvement is gained through the immediate processing of the first arriving packet.

Switched Combining Two-way Radio Example

In land-mobile radio, where vehicle-mounted and hand-held radios communicate with a base station radio over a single frequency, space diversity is achieved by having several receivers at dif-ferent sites. Diversity combining, or voting, in two-way radio systems is a method for improving talk-back range from walkie-talkie and vehicular mobile radios.

The receivers are connected to a device referred to as a voting comparator or voter.

The voting comparator performs an evaluation of all received signals and picks the most usable received signal. In repeater systems, the voted signal is retransmitted. In simplex systems, it goes to the console speaker at the base station. Audio from a receiver that is not voted is ignored. Voting comparators in analog FM systems can switch between receivers in tenths- or hundredths-of-a-second, (faster than one syllable). So long as an intelligible signal gets to a single receiver in the system, the repeated audio, or audio sent to the console speaker, would be intelligible.

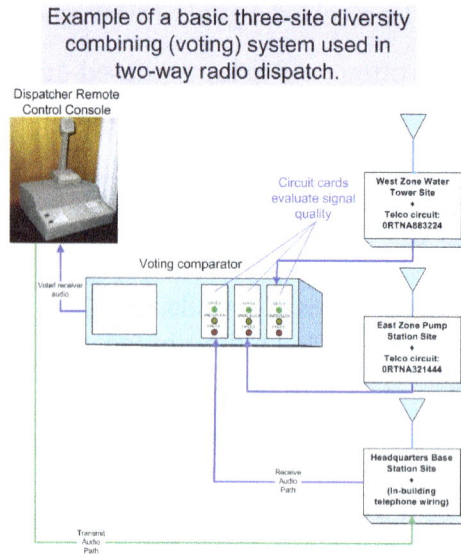

Example of a basic three-site diversity combining (voting) system used in two-way radio dispatch.

System diagram for a water utility two-way radio system with voting.

Corresponding Receiver Locations

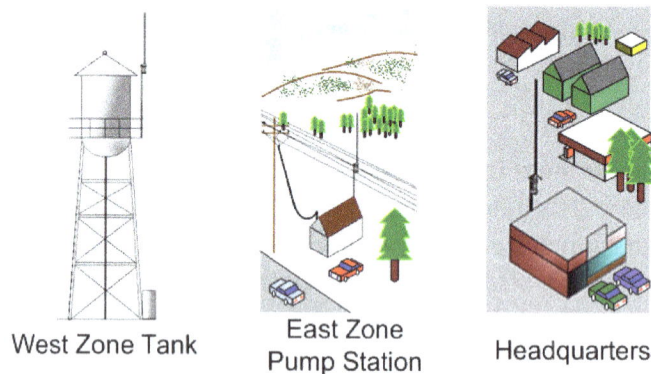

West Zone Tank

East Zone Pump Station

Headquarters

Water utility equipment sites.

In this arrangement, receivers at remote sites are connected to the voting comparator by private telephone lines, a channel in a D4 channel bank on a DS1, or an analog microwave baseband channel.

How Signals are Evaluated

The earliest voting comparators relied on a tone encoded on a separate audio path, requiring each receiver site to have a 4-wire circuit or two audio paths. The pitch of the tone changed to represent the received signal level, or FM receiver limiter voltage, at the remote receiver. This worked poorly because it did not account for microwave baseband noise or noisy telephone company circuits.

Newer voting comparators compare signal-to-noise ratio at the voting comparator, accounting for end-to-end noise, bad phone lines, poor level discipline, as well as the best diversity reception path.

Walkie-talkie Talk-back Range

When communicating with hand-held radios (walkie-talkies), base stations generally talk out further than they can receive. Voting among several receivers at different sites increases the probability that one of the receivers will acquire a usable signal from two-way radios in a system.

Interference Reduction

Diversity combining reduces one possible single-point failure: any single receiver failure, or local interference to a single receiver, will not block reception on the entire system. Equipment sites can host many radio transmitters and receivers. A single site is subject to local, site-specific interfering signals. These interfering signals may come and go as transmitters switch on and off.

A potential problem with receivers located at high-elevation receiver sites is that they may acquire signals from distant counties, prefectures, or other provinces. These unwanted, distant signals can be stronger than desired signals from local walkie-talkies. The distant signals may block local weak signals in some cases. Having several receive sites increases the probability that one of the sites will receive the local signal in the presence of a distant, undesired one. Selective calling can eliminate users having to listen to the audio of distant signals even though the distant signals are within receive range of one or more receivers.

Coverage

A minimum of 95% coverage is cited in literature for critical or emergency service two-way radio systems. One definition of system coverage is Telecommunications Industry Association (TIA) TSB-88A standard.

Vote-lock or Vote-and-hold Option

The majority of installations using diversity combining equipment continually evaluate the best received signal against all other signals. Throughout the length of a received transmission, the comparator may switch receivers as often as every few tenths of a second. As a walkie-talkie user causes a signal fade by turning their head, or a passing tractor-trailer rig blocks their signal at the voted receive site, the combining unit rapidly changes to a different receiver.

In some installations, diversity combining equipment is configured to lock on a receiver. For example, in some rural, regional coverage systems, the receivers each cover a unique geographic area. There is not much overlap. If the system consisted of two sites, north and south, it would pick the better of the two and remain locked on that receiver until the transmission ended. This works better with mobile radios because their signal strengths tend to be steady.

In some cases this vote-and-hold is used to steer transmitter selection. Consider the case of a regional system with two base stations: north and south. If the diversity combining equipment votes "north," the next time the dispatcher presses the transmit button, the north transmitter will key. Called *transmitter steering*, this is supposed to automate transmitter selection in systems where more than one transmitter site is available. In some instances it doesn't work well.

In mobile data systems, the vote lock option is preferred because constant switching between re-

ceivers causes lost data packets. The diversity combining equipment switches fast enough that syllables are not lost but not fast enough that bits are not lost. Mobile data systems usually originate
with modems in mobile radios. Mobile radios usually produce solid signals into more than one
receive site, so the signal strengths are strong enough for vote locking to work well.

Phantom Circuit

In telecommunication and electrical engineering, a phantom circuit is an electrical circuit derived
from suitably arranged wires with one or more conductive paths being a circuit in itself and at the
same time acting as one conductor of another circuit.

Phantom Group

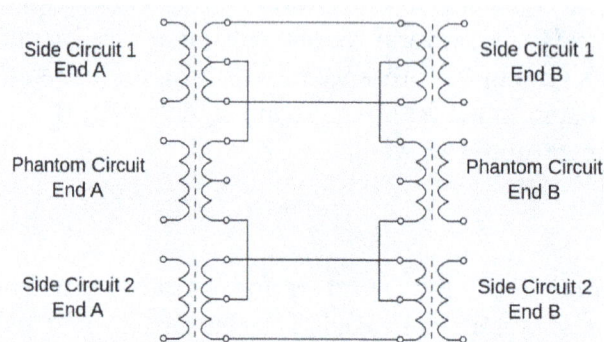

Phantom circuit derived from two subscriber circuits

A phantom group is composed of three circuits that are derived from two single-channel circuits
to form a *phantom circuit*. Here the phantom circuit is a third circuit derived from two suitably
arranged pairs of wires, called side circuits, with each pair of wires being a circuit in itself and at the
same time acting as one conductor of the third circuit. The "side circuits" within phantom circuits can
be coupled to their respective voltage drops by center-tapped transformers, usually called "repeating
coils". The center taps are on the line side of the side circuits. Current from the phantom circuit is
split evenly by the center taps. This cancels crosstalk from the phantom circuit to the side circuits.

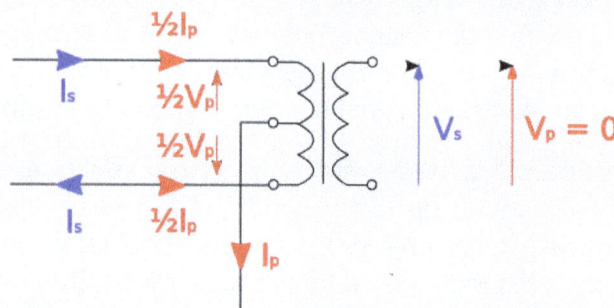

Diagram showing how the phantom currents (red) cancel in the transformer. Side circuit currents (blue) do not cancel
and are transmitted through the transformer.

Phantom working increased the number of circuits on long distance routes in the early 20th century without putting up more wires. Phantoming declined with the adoption of carrier systems.

It is theoretically possible to create a phantom circuit from two other phantom circuits and so on up in a pyramid with a maximum 2n-1 circuits being derived from n original circuits. However, more than one level of phantoming is usually impractical. Isolation between the phantom circuit and the side circuits relies on accurate balance of the line and transformers. Imperfect balance results in crosstalk between the phantom and side circuits and this effect accumulates as each level of phantoms is added. Even small levels of crosstalk are unacceptable on analogue telecommunications circuits since speech crosstalk is still intelligible down to quite low levels.

Phantom Microphone Powering

Condenser microphones have impedance converter (current amplifier) circuitry that requires powering; in addition, the capsule of any non-electret, non-RF condenser microphone requires a polarizing voltage to be applied. Since the mid- to late 1960s most balanced, professional condenser microphones for recording and broadcast have used phantom powering. It can be provided by outboard AC or battery supplies, but nowadays is most often built into the mixing console, recorder or microphone preamplifier to which the microphones are connected.

By far the most common circuit uses +48 VDC fed through a matched pair of 6.8 kOhm resistors for each input channel. This arrangement has been standardized by the IEC and ISO, along with a less-commonly-used arrangement with +12 VDC and 680 Ohm feed resistors.

As a practical matter, phantom powering allows the same two-conductor shielded cables to be used for both dynamic microphones and condenser microphones, while being harmless to balanced microphones that aren't designed to consume it, since the circuit balance prevents any substantial DC from flowing through the output circuit of those microphones.

DC Phantom

Simple DC signalling can be achieved on a telecommunications line in a similar way to phantom powering of microphones. A switch connected to the transformer centre-tap at one end of the line can operate a similarly connected relay at the other end. The return path is through the ground connection. This arrangement can be used for remotely controlling equipment.

Carrier Circuit Phantoms

From the 1950s to around the 1980s, using phantoms on star-quad trunk carrier circuits was a popular method of deriving a high quality broadcast audio circuit. The multiplexed FDM telecommunications carrier system usually did not use the baseband of the cable because it was inconvenient to separate low frequencies with filters. On the other hand, a one-way audio phantom could be formed from the two pairs (go and return signals) making up the star-quad cable.

Unloaded Phantom

Unloaded phantom is a phantom configuration of loaded lines (a circuit fitted with loading coils). The idea here is not to create additional circuits. Rather, the purpose is to cancel or greatly reduce the effect of the loading coils fitted to a line. The reason for doing this is that loaded lines have a definite cut-off frequency and it may be desired to equalise the line to a frequency which is higher than this, for example to make a circuit suitable for use by a broadcaster. Ideally, the loading would be removed or reduced for a permanent connection, but this is not feasible for temporary arrangements such as a requirement for outside broadcast. Instead, two circuits in a phantom configuration can be used to greatly reduce the inductance being inserted by the loading coils, and hence the loading effect.

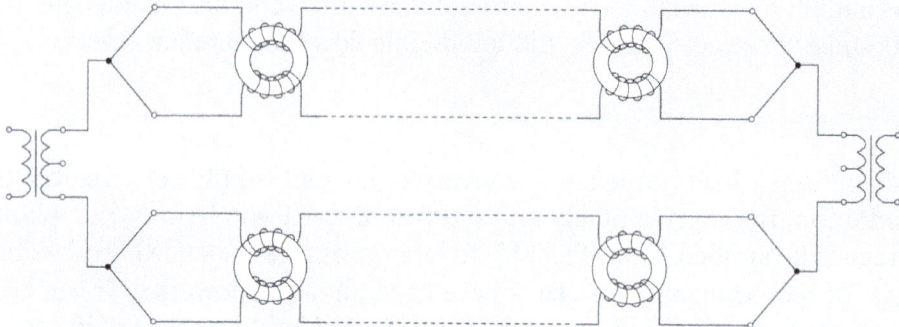

Unloaded phantom configuration. The windings of the loading coil are wound such that the magnetic flux induced in the core is normally in the same direction for both windings. However, in the phantom configuration the flux cancels.

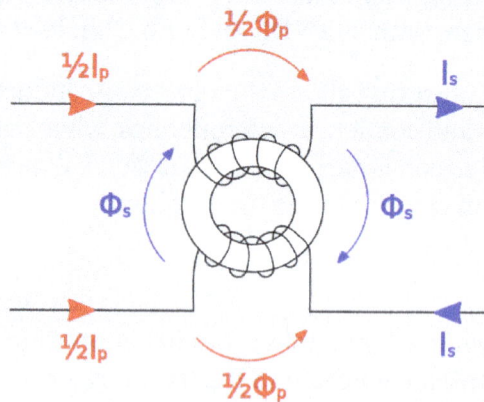

Diagram showing how the flux due to the phantom currents (red) is cancelled in the load coil. Flux due to normal line currents (blue) is additive.

It works because the loading coils used on balanced lines have two windings, one for each leg of the circuit. They are both wound on a common core and the windings are so arranged that the magnetic flux induced by both of them is in the same direction. Both windings induce an emf in each other as well as their own self-induction. This effect greatly increases the inductance of the coil and hence its loading effectiveness. By contrast, when the circuit is in the phantom configuration the currents in the two wires of each pair are in the same direction and the magnetic flux is being cancelled. This has precisely the opposite effect and the inductance is greatly reduced.

This configuration is most commonly used on the two pairs of a star-quad cable. It is not so successful with other pairs of wires. The difference in the path of the two pairs can easily destroy the balance and results in crosstalk and interference.

This configuration can also be called "bunched pairs". However, "bunched pairs" can also refer to the straightforward connection of two lines in parallel which is not a phantom circuit and will not reduce the loading.

Through-the-Earth Mine Communications

Through-the-Earth (TTE) signalling is a type of radio signalling used in underground mines and caves that uses low-frequency waves to penetrate dirt and rock, which are opaque to higher-frequency conventional radio signals.

In mining, these higher-frequency signals can be relayed underground through various antennas, repeater or mesh configurations, but communication is restricted to line of sight to these antenna and repeaters systems.

Overview

Through-the-Earth transmission can overcome these restrictions by using ultra-low frequency (300–3000 Hz) signals, which can travel through several hundred feet of rock strata. The antenna cable can be located on the surface only at a mine site, and provide signal coverage to all parts of the underground mine. The antenna may be placed in a "loop" formation around the perimeter of the mine site (or wherever coverage is needed) for systems using magnetic fields to carry signals. Systems that use electric fields as the signal carrier are not subject to this limitation. Transmissions propagate through rock strata which is used as the medium to carry the ultra-low-frequency signals. This is important in mining applications, particularly after any significant incident, such as fire or explosion, which would destroy much of the fixed communication infrastructure underground.

If the terrain makes a loop surface antenna impractical to install, then the antenna can be installed underground or a non-magnetic field type carrier may be used. But because the signal travels through rock, the antenna does not need to run into all parts of the mine to achieve mine wide signal coverage, thus minimizing the risk of damage during an incident.

Cave Radios

Portable magnetic-loop cave radios have been used by cavers for two-way communication and cave surveying since the 1960s. In a typical setup the transmitting loop, consisting of a many turns of copper wire, is oriented horizontally within the cave using a spirit level, and driven at a few kHz. Though such a small antenna is a very poor radiator of propagating radio waves at this low frequency, its local AC magnetic field is strong enough to be detected by a similar receiving antenna up to a few hundred meters away. The received signal's strength and its dependence on orientation of the receiving coil yields approximate distance and directional information.

Personal Emergency Device

There are several systems that have been recently developed. One system is known as the PED System, where PED is an acronym for Personal Emergency Device. Initially developed after a mining disaster in Australia at Moura No. 4 Coal Mine in 1986, and further developed after the Moura No. 2 Coal Mine explosion in 1994 where the need for a communication system to survive major incidents underground was identified in the inquiries into the disasters.

PED is a one-way text paging device, with wide use in Australia, as well as installations in the United States, China, Canada, Mongolia, Chile, Tanzania, and Sweden. Australian company Mine Site Technologies began the development of PED in 1987, and it became commercially available and Mine Safety & Health Administration (MSHA) approved in 1991. The best documented use of PED during a mine emergency is from the Willow Creek Mine Fire in 1998 in Utah, where it was able to quickly alert miners underground of the need to evacuate before toxic fumes from the fire filled the mine. Reports of this use can be seen on the Mine Safety & Health Administration (MSHA) web site at and.

Development

On-going developments based on initial research undertaken in Australia by the Commonwealth Scientific and Industrial Research Organisation (CSIRO) after the 1994 mine accident and given further impetus by the investigations into the Sago Mine disaster the PED System is undergoing further development to provide two-way through-the-Earth communication capability.

Emerging technologies have recently been developed such as the Rescue Dog Emergency Through the Earth Communication System developed by E-Spectrum Technologies. The Rescue Dog is a two-way extended-range portable through-the-Earth solution that was developed in the US in cooperation with The National Institute for Occupational Safety and Health (or NIOSH) which does not rely on large loop surface antennas for signal transmission. New non-portable systems have also been developed by companies such as Lockheed Martin for use in emergency chambers to provide post-accident, two-way, emergency voice and text communications independent of surface or in-mine infrastructure.

New Technologies

A new wireless "Miner Lifeline" telecommunication technology is being tested in 2012 at the West Virginia Robinson Run mine (recent production 6,000,000 short tons (5,400,000 t) per year of coal using 600 miners). The system supports voice, text, or SOS sent on a "bubble" of magnetic waves, and "can move more than 1,500 feet (460 m) up or down and 2,000 feet (610 m) laterally, arriving in less than a minute."

Two-way Communication

Two-way communication is a form of transmission in which both parties involved transmit information. Two-Way communication has also been referred to as interpersonal communication. Common forms of two-way communication are:

- Amateur Radio, CB or FRS radio contacts.

- Chatrooms and Instant Messaging.

- Computer networks.

- In-person communication.

- Telephone conversations.

A cycle of communication and two-way communication are actually two different things. If we examine closely the anatomy of communication – the actual structure and parts – we will discover that a cycle of communication is not a two-way communication in its entirety. Meaning, two way communication is not as simple as one may infer. One can improve two-way or interpersonal communication by focusing on the eyes of the person speaking, making eye contact, watching body language, responding appropriately with comments, questions, and paraphrasing, and summarizing to confirm main points and an accurate understanding.

Two-way communication involves feedback from the receiver to the sender. This allows the sender to know the message was received accurately by the receiver. This chart demonstrates two-way communication and feedback.

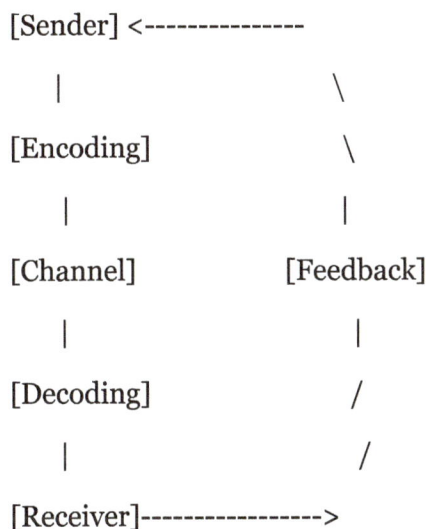

```
        [Sender] <---------------

           |                    \

        [Encoding]               \

           |                      |

        [Channel]            [Feedback]

           |                      |

        [Decoding]               /

           |                     /

        [Receiver]----------------->
```

Amateur Radio, Citizen Band Radio, and Family Radio Service

Amateur Radio is used for entertainment and as a hobby by many groups of people. These individuals label themselves as "Hams." Amateur Radios are also known to be a reliable means of communication when all other forms are not operating. In times of disaster, communication through Amateur radios has led to lives being saved. Citizen Band Radio, CB radio, can be used by anyone who is not a member of a foreign government. It is meant for short range communication using devices that mimic walkie-talkies. Family Radio Service, FRS, is also meant for short range communication using devices that mimic walkie-talkies. Like the CB radio, the FRS does not require a license and can be used by anyone who is not a member of a foreign government.

Chat Rooms and Instant Messaging

Instant messaging became wildly popular around 1996 and spread even more with AOL in 1997.

The concept behind IM is that it is a way of quick communication between two people due to tools such as knowing when messages are seen or knowing when others are online. Many social media sites have integrated IM into their sites as ways to spread communication. Chat Rooms are very similar, only they are messages to a group of people. Chat rooms are often public; meaning that you can send a message and anyone can freely join the "room" and view the message as well as respond.

In-person Communication

As it relates to business, 75% of people believe in-person communication is critical. In-person interaction is useful for resolving problems more efficiently, generating long-term relationships, and resolving a problem or creating an opportunity quickly. 4 out of 6 of the most important attributes of building a relationship cannot be achieved without the power of in-person, which requires a rich communication environment. Business executives believe in-person collaboration is critical for more than 50 percent of key business strategic and tactical business processes when engaging with colleagues, customers, or partners.

Telephone Conversations

The telephone is a device that is relatively easy to understand and use. In fact, the telephone connections used today have remained remarkably unchanged compared to those used almost a century ago. In addition, your connection to the phone company is rather straightforward as well. The telephone makes it easy to connect instantly with others from all over the world, making it simple to have a two-way conversation with a neighbor or with someone many miles away. Phones have undergone some changes over the years. Today, for example, phones use electronic switches instead of operators. The switch uses a dial tone so that when you pick up the phone you are aware that both the switch and the phone are functioning properly.

Computer Networks

Computer Networks are used to have two-way communication by having computers exchange data. Ways that this is possible is wired interconnects and wireless interconnects. Types of wired interconnects are Ethernets and fiber optic cables. Ethernets connect local devices through Ethernet cables. Fiber runs underground for long distances and is the main source of Internet in most homes and businesses. Types of wireless interconnects include Wi-Fi and Bluetooth. The problem with these networks is that they don't have unlimited connection span. To expand the reach there are wide area interconnects such as satellite and cellular networks. Also, there are long distance interconnects which need *backhaul* to move the data back and forth and *last mile* to connect the provider to the network.

Underwater Acoustic Communication

Underwater acoustic communication is a technique of sending and receiving messages below water. There are several ways of employing such communication but the most common is by using hydrophones. Underwater communication is difficult due to factors such as multi-path propagation, time variations of the channel, small available bandwidth and strong signal attenuation, especially over long ranges. Compared to terrestrial communication, underwater communication has low data rates because it uses acoustic waves instead of electromagnetic waves.

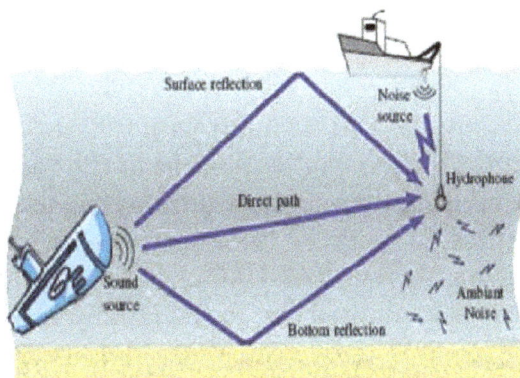

Example of multi-path propagation

At the beginning of the 20th century, some ships communicated by underwater bells, the system being competitive with the primitive Maritime radionavigation service of the time. The later Fessenden oscillator allowed communication with submarines.

Types of Modulation Used for Underwater Acoustic Communications

In general the modulation methods developed for radio communications can be adapted for underwater acoustic communications (UAC). However some of the modulation schemes are more suited to the unique underwater acoustic communication channel than others. Some of the modulation methods used for UAC are as follows:

- Frequency Shift Keying (FSK)

- Phase Shift Keying (PSK)

- Frequency Hopped Spread Spectrum (FHSS)

- Direct Sequence Spread Spectrum (DSSS)

- Frequency and Pulse-position modulation (FPPM and PPM)

- Multiple Frequency Shift Keying (MFSK)

- Orthogonal Frequency-Division Multiplexing (OFDM)

The following is a discussion on the different types of modulation and their utility to UAC.

Frequency Shift Keying as Applied to UAC

FSK is the earliest form of modulation used for more advanced forms of UAC by acoustic modems. The earliest forms of UAC prior to FSK has been by percussion of different objects underwater and this method has been used to measure the speed of sound in water.

FSK usually employs two distinct frequencies to modulate data. For example Frequency F1 to indicate bit 0 and frequency F2 to indicate bit 1. Hence a binary string can be transmitted by alternating these two frequencies depending on whether it is a 0 or 1. The receiver can be as simple as having analogue matched filters to the two frequencies and a level detector to decide if a 1 or 0 was received. This is a relatively easy form of modulation and therefore used in the earliest acoustic modems. However more sophisticated Demodulator using Digital Signal Processors (DSP) can be used in the present day.

The biggest challenge FSK faces in the UAC is multi-path reflections. With multi-path (particularly in UAC) several strong reflections can be present at the receiving hydrophone and the threshold detectors become confused, thus severely limiting the use of this type of UAC to vertical channels. Adaptive equalization methods have been tried with limited success. Adaptive equalization tries to model the highly reflective UAC channel and subtract the effects from the received signal. The success has been limited due to the rapidly varying conditions and the difficulty to adapt in time.

Phase Shift Keying

Phase-shift keying (PSK) is a digital modulation scheme that conveys data by changing (modulating) the phase of a reference signal (the carrier wave).The signal is impressed into the magnetic field x,y area by varying the sine and cosine inputs at a precise time. It is widely used for wireless LANs, RFID and Bluetooth communication.

Any digital modulation scheme uses a finite number of distinct signals to represent digital data. PSK uses a finite number of phases, each assigned a unique pattern of binary digits. Usually, each phase encodes an equal number of bits. Each pattern of bits forms the symbol that is represented by the particular phase. The demodulator, which is designed specifically for the symbol-set used by the modulator, determines the phase of the received signal and maps it back to the symbol it represents, thus recovering the original data. This requires the receiver to be able to compare the phase of the received signal to a reference signal — such a system is termed coherent (and referred to as CPSK).

Alternatively, instead of operating with respect to a constant reference wave, the broadcast can operate with respect to itself. Changes in phase of a single broadcast waveform can be considered the significant items. In this system, the demodulator determines the changes in the phase of the received signal rather than the phase (relative to a reference wave) itself. Since this scheme depends on the difference between successive phases, it is termed differential phase-shift keying (DPSK). DPSK can be significantly simpler to implement than ordinary PSK since there is no need for the demodulator to have a copy of the reference signal to determine the exact phase of the received signal (it is a non-coherent scheme). In exchange, it produces more erroneous demodulation.

Orthogonal Frequency-Division Multiplexing

Orthogonal Frequency-Division Multiplexing (OFDM) is a digital multi-carrier modulation scheme. OFDM conveys data on several parallel data channel by incorporating closely spaced orthogonal sub-carrier signals

OFDM is a favorable communication scheme in underwater acoustic communications thanks to its resilience against frequency selective channels with long delay spreads.

Use of Vector Sensor Receivers

A vector sensor is capable of measuring important non-scalar components of the acoustic field such as the wave velocity, which cannot be obtained by a single scalar pressure sensor.

In recent decades, extensive research has been conducted on the theory and design of vector sensors. Many vector sensor signal processing algorithms have been designed. They have been mainly used for underwater target localization and sonar applications.

Earlier underwater acoustic communication systems have been relying on scalar sensors only, which measure the pressure of the acoustic field. Vector sensors measure the scalar and vector components of the acoustic field in a single point in space, therefore can serve as a compact multichannel receiver. This is different from the existing multichannel underwater receivers, which are composed of spatially separated pressure-only sensors, which may result in large-size arrays.

In general, there are two types of vector sensors: inertial and gradient. Inertial sensors truly measure the velocity or acceleration by responding to the acoustic medium motion, whereas gradient sensors employ a finite-difference approximation to estimate the gradients of the acoustic field such as velocity and acceleration.

In the example of vector sensor communications shown, there is one transmitter pressure transducer, shown by a black dot, whereas for reception we use a vector sensor, shown by a black square, which measures the pressure and the y and z components of the velocity. This is a 1×3 single-input multiple-output (SIMO) system. With more pressure transmitters, one can have a multiple-input multiple-output (MIMO) system also.

References

- Daniel Minoli (3 February 2009). Satellite Systems Engineering in an IPv6 Environment. CRC Press. pp. 136–. ISBN 978-1-4200-7868-8. Retrieved 1 June 2012.

- Halit Eren (Nov 16, 2005). Wireless Sensors and Instruments: Networks, Design, and Applications. CRC Press. p. 112. ISBN 9781420037401.

- Guowang Miao; Jens Zander; Ki Won Sung; Ben Slimane (2016). Fundamentals of Mobile Data Networks. Cambridge University Press. ISBN 1107143217.

- Martin, Donald; Anderson, Paul; Bartamian, Lucy (March 16, 2007). "Communications Satellites" (5th ed.). AIAA. ISBN 978-1884989193.

- Kornel Terplan (2000). The Telecommunications Handbook. CRC Press. pp. 266–. ISBN 978-0-8493-3137-4. Retrieved 1 June 2012.

- Guowang Miao; Jens Zander; Ki Won Sung; Ben Slimane (2016). Fundamentals of Mobile Data Networks. Cambridge University Press. ISBN 1107143217.

- "Military Satellite Communications Fundamentals | The Aerospace Corporation". Aerospace. 2010-04-01. Retrieved 2016-02-10.

- "Arthur C. Clarke, inventor of satellite, visionary in technology, dead at 90". Engadget.com. 2008-03-18. Retrieved 2016-02-10.

- "Communication: How the rover can communicate through Mars-orbiting spacecraft". Jet Propulsion Laboratory. Retrieved 21 January 2016.

- "Fundamentals of Communications Access Technologies: FDMA, TDMA, CDMA, OFDMA, AND SDMA". Electronic Design. 2013-01-22. Retrieved 2014-08-28.

- "DIRECTV's Spaceway F1 Satellite Launches New Era in High-Definition Programming; Next Generation Satellite Will Initiate Historic Expansion of DIRECTV". SpaceRef. Retrieved 2012-05-11.

- Rentschler, M.; Laukemann, P., "Performance analysis of parallel redundant WLAN," Emerging Technologies & Factory Automation (ETFA), 2012 IEEE 17th Conference on , vol., no., pp.1,8, 17-21 Sept. 2012

- Mashaghi et al. Noise reduction by signal combination in Fourier space applied to drift correction in optical tweezers, Rev. Sci. Instrum. 82, 115103 (2011)

Understanding Modulation

Modulation is the procedure of fluctuating one or more properties of a periodic waveform. A modulator helps in the performance of modulation. There are a number of modulations such as amplitude modulation, quadrature modulation, frequency modulation and demodulation. The major components of modulation are discussed in this section.

Modulation

In electronics and telecommunications, modulation is the process of varying one or more properties of a periodic waveform, called the *carrier signal*, with a modulating signal that typically contains information to be transmitted.

In telecommunications, modulation is the process of conveying a message signal, for example a digital bit stream or an analog audio signal, inside another signal that can be physically transmitted. Modulation of a sine waveform transforms a baseband message signal into a passband signal.

A modulator is a device that performs modulation. A demodulator (sometimes *detector* or *demod*) is a device that performs demodulation, the inverse of modulation. A modem (from modulator–demodulator) can perform both operations.

The aim of analog modulation is to transfer an analog baseband (or lowpass) signal, for example an audio signal or TV signal, over an analog bandpass channel at a different frequency, for example over a limited radio frequency band or a cable TV network channel.

The aim of digital modulation is to transfer a digital bit stream over an analog bandpass channel, for example over the public switched telephone network (where a bandpass filter limits the frequency range to 300–3400 Hz) or over a limited radio frequency band.

Analog and digital modulation facilitate frequency division multiplexing (FDM), where several low pass information signals are transferred simultaneously over the same shared physical medium, using separate passband channels (several different carrier frequencies).

The aim of digital baseband modulation methods, also known as line coding, is to transfer a digital bit stream over a baseband channel, typically a non-filtered copper wire such as a serial bus or a wired local area network.

The aim of pulse modulation methods is to transfer a narrowband analog signal, for example a phone call over a wideband baseband channel or, in some of the schemes, as a bit stream over another digital transmission system.

In music synthesizers, modulation may be used to synthesise waveforms with an extensive overtone spectrum using a small number of oscillators. In this case the carrier frequency is typically in the same order or much lower than the modulating waveform.

Analog Modulation Methods

In analog modulation, the modulation is applied continuously in response to the analog information signal.

List of Analog Modulation Techniques

Common analog modulation techniques are:

- Amplitude modulation (AM) (here the amplitude of the carrier signal is varied in accordance to the instantaneous amplitude of the modulating signal)
    - Double-sideband modulation (DSB)
        - Double-sideband modulation with carrier (DSB-WC) (used on the AM radio broadcasting band)
        - Double-sideband suppressed-carrier transmission (DSB-SC)
        - Double-sideband reduced carrier transmission (DSB-RC)
    - Single-sideband modulation (SSB, or SSB-AM)
        - Single-sideband modulation with carrier (SSB-WC)
        - Single-sideband modulation suppressed carrier modulation (SSB-SC)
    - Vestigial sideband modulation (VSB, or VSB-AM)
    - Quadrature amplitude modulation (QAM)
- Angle modulation, which is approximately constant envelope
    - Frequency modulation (FM) (here the frequency of the carrier signal is varied in accordance to the instantaneous amplitude of the modulating signal)
    - Phase modulation (PM) (here the phase shift of the carrier signal is varied in accordance with the instantaneous amplitude of the modulating signal)
    - Transpositional Modulation (TM), in which the waveform inflection is modified resulting in a signal where each quarter cycle is transposed in the modulation process. TM is a pesudo-analog modulation (AM). Where an AM carrier also carries a phase variable phase f(ϕ). TM is f(AM,ϕ)

Digital Modulation Methods

In digital modulation, an analog carrier signal is modulated by a discrete signal. Digital modulation methods can be considered as digital-to-analog conversion, and the corresponding demodulation or detection as analog-to-digital conversion. The changes in the carrier signal are chosen from a finite number of M alternative symbols (the *modulation alphabet*).

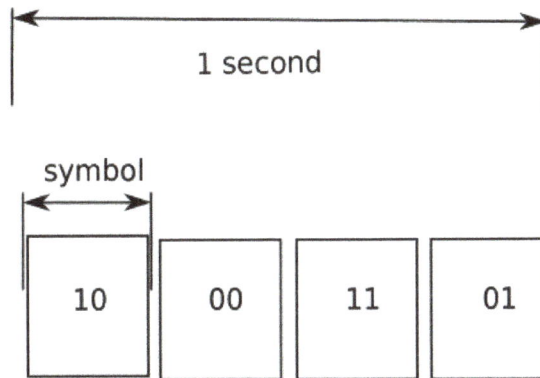

Schematic of 4 baud (8 bit/s) data link containing arbitrarily chosen values.

A simple example: A telephone line is designed for transferring audible sounds, for example tones, and not digital bits (zeros and ones). Computers may however communicate over a telephone line by means of modems, which are representing the digital bits by tones, called symbols. If there are four alternative symbols (corresponding to a musical instrument that can generate four different tones, one at a time), the first symbol may represent the bit sequence 00, the second 01, the third 10 and the fourth 11. If the modem plays a melody consisting of 1000 tones per second, the symbol rate is 1000 symbols/second, or baud. Since each tone (i.e., symbol) represents a message consisting of two digital bits in this example, the bit rate is twice the symbol rate, i.e. 2000 bits per second. This is similar to the technique used by dialup modems as opposed to DSL modems.

According to one definition of digital signal, the modulated signal is a digital signal. According to another definition, the modulation is a form of digital-to-analog conversion. Most textbooks would consider digital modulation schemes as a form of digital transmission, synonymous to data transmission; very few would consider it as analog transmission.

Fundamental Digital Modulation Methods

The most fundamental digital modulation techniques are based on keying:

- PSK (phase-shift keying): a finite number of phases are used.

- FSK (frequency-shift keying): a finite number of frequencies are used.

- ASK (amplitude-shift keying): a finite number of amplitudes are used.

- QAM (quadrature amplitude modulation): a finite number of at least two phases and at least two amplitudes are used.

In QAM, an inphase signal (or I, with one example being a cosine waveform) and a quadrature phase signal (or Q, with an example being a sine wave) are amplitude modulated with a finite number of amplitudes, and then summed. It can be seen as a two-channel system, each channel using ASK. The resulting signal is equivalent to a combination of PSK and ASK.

In all of the above methods, each of these phases, frequencies or amplitudes are assigned a unique pattern of binary bits. Usually, each phase, frequency or amplitude encodes an equal number of bits. This number of bits comprises the *symbol* that is represented by the particular phase, frequency or amplitude.

If the alphabet consists of $M = 2^N$ alternative symbols, each symbol represents a message consisting of N bits. If the symbol rate (also known as the baud rate) is f_s symbols/second (or baud), the data rate is Nf_s bit/second.

For example, with an alphabet consisting of 16 alternative symbols, each symbol represents 4 bits. Thus, the data rate is four times the baud rate.

In the case of PSK, ASK or QAM, where the carrier frequency of the modulated signal is constant, the modulation alphabet is often conveniently represented on a constellation diagram, showing the amplitude of the I signal at the x-axis, and the amplitude of the Q signal at the y-axis, for each symbol.

Modulator and Detector Principles of Operation

PSK and ASK, and sometimes also FSK, are often generated and detected using the principle of QAM. The I and Q signals can be combined into a complex-valued signal $I+jQ$ (where j is the imaginary unit). The resulting so called equivalent lowpass signal or equivalent baseband signal is a complex-valued representation of the real-valued modulated physical signal (the so-called passband signal or RF signal).

These are the general steps used by the modulator to transmit data:

1. Group the incoming data bits into codewords, one for each symbol that will be transmitted.

2. Map the codewords to attributes, for example amplitudes of the I and Q signals (the equivalent low pass signal), or frequency or phase values.

3. Adapt pulse shaping or some other filtering to limit the bandwidth and form the spectrum of the equivalent low pass signal, typically using digital signal processing.

4. Perform digital to analog conversion (DAC) of the I and Q signals (since today all of the above is normally achieved using digital signal processing, DSP).

5. Generate a high frequency sine carrier waveform, and perhaps also a cosine quadrature component. Carry out the modulation, for example by multiplying the sine and cosine waveform with the I and Q signals, resulting in the equivalent low pass signal being frequency shifted to the modulated passband signal or RF signal. Sometimes this is achieved using DSP technology, for example direct digital synthesis using a waveform table, instead of analog signal processing. In that case the above DAC step should be done after this step.

6. Amplification and analog bandpass filtering to avoid harmonic distortion and periodic spectrum.

At the receiver side, the demodulator typically performs:

1. Bandpass filtering.

2. Automatic gain control, AGC (to compensate for attenuation, for example fading).

3. Frequency shifting of the RF signal to the equivalent baseband I and Q signals, or to an intermediate frequency (IF) signal, by multiplying the RF signal with a local oscillator sine-wave and cosine wave frequency.

4. Sampling and analog-to-digital conversion (ADC) (sometimes before or instead of the above point, for example by means of undersampling).

5. Equalization filtering, for example a matched filter, compensation for multipath propagation, time spreading, phase distortion and frequency selective fading, to avoid intersymbol interference and symbol distortion.

6. Detection of the amplitudes of the I and Q signals, or the frequency or phase of the IF signal.

7. Quantization of the amplitudes, frequencies or phases to the nearest allowed symbol values.

8. Mapping of the quantized amplitudes, frequencies or phases to codewords (bit groups).

9. Parallel-to-serial conversion of the codewords into a bit stream.

10. Pass the resultant bit stream on for further processing such as removal of any error-correcting codes.

As is common to all digital communication systems, the design of both the modulator and demodulator must be done simultaneously. Digital modulation schemes are possible because the transmitter-receiver pair have prior knowledge of how data is encoded and represented in the communications system. In all digital communication systems, both the modulator at the transmitter and the demodulator at the receiver are structured so that they perform inverse operations.

Non-coherent modulation methods do not require a receiver reference clock signal that is phase synchronized with the sender carrier signal. In this case, modulation symbols (rather than bits, characters, or data packets) are asynchronously transferred. The opposite is coherent modulation.

List of Common Digital Modulation Techniques

The most common digital modulation techniques are:

- Phase-shift keying (PSK)

    o Binary PSK (BPSK), using M=2 symbols

    o Quadrature PSK (QPSK), using M=4 symbols

    o 8PSK, using M=8 symbols

    o 16PSK, using M=16 symbols

    o Differential PSK (DPSK)

    o Differential QPSK (DQPSK)

- o   Offset QPSK (OQPSK)

- o $\pi/4$–QPSK

- Frequency-shift keying (FSK)

  - o   Audio frequency-shift keying (AFSK)

  - o   Multi-frequency shift keying (M-ary FSK or MFSK)

  - o   Dual-tone multi-frequency (DTMF)

- Amplitude-shift keying (ASK)

- On-off keying (OOK), the most common ASK form

  - o   M-ary vestigial sideband modulation, for example 8VSB

- Quadrature amplitude modulation (QAM), a combination of PSK and ASK

  - o   Polar modulation like QAM a combination of PSK and ASK

- Continuous phase modulation (CPM) methods

  - o   Minimum-shift keying (MSK)

  - o   Gaussian minimum-shift keying (GMSK)

  - o   Continuous-phase frequency-shift keying (CPFSK)

- Orthogonal frequency-division multiplexing (OFDM) modulation

  - o   Discrete multitone (DMT), including adaptive modulation and bit-loading

- Wavelet modulation

- Trellis coded modulation (TCM), also known as Trellis modulation

- Spread-spectrum techniques

  - o   Direct-sequence spread spectrum (DSSS)

  - o   Chirp spread spectrum (CSS) according to IEEE 802.15.4a CSS uses pseudo-stochastic coding

  - o   Frequency-hopping spread spectrum (FHSS) applies a special scheme for channel release

MSK and GMSK are particular cases of continuous phase modulation. Indeed, MSK is a particular case of the sub-family of CPM known as continuous-phase frequency-shift keying (CPFSK) which is defined by a rectangular frequency pulse (i.e. a linearly increasing phase pulse) of one symbol-time duration (total response signaling).

OFDM is based on the idea of frequency-division multiplexing (FDM), but the multiplexed streams are all parts of a single original stream. The bit stream is split into several parallel data streams, each transferred over its own sub-carrier using some conventional digital modulation scheme. The modulated sub-carriers are summed to form an OFDM signal. This dividing and recombining helps with handling channel impairments. OFDM is considered as a modulation technique rather than a multiplex technique, since it transfers one bit stream over one communication channel using one sequence of so-called OFDM symbols. OFDM can be extended to multi-user channel access method in the orthogonal frequency-division multiple access (OFDMA) and multi-carrier code division multiple access (MC-CDMA) schemes, allowing several users to share the same physical medium by giving different sub-carriers or spreading codes to different users.

Of the two kinds of RF power amplifier, switching amplifiers (Class D amplifiers) cost less and use less battery power than linear amplifiers of the same output power. However, they only work with relatively constant-amplitude-modulation signals such as angle modulation (FSK or PSK) and CDMA, but not with QAM and OFDM. Nevertheless, even though switching amplifiers are completely unsuitable for normal QAM constellations, often the QAM modulation principle are used to drive switching amplifiers with these FM and other waveforms, and sometimes QAM demodulators are used to receive the signals put out by these switching amplifiers.

Automatic Digital Modulation Recognition (ADMR)

Automatic digital modulation recognition in intelligent communication systems is one of the most important issues in software defined radio and cognitive radio. According to incremental expanse of intelligent receivers, automatic modulation recognition becomes a challenging topic in telecommunication systems and computer engineering. Such systems have many civil and military applications. Moreover, blind recognition of modulation type is an important problem in commercial systems, especially in software defined radio. Usually in such systems, there are some extra information for system configuration, but considering blind approaches in intelligent receivers, we can reduce information overload and increase transmission performance. Obviously, with no knowledge of the transmitted data and many unknown parameters at the receiver, such as the signal power, carrier frequency and phase offsets, timing information, etc., blind identification of the modulation is a difficult task. This becomes even more challenging in real-world scenarios with multipath fading, frequency-selective and time-varying channels.

There are two main approaches to automatic modulation recognition. The first approach uses likelihood-based methods to assign an input signal to a proper class. Another recent approach is based on feature extraction.

Digital Baseband Modulation or Line Coding

The term digital baseband modulation (or digital baseband transmission) is synonymous to line codes. These are methods to transfer a digital bit stream over an analog baseband channel (a.k.a. lowpass channel) using a pulse train, i.e. a discrete number of signal levels, by directly modulating the voltage or current on a cable. Common examples are unipolar, non-return-to-zero (NRZ), Manchester and alternate mark inversion (AMI) codings.

Amplitude Modulation

Amplitude modulation (AM) is a modulation technique used in electronic communication, most commonly for transmitting information via a radio carrier wave. In amplitude modulation, the amplitude (signal strength) of the carrier wave is varied in proportion to the waveform being transmitted. That waveform may, for instance, correspond to the sounds to be reproduced by a loudspeaker, or the light intensity of television pixels. This technique contrasts with frequency modulation, in which the frequency of the carrier signal is varied, and phase modulation, in which its phase is varied.

AM was the earliest modulation method used to transmit voice by radio. It was developed during the first two decades of the 20th century beginning with Roberto Landell De Moura and Reginald Fessenden's radiotelephone experiments in 1900. It remains in use today in many forms of communication; for example it is used in portable two way radios, VHF aircraft radio, Citizen's Band Radio, and in computer modems (in the form of QAM). "AM" is often used to refer to mediumwave AM radio broadcasting.

Forms of Amplitude Modulation

In electronics and telecommunications, modulation means varying some aspect of a higher frequency continuous wave carrier signal with an information-bearing modulation waveform, such as an audio signal which represents sound, or a video signal which represents images, so the carrier will "carry" the information. When it reaches its destination, the information signal is extracted from the modulated carrier by demodulation.

In amplitude modulation, the amplitude or "strength" of the carrier oscillations is what is varied. For example, in AM radio communication, a continuous wave radio-frequency signal (a sinusoidal carrier wave) has its amplitude modulated by an audio waveform before transmission. The audio waveform modifies the amplitude of the carrier wave and determines the *envelope* of the waveform. In the frequency domain, amplitude modulation produces a signal with power concentrated at the carrier frequency and two adjacent sidebands. Each sideband is equal in bandwidth to that of the modulating signal, and is a mirror image of the other. Standard AM is thus sometimes called "double-sideband amplitude modulation" (DSB-AM) to distinguish it from more sophisticated modulation methods also based on AM.

One disadvantage of all amplitude modulation techniques (not only standard AM) is that the receiver amplifies and detects noise and electromagnetic interference in equal proportion to the signal. Increasing the received signal to noise ratio, say, by a factor of 10 (a 10 decibel improvement), thus would require increasing the transmitter power by a factor of 10. This is in contrast to frequency modulation (FM) and digital radio where the effect of such noise following demodulation is strongly reduced so long as the received signal is well above the threshold for reception. For this reason AM broadcast is not favored for music and high fidelity broadcasting, but rather for voice communications and broadcasts (sports, news, talk radio etc.).

Another disadvantage of AM is that it is inefficient in power usage; at least two-thirds of the power is concentrated in the carrier signal. The carrier signal contains none of the original information being transmitted (voice, video, data, etc.). However its presence provides a simple means of demodulation using envelope detection, providing a frequency and phase reference to extract the modulation from the sidebands. In some modulation systems based on AM, a lower transmitter power is required through partial or total elimination of the carrier component, however receivers for these signals are more complex and costly. The receiver may regenerate a copy of the carrier frequency (usually as shifted to the intermediate frequency) from a greatly reduced "pilot" carrier (in reduced-carrier transmission or DSB-RC) to use in the demodulation process. Even with the carrier totally eliminated in double-sideband suppressed-carrier transmission, carrier regeneration is possible using a Costas phase-locked loop. This doesn't work however for single-sideband suppressed-carrier transmission (SSB-SC), leading to the characteristic "Donald Duck" sound from such receivers when slightly detuned. Single sideband is nevertheless used widely in amateur radio and other voice communications both due to its power efficiency and bandwidth efficiency (cutting the RF bandwidth in half compared to standard AM). On the other hand, in medium wave and short wave broadcasting, standard AM with the full carrier allows for reception using inexpensive receivers. The broadcaster absorbs the extra power cost to greatly increase potential audience.

An additional function provided by the carrier in standard AM, but which is lost in either single or double-sideband suppressed-carrier transmission, is that it provides an amplitude reference. In the receiver, the automatic gain control (AGC) responds to the carrier so that the reproduced audio level stays in a fixed proportion to the original modulation. On the other hand, with suppressed-carrier transmissions there is *no* transmitted power during pauses in the modulation, so the AGC must respond to peaks of the transmitted power during peaks in the modulation. This typically involves a so-called *fast attack, slow decay* circuit which holds the AGC level for a second or more following such peaks, in between syllables or short pauses in the program. This is very acceptable for communications radios, where compression of the audio aids intelligibility. However it is absolutely undesired for music or normal broadcast programming, where a faithful reproduction of the original program, including its varying modulation levels, is expected.

A trivial form of AM which can be used for transmitting binary data is on-off keying, the simplest form of *amplitude-shift keying*, in which ones and zeros are represented by the presence or absence of a carrier. On-off keying is likewise used by radio amateurs to transmit Morse code where it is known as continuous wave (CW) operation, even though the transmission is not strictly "continuous." A more complex form of AM, Quadrature amplitude modulation is now more commonly used with digital data, while making more efficient use of the available bandwidth.

ITU Designations

In 1982, the International Telecommunication Union (ITU) designated the types of amplitude modulation:

Designation	Description
A3E	double-sideband a full-carrier - the basic Amplitude modulation scheme
R3E	single-sideband reduced-carrier
H3E	single-sideband full-carrier
J3E	single-sideband suppressed-carrier
B8E	independent-sideband emission
C3F	vestigial-sideband
Lincompex	linked compressor and expander

History

One of the crude pre-vacuum tube AM transmitters, a Telefunken arc transmitter from 1906. The carrier wave is generated by 6 electric arcs in the vertical tubes, connected to a tuned circuit. Modulation is done by the large carbon microphone *(cone shape)* in the antenna lead.

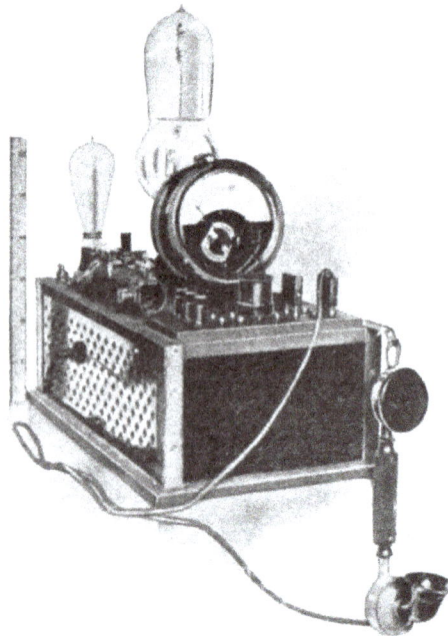

One of the f rst vacuum tube AM radio transmitters, built by Meissner in 1913 with an early triode tube by Robert von Lieben. He used it in a historic 36 km (24 mi) voice transmission from Berlin to Nauen, Germany. Compare its small size with above transmitter.

Although AM was used in a few crude experiments in multiplex telegraph and telephone transmission in the late 1800s, the practical development of amplitude modulation is synonymous with the development between 1900 and 1920 of "radiotelephone" transmission, that is, the effort to send sound (audio) by radio waves. The first radio transmitters, called spark gap transmitters, transmitted information by wireless telegraphy, using different length pulses of carrier wave to spell out text messages in Morse code. They couldn't transmit audio because the carrier consisted of strings of damped waves, pulses of radio waves that declined to zero, that sounded like a buzz in receivers. In effect they were already amplitude modulated.

Continuous Waves

The first AM transmission was made by Canadian researcher Reginald Fessenden on 23 December 1900 using a spark gap transmitter with a specially designed high frequency 10 kHz interrupter, over a distance of 1 mile (1.6 km) at Cobb Island, Maryland, USA. His first transmitted words were, "Hello. One, two, three, four. Is it snowing where you are, Mr. Thiessen?". The words were barely intelligible above the background buzz of the spark.

Fessenden was a significant figure in the development of AM radio. He was one of the first researchers to realize, from experiments like the above, that the existing technology for producing radio waves, the spark transmitter, was not usable for amplitude modulation, and that a new kind of transmitter, one that produced sinusoidal *continuous waves*, was needed. This was a radical idea at the time, because experts believed the impulsive spark was necessary to produce radio frequency waves, and Fessenden was ridiculed. He invented and helped develop one of the first continuous wave transmitters - the Alexanderson alternator, with which he made what is considered the first AM public entertainment broadcast on Christmas Eve, 1906. He also discovered the principle on which AM modulation is based, heterodyning, and invented one of the first detectors able to rectify and receive AM, the electrolytic detector or "liquid baretter", in 1902. Other radio detectors invented for wireless telegraphy, such as the Fleming valve (1904) and the crystal detector (1906) also proved able to rectify AM signals, so the technological hurdle was generating AM waves; receiving them was not a problem.

Early Technologies

Early experiments in AM radio transmission, conducted by Fessenden, Valdamar Poulsen, Ernst Ruhmer, Quirino Majorana, Charles Harrold, and Lee De Forest, were hampered by the lack of a technology for amplification. The first practical continuous wave AM transmitters were based on either the huge, expensive Alexanderson alternator, developed 1906-1910, or versions of the Poulsen arc transmitter (arc converter), invented in 1903. The modifications necessary to transmit AM were clumsy and resulted in very low quality audio. Modulation was usually accomplished by a carbon microphone inserted directly in the antenna or ground wire; its varying resistance varied the current to the antenna. The limited power handling ability of the microphone severely limited the power of the first radiotelephones; many of the microphones were water-cooled.

Vacuum Tubes

The discovery in 1912 of the amplifying ability of the Audion vacuum tube, invented in 1906 by Lee De Forest, solved these problems. The vacuum tube feedback oscillator, invented in 1912 by Edwin Armstrong and Alexander Meissner, was a cheap source of continuous waves and could be easily modulated to make an AM transmitter. Modulation did not have to be done at the output but could be applied to the signal before the final amplifier tube, so the microphone or other audio source didn't have to handle high power. Wartime research greatly advanced the art of AM modulation, and after the war the availability of cheap tubes sparked a great increase in the number of radio stations experimenting with AM transmission of news or music. The vacuum tube was responsible for the rise of AM radio broadcasting around 1920, the first electronic mass entertainment medium. Amplitude modulation was virtually the only type used for radio broadcasting until FM broadcasting began after World War 2.

At the same time as AM radio began, telephone companies such as AT&T were developing the other large application for AM: sending multiple telephone calls through a single wire by modulating them on separate carrier frequencies, called *frequency division multiplexing*.

Single-sideband

John Renshaw Carson in 1915 did the first mathematical analysis of amplitude modulation, showing that a signal and carrier frequency combined in a nonlinear device would create two sidebands on either side of the carrier frequency, and passing the modulated signal through another nonlinear device would extract the original baseband signal. His analysis also showed only one sideband was necessary to transmit the audio signal, and Carson patented single-sideband modulation (SSB) on 1 December 1915. This more advanced variant of amplitude modulation was adopted by AT&T for longwave transatlantic telephone service beginning 7 January 1927. After WW2 it was developed by the military for aircraft communication.

Simplified Analysis of Standard AM

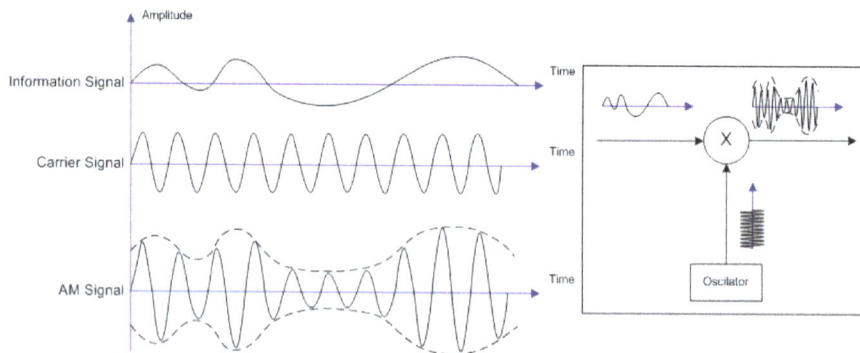

Illustration of Amplitude Modulation

Consider a carrier wave (sine wave) of frequency f_c and amplitude A given by:

$$c(t) = A \cdot \sin(2\pi f_c t).$$

Let $m(t)$ represent the modulation waveform. For this example we shall take the modulation to be simply a sine wave of a frequency f_m, a much lower frequency (such as an audio frequency) than f_c:

$$m(t) = M \cdot \cos(2\pi f_m t + \phi),$$

where M is the amplitude of the modulation. We shall insist that $M<1$ so that *(1+m(t))* is always positive. If $M>1$ then overmodulation occurs and reconstruction of message signal from the transmitted signal would lead in loss of original signal. Amplitude modulation results when the carrier $c(t)$ is multiplied by the positive quantity *(1+m(t))*:

$$y(t)=[1+m(t)]\cdot c(t)$$

$$=[1+M\cdot\cos(2\pi f_m t+\phi)]\cdot A\cdot\sin(2\pi f_c t)$$

In this simple case M is identical to the modulation index, discussed below. With $M=0.5$ the amplitude modulated signal $y(t)$ thus corresponds to the top graph (labelled "50% Modulation") in Figure 4.

Using prosthaphaeresis identities, $y(t)$ can be shown to be the sum of three sine waves:

$$y(t) = A\cdot\sin(2\pi f_c t)+\frac{AM}{2}\left[\sin(2\pi(f_c + f_m)t+\phi)+\sin(2\pi(f_c - f_m)t-\phi)\right].$$

Therefore, the modulated signal has three components: the carrier wave $c(t)$ which is unchanged, and two pure sine waves (known as sidebands) with frequencies slightly above and below the carrier frequency f_c.

Spectrum

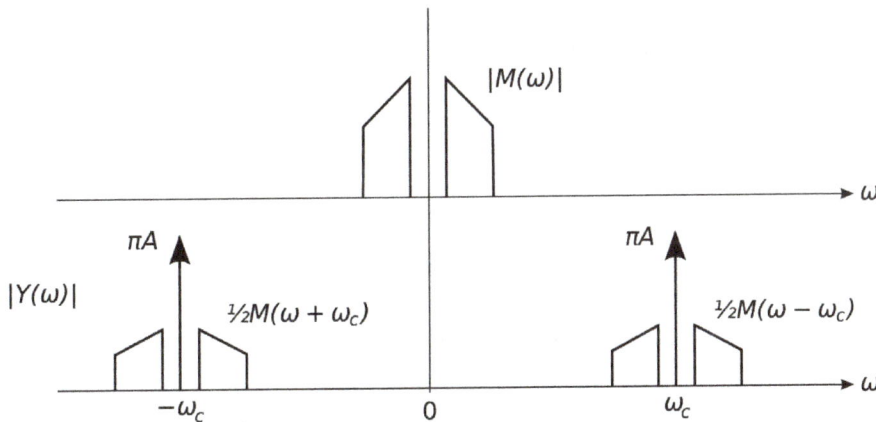

Fig 1: Double-sided spectra of baseband and AM signals.

Of course a useful modulation signal $m(t)$ will generally not consist of a single sine wave, as treated above. However, by the principle of Fourier decomposition, $m(t)$ can be expressed as the sum of a number of sine waves of various frequencies, amplitudes, and phases. Carrying out the multiplication of $1+m(t)$ with $c(t)$ as above then yields a result consisting of a sum of sine waves. Again the carrier $c(t)$ is present unchanged, but for each frequency component of m at f_i there are two sidebands at frequencies $f_c + f_i$ and $f_c - f_i$. The collection of the former frequencies above the carrier frequency is known as the upper sideband, and those below constitute the lower sideband. In a slightly different way of looking at it, we can consider the modulation $m(t)$ to consist of an equal mix of positive and negative frequency components (as results from a formal Fourier transform of a real valued quantity) as shown in the top of Fig. 2. Then one can view the sidebands as that modulation $m(t)$ having simply been shifted in frequency by f_c as depicted at the bottom right of Fig. 2 (formally, the modulated signal also contains identical components at negative frequencies, shown at the bottom left of Fig. 2 for completeness).

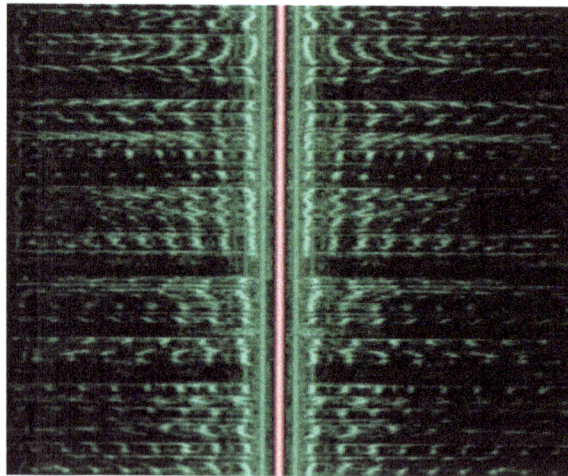

Fig 2: The spectrogram of an AM voice broadcast shows the two sidebands (green) on either side of the carrier (red) with time proceeding in the vertical direction.

If we just look at the short-term spectrum of modulation, changing as it would for a human voice for instance, then we can plot the frequency content (horizontal axis) as a function of time (vertical axis) as in Fig. 3. It can again be seen that as the modulation frequency content varies, at any point in time there is an upper sideband generated according to those frequencies shifted *above* the carrier frequency, and the same content mirror-imaged in the lower sideband below the carrier frequency. At all times, the carrier itself remains constant, and of greater power than the total sideband power.

Power and Spectrum Efficiency

The RF bandwidth of an AM transmission (refer to Figure 2, but only considering positive frequencies) is twice the bandwidth of the modulating (or "baseband") signal, since the upper and lower sidebands around the carrier frequency each have a bandwidth as wide as the highest modulating frequency. Although the bandwidth of an AM signal is narrower than one using frequency modulation (FM), it is twice as wide as single-sideband techniques; it thus may be viewed as spectrally inefficient. Within a frequency band, only half as many transmissions (or "channels") can thus be accommodated. For this reason television employs a variant of single-sideband (known as vestigial sideband, somewhat of a compromise in terms of bandwidth) in order to reduce the required channel spacing.

Another improvement over standard AM is obtained through reduction or suppression of the carrier component of the modulated spectrum. In Figure 2 this is the spike in between the sidebands; even with full (100%) sine wave modulation, the power in the carrier component is twice that in the sidebands, yet it carries no unique information. Thus there is a great advantage in efficiency in reducing or totally suppressing the carrier, either in conjunction with elimination of one sideband (single-sideband suppressed-carrier transmission) or with both sidebands remaining (double sideband suppressed carrier). While these suppressed carrier transmissions are efficient in terms of transmitter power, they require more sophisticated receivers employing synchronous detection and regeneration of the carrier frequency. For that reason, standard AM continues to be widely used, especially in broadcast transmission, to allow for the use of inexpensive receivers using envelope detection. Even (analog) television, with a (largely) suppressed lower sideband, includes sufficient carrier power for use of envelope detection. But for communications systems where both transmitters and receivers can be optimized, suppression of both one sideband and the carrier represent a net advantage and are frequently employed.

Modulation Index

The AM modulation index is a measure based on the ratio of the modulation excursions of the RF signal to the level of the unmodulated carrier. It is thus defined as:

$$h = \frac{\text{peak value of } m(t)}{A} = \frac{M}{A}$$

where M and   are the modulation amplitude and carrier amplitude, respectively; the modulation amplitude is the peak (positive or negative) change in the RF amplitude from its unmodulated value. Modulation index is normally expressed as a percentage, and may be displayed on a meter connected to an AM transmitter.

So if $h = 0.5$, carrier amplitude varies by 50% above (and below) its unmodulated level, as is shown in the first waveform, below. For $h = 1.0$, it varies by 100% as shown in the illustration below it. With 100% modulation the wave amplitude sometimes reaches zero, and this represents full modulation using standard AM and is often a target (in order to obtain the highest possible signal to noise ratio) but mustn't be exceeded. Increasing the modulating signal beyond that point, known as overmodulation, causes a standard AM modulator to fail, as the negative excursions of the wave envelope cannot become less than zero, resulting in distortion ("clipping") of the received modulation. Transmitters typically incorporate a limiter circuit to avoid overmodulation, and/or a compressor circuit (especially for voice communications) in order to still approach 100% modulation for maximum intelligibility above the noise. Such circuits are sometimes referred to as a vogad.

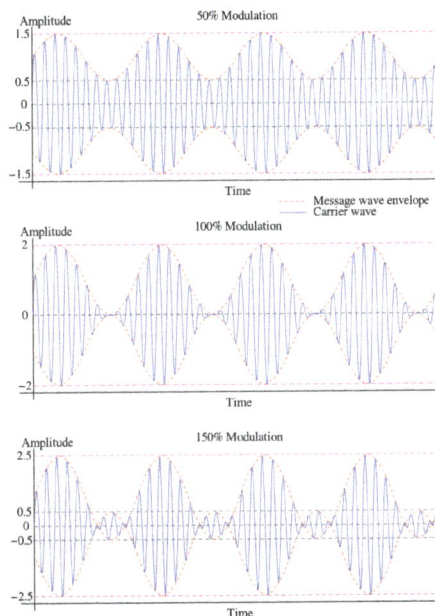

Fig 3: Modulation depth. In the diagram, the unmodulated carrier has an amplitude of 1.

However it is possible to talk about a modulation index exceeding 100%, without introducing distortion, in the case of double-sideband reduced-carrier transmission. In that case, negative excursions beyond zero entail a reversal of the carrier phase, as shown in the third waveform below. This cannot be produced using the efficient high-level (output stage) modulation techniques which are widely used especially in high power broadcast transmitters. Rather, a special modulator produces such a waveform at a low level followed by a linear amplifier. What's more, a standard AM receiver using an envelope detector is incapable of properly demodulating such a signal. Rather, synchronous detection is required. Thus double-sideband transmission is generally *not* referred to as "AM" even though it generates an identical RF waveform as standard AM as long as the modulation index is below 100%. Such systems more often attempt a radical reduction of the carrier level compared to the sidebands (where the useful information is present) to the point of double-sideband suppressed-carrier transmission where the carrier is (ideally) reduced to zero. In all such cases the term "modulation index" loses its value as it refers to the ratio of the modulation amplitude to a rather small (or zero) remaining carrier amplitude.

Modulation Methods

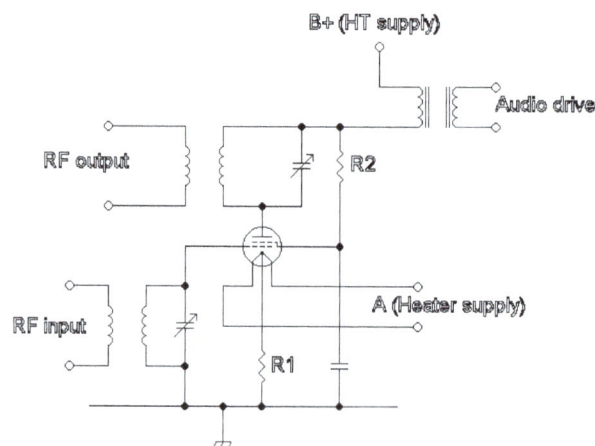

Anode (plate) modulation. A tetrode's plate and screen grid voltage is modulated via an audio transformer. The resistor R1 sets the grid bias; both the input and output are tuned circuits with inductive coupling.

Modulation circuit designs may be classified as low- or high-level (depending on whether they modulate in a low-power domain—followed by amplification for transmission—or in the high-power domain of the transmitted signal).

Low-level Generation

In modern radio systems, modulated signals are generated via digital signal processing (DSP). With DSP many types of AM are possible with software control (including DSB with carrier, SSB suppressed-carrier and independent sideband, or ISB). Calculated digital samples are converted to voltages with a digital to analog converter, typically at a frequency less than the desired RF-output frequency. The analog signal must then be shifted in frequency and linearly amplified to the desired frequency and power level (linear amplification must be used to prevent modulation distortion). This low-level method for AM is used in many Amateur Radio transceivers.

AM may also be generated at a low level, using analog methods described in the next section.

High-level Generation

High-power AM transmitters (such as those used for AM broadcasting) are based on high-efficiency class-D and class-E power amplifier stages, modulated by varying the supply voltage.

Older designs (for broadcast and amateur radio) also generate AM by controlling the gain of the transmitter's final amplifier (generally class-C, for efficiency). The following types are for vacuum tube transmitters (but similar options are available with transistors):

- Plate modulation: In plate modulation, the plate voltage of the RF amplifier is modulated with the audio signal. The audio power requirement is 50 percent of the RF-carrier power.

- Heising (constant-current) modulation: RF amplifier plate voltage is fed through a "choke" (high-value inductor). The AM modulation tube plate is fed through the same inductor, so the modulator tube diverts current from the RF amplifier. The choke acts as a constant current source in the audio range. This system has a low power efficiency.

- Control grid modulation: The operating bias and gain of the final RF amplifier can be controlled by varying the voltage of the control grid. This method requires little audio power, but care must be taken to reduce distortion.

- Clamp tube (screen grid) modulation: The screen-grid bias may be controlled through a "clamp tube", which reduces voltage according to the modulation signal. It is difficult to approach 100-percent modulation while maintaining low distortion with this system.

- Doherty modulation: One tube provides the power under carrier conditions and another operates only for positive modulation peaks. Overall efficiency is good, and distortion is low.

- Outphasing modulation: Two tubes are operated in parallel, but partially out of phase with each other. As they are differentially phase modulated their combined amplitude is greater or smaller. Efficiency is good and distortion low when properly adjusted.

- Pulse width modulation (PWM) or Pulse duration modulation (PDM): A highly efficient high voltage power supply is applied to the tube plate. The output voltage of this supply is varied at an audio rate to follow the program. This system was pioneered by Hilmer Swanson and has a number of variations, all of which achieve high efficiency and sound quality.

Demodulation Methods

The simplest form of AM demodulator consists of a diode which is configured to act as envelope detector. Another type of demodulator, the product detector, can provide better-quality demodulation with additional circuit complexity.

Quadrature Amplitude Modulation

Quadrature amplitude modulation (QAM) is both an analog and a digital modulation scheme. It conveys two analog message signals, or two digital bit streams, by changing (*modulating*) the amplitudes of two carrier waves, using the amplitude-shift keying (ASK) digital modulation scheme or amplitude modulation (AM) analog modulation scheme. The two carrier waves of the same frequency, usually sinusoids, are out of phase with each other by 90° and are thus called quadrature carriers or quadrature components — hence the name of the scheme. The modulated waves are summed, and the final waveform is a combination of both phase-shift keying (PSK) and amplitude-shift keying (ASK), or, in the analog case, of phase modulation (PM) and amplitude modulation. In the digital QAM case, a finite number of at least two phases and at least two amplitudes are used. PSK modulators are often designed using the QAM principle, but are not considered as QAM since the amplitude of the modulated carrier signal is constant. QAM is used extensively as a modulation scheme for digital telecommunication systems. Arbitrarily high spectral efficiencies can be achieved with QAM by setting a suitable constellation size, limited only by the noise level and linearity of the communications channel.

QAM is being used in optical fiber systems as bit rates increase; QAM16 and QAM64 can be optically emulated with a 3-path interferometer.

Introduction

Like all modulation schemes, QAM conveys data by changing some aspect of a carrier signal, or the carrier wave, (usually a sinusoid) in response to a data signal. In the case of QAM, the amplitude of two waves of the same frequency, 90° out-of-phase with each other (in quadrature) are changed (*modulated* or *keyed*) to represent the data signal. Amplitude modulating two carriers in quadrature can be equivalently viewed as both amplitude modulating and phase modulating a single carrier.

Phase modulation (analog PM) and phase-shift keying (digital PSK) can be regarded as a special case of QAM, where the magnitude of the modulating signal is a constant, with only the phase varying. This can also be extended to frequency modulation (FM) and frequency-shift keying (FSK), for these can be regarded as a special case of phase modulation.

Analog QAM

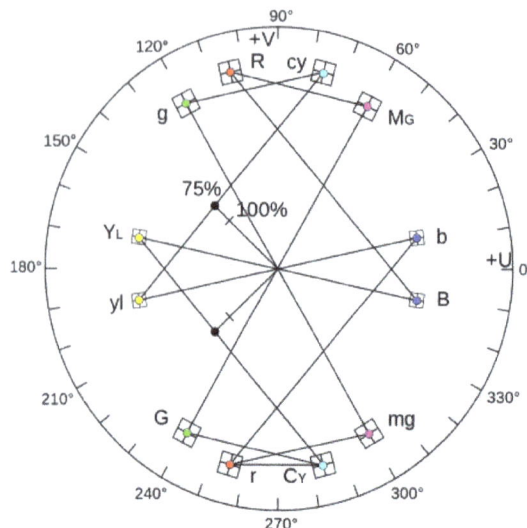

Analog QAM: measured PAL colour bar signal on a vector analyser screen.

When transmitting two signals by modulating them with QAM, the transmitted signal will be of the form:

$$s(t) = Re\left\{[I(t)+iQ(t)]e^{i2\pi f_0 t}\right\}$$

$$= I(t)\cos(2\pi f_0 t) - Q(t)\sin(2\pi f_0 t)$$

where $i^2 = -1$, $I(t)$ and $Q(t)$ are the modulating signals, f_0 is the carrier frequency and $Re\{\}$ is the real part.

At the receiver, these two modulating signals can be demodulated using a coherent demodulator. Such a receiver multiplies the received signal separately with both a cosine and sine signal to produce the received estimates of $I(t)$ and $Q(t)$ respectively. Because of the orthogonality property of the carrier signals, it is possible to detect the modulating signals independently.

In the ideal case $I(t)$ is demodulated by multiplying the transmitted signal with a cosine signal:

$$r(t) = s(t)\cos(2\pi f_0 t)$$

$$= I(t)\cos(2\pi f_0 t)\cos(2\pi f_0 t) - Q(t)\sin(2\pi f_0 t)\cos(2\pi f_0 t)$$

Using standard trigonometric identities, we can write it as:

$$r(t) = \frac{1}{2}I(t)[1+\cos(4\pi f_0 t)] - \frac{1}{2}Q(t)\sin(4\pi f_0 t)$$

$$= \frac{1}{2}I(t) + \frac{1}{2}[I(t)\cos(4\pi f_0 t) - Q(t)\sin(4\pi f_0 t)]$$

Low-pass filtering $r(t)$ removes the high frequency terms (containing $4\pi f_0 t$), leaving only the $I(t)$ term. This filtered signal is unaffected by $Q(t)$, showing that the in-phase component can be received independently of the quadrature component. Similarly, we may multiply $s(t)$ by a sine wave and then low-pass filter to extract $Q(t)$.

Analog QAM suffers from the same problem as Single-sideband modulation: the exact phase of the carrier is required for correct demodulation at the receiver. If the demodulating phase is even a little off, it results in crosstalk between the modulated signals. This issue of carrier synchronization at the receiver must be handled somehow in QAM systems. The coherent demodulator needs to be exactly in phase with the received signal, or otherwise the modulated signals cannot be independently received. This is achieved typically by transmitting a burst subcarrier or a Pilot signal.

Analog QAM is used in:

- NTSC and PAL analog Color television systems, where the I- and Q-signals carry the components of chroma (colour) information. The QAM carrier phase is recovered from a special Colorburst transmitted at the beginning of each scan line.

- C-QUAM ("Compatible QAM") is used in AM stereo radio to carry the stereo difference information.

Fourier Analysis of QAM

In the frequency domain, QAM has a similar spectral pattern to DSB-SC modulation. Using the properties of the Fourier transform, we find that:

$$S(f) = \frac{1}{2}\left[M_I(f - f_0) + M_I(f + f_0)\right] + \frac{i}{2}\left[M_Q(f - f_0) - M_Q(f + f_0)\right]$$

where $S(f)$, $M_I(f)$ and $M_Q(f)$ are the Fourier transforms (frequency-domain representations) of $s(t)$, $I(t)$ and $Q(t)$, respectively.

Quantized QAM

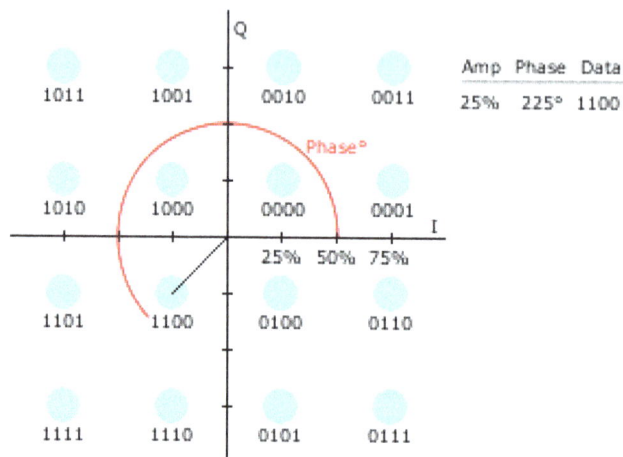

Digital 16-QAM with example constellation points

As in many digital modulation schemes, the constellation diagram is useful for QAM. In QAM, the constellation points are usually arranged in a square grid with equal vertical and horizontal spacing, although other configurations are possible (e.g. Cross-QAM). Since in digital telecommunications the data are usually binary, the number of points in the grid is usually a power of 2 (2, 4, 8, ...). Since QAM is usually square, some of these are rare—the most common forms are 16-QAM, 64-QAM and 256-QAM. By moving to a higher-order constellation, it is possible to transmit more bits per symbol. However, if the mean energy of the constellation is to remain the same (by way of making a fair comparison), the points must be closer together and are thus more susceptible to noise and other corruption; this results in a higher bit error rate and so higher-order QAM can deliver more data less reliably than lower-order QAM, for constant mean constellation energy. Using higher-order QAM without increasing the bit error rate requires a higher signal-to-noise ratio (SNR) by increasing signal energy, reducing noise, or both.

If data-rates beyond those offered by 8-PSK are required, it is more usual to move to QAM since it achieves a greater distance between adjacent points in the I-Q plane by distributing the points more evenly. The complicating factor is that the points are no longer all the same amplitude and so the demodulator must now correctly detect both phase and amplitude, rather than just phase.

64-QAM and 256-QAM are often used in digital cable television and cable modem applications. In the United States, 64-QAM and 256-QAM are the mandated modulation schemes for digital cable as standardised by the SCTE in the standard ANSI/SCTE 07 2013. Note that many marketing people will refer to these as QAM-64 and QAM-256. In the UK, 64-QAM is used for digital terrestrial television (Freeview) whilst 256-QAM is used for Freeview-HD.

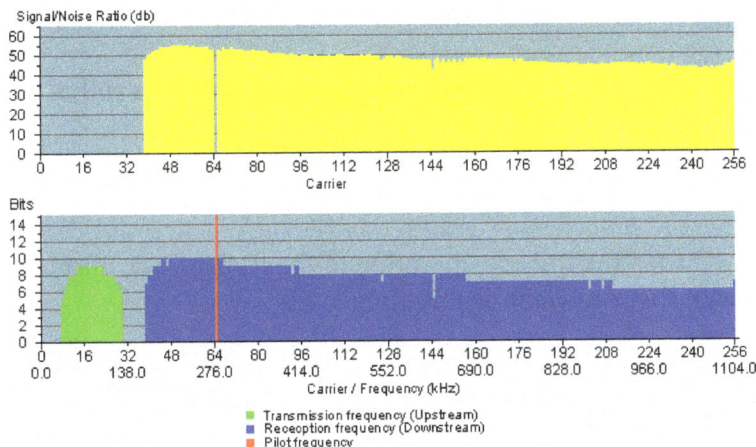

Bit-loading (bits per QAM constellation) on an ADSL line

Communication systems designed to achieve very high levels of spectral efficiency usually employ very dense QAM constellations. For example, current Homeplug AV2 500-Mbit powerline Ethernet devices use 1024-QAM and 4096-QAM, as well as future devices using ITU-T G.hn standard for networking over existing home wiring (coaxial cable, phone lines and power lines); 4096-QAM provides 12 bits/symbol. Another example is ADSL technology for copper twisted pairs, whose constellation size goes up to 32768-QAM (in ADSL terminology this is referred to as bit-loading, or bit per tone, 32768-QAM being equivalent to 15 bits per tone).

Ultra-high capacity Microwave Backhaul Systems also use 1024-QAM. With 1024-QAM, Adaptive Coding and Modulation (ACM), and XPIC, Vendors can obtain Gigabit capacity in a single 56 MHz channel.

Ideal Structure

Transmitter

The following picture shows the ideal structure of a QAM transmitter, with a carrier center frequency and the frequency response of the transmitter's filter :

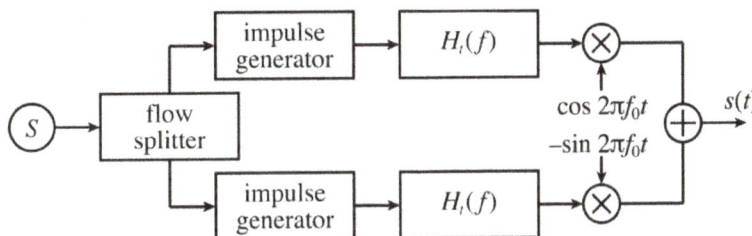

First the flow of bits to be transmitted is split into two equal parts: this process generates two independent signals to be transmitted. They are encoded separately just like they were in an amplitude-shift keying (ASK) modulator. Then one channel (the one "in phase") is multiplied by a cosine, while the other channel (in "quadrature") is multiplied by a sine. This way there is a phase of 90° between them. They are simply added one to the other and sent through the real channel.

The sent signal can be expressed in the form:

$$s(t) = \sum_{n=-\infty}^{\infty} \left[v_c[n] \cdot h_t(t - nT_s) \cos(2\pi f_0 t) - v_s[n] \cdot h_t(t - nT_s) \sin(2\pi f_0 t) \right]$$

where $v_c[n]$ and $v_s[n]$ are the voltages applied in response to the nth symbol to the cosine and sine waves respectively.

Receiver

The receiver simply performs the inverse operation of the transmitter. Its ideal structure is shown in the picture below with the receive filter's frequency response :

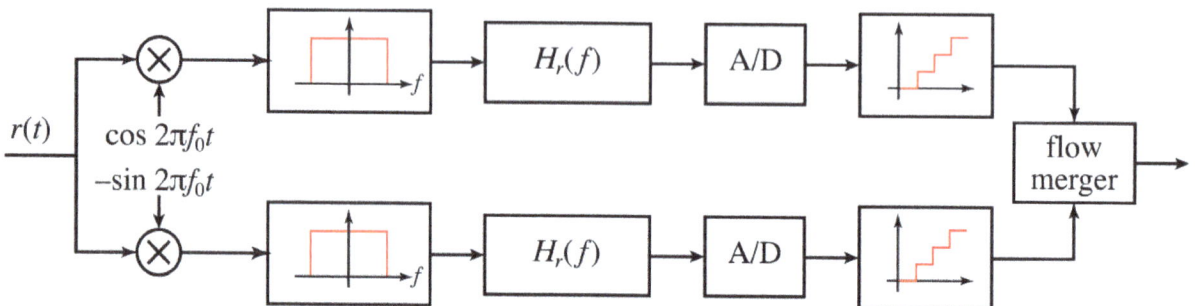

Multiplying by a cosine (or a sine) and by a low-pass filter it is possible to extract the component in phase (or in quadrature). Then there is only an ASK demodulator and the two flows of data are merged back.

In practice, there is an unknown phase delay between the transmitter and receiver that must be compensated by *synchronization* of the receivers local oscillator (i.e. the sine and cosine functions in the above figure). In mobile applications, there will often be an offset in the relative *frequency* as well, due to the possible presence of a Doppler shift proportional to the relative velocity of the transmitter and receiver. Both the phase and frequency variations introduced by the channel must be compensated by properly tuning the sine and cosine components, which requires a *phase reference*, and is typically accomplished using a Phase-Locked Loop (PLL).

In any application, the low-pass filter and the receive H_r filter will be implemented as a single combined filter. Here they are shown as separate just to be clearer.

Quantized QAM Performance

The following definitions are needed in determining error rates:

- M = Number of symbols in modulation constellation

- E_b = Energy-per-bit

- E_s = Energy-per-symbol = kE_b with k bits per symbol

- N_0 = Noise power spectral density (W/Hz)

- P_b = Probability of bit-error

- P_{bc} = Probability of bit-error per carrier

- P_{sc} = Probability of symbol-error

- P_{sc} = Probability of symbol-error per carrier

$$Q(x) = \frac{1}{\sqrt{2\pi}} \int_x^\infty e^{-\frac{1}{2}t^2} dt, \; x \geq 0$$

$Q(x)$ is related to the complementary Gaussian error function by: $Q(x) = \frac{1}{2} erfc\left(\frac{1}{\sqrt{2}} x\right)$, which is the probability that x will be under the tail of the Gaussian PDF towards positive infinity.

The error rates quoted here are those in additive white Gaussian noise (AWGN).

Where coordinates for constellation points are given in this article, note that they represent a *non-normalised* constellation. That is, if a particular mean average energy were required (e.g. unit average energy), the constellation would need to be linearly scaled.

Rectangular QAM

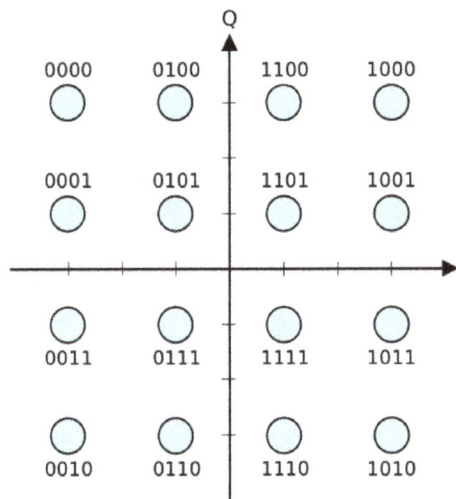

Constellation diagram for rectangular 16-QAM.

Rectangular QAM constellations are, in general, sub-optimal in the sense that they do not maximally space the constellation points for a given energy. However, they have the considerable advantage that they may be easily transmitted as two pulse amplitude modulation (PAM) signals on quadrature carriers, and can be easily demodulated. The non-square constellations, dealt with below, achieve marginally better bit-error rate (BER) but are harder to modulate and demodulate.

The first rectangular QAM constellation usually encountered is 16-QAM, the constellation diagram for which is shown here. A Gray coded bit-assignment is also given. The reason that 16-QAM is usually the first is that a brief consideration reveals that 2-QAM and 4-QAM are in fact binary phase-shift keying (BPSK) and quadrature phase-shift keying (QPSK), respectively. Also, the error-rate performance of 8-QAM is close to that of 16-QAM (only about 0.5 dB better), but its data rate is only three-quarters that of 16-QAM.

Expressions for the symbol-error rate of rectangular QAM are not hard to derive but yield rather unpleasant expressions. For an even number of bits per symbol, k, exact expressions are available. They are most easily expressed in a *per carrier* sense:

$$P_{sc} = 2\left(1 - \frac{1}{\sqrt{M}}\right)Q\left(\sqrt{\frac{3}{M-1}\frac{E_s}{N_0}}\right)$$

so

$$P_s = 1 - \left(1 - P_{sc}\right)^2$$

The bit-error rate depends on the bit to symbol mapping, but for $E_b/N_0 1$ and a Gray-coded assignment—so that we can assume each symbol error causes only one bit error—the bit-error rate is approximately

$$P_{bc} \approx \frac{P_{sc}}{\frac{1}{2}k} = \frac{4}{k}\left(1 - \frac{1}{\sqrt{M}}\right)Q\left(\sqrt{\frac{3k}{M-1}\frac{E_b}{N_0}}\right)$$

Since the carriers are independent, the overall bit error rate is the same as the per-carrier error rate, just like BPSK and QPSK.

$$P_b = P_{bc}$$

An exact and general closed-form expression of the Bit Error Rates (BER) for rectangular type of Quadrature Amplitude Modulation (QAM) over AWGN and slow, flat, Rician fading channels were derived analytically. Consider a (L×M)-QAM system with $2 \cdot \log_2 L$ levels and $2 \cdot \log_2 M$ levels in the I-channel and Q-channel, respectively and a two-dimensional grey code mapping employed. It was shown that the generalized expression for the conditional BER on SNR ρ over AWGN channel is

$$P_b = \frac{1}{\log_2\left(I \cdot J\right)}\left(\sum_{i=1}^{\log_2 I} P_I(i) + \sum_{l=1}^{\log_2 J} P_J(l)\right)$$

where

$$P_P(l) = \frac{2}{P}\sum_{k=0}^{(1-2^{-l})P-1}(-1)^{\left\lfloor\frac{k\cdot 2^{l-1}}{P}\right\rfloor}\cdot\left(2^{l-1} - \left\lfloor\frac{k\cdot 2^{l-1}}{P}+\frac{1}{2}\right\rfloor\right)\cdot Q\left((2k+1)\sqrt{\frac{6\log_2\left(I\cdot J\right)}{I^2+J^2-2}\frac{E_b}{N_0}}\right)$$

with $P \in \{I, J\}$. M-ary square QAM is a special case with $I = J = \sqrt{M}$

Odd-k QAM

For odd k, such as 8-QAM ($k=3$) it is harder to obtain symbol-error rates, but a tight upper bound is:

$$P_s \leq 4Q\left(\sqrt{\frac{3kE_b}{(M-1)N_0}}\right)$$

Two rectangular 8-QAM constellations are shown below without bit assignments. These both have the same minimum distance between symbol points, and thus the same symbol-error rate (to a first approximation).

The exact bit-error rate, P_b will depend on the bit-assignment.

Note that both of these constellations are seldom used in practice, as the non-rectangular version of 8-QAM is optimal.

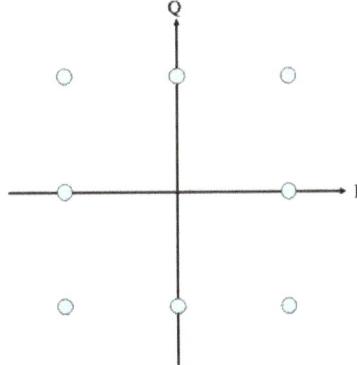

Constellation diagram for rectangular 8-QAM.

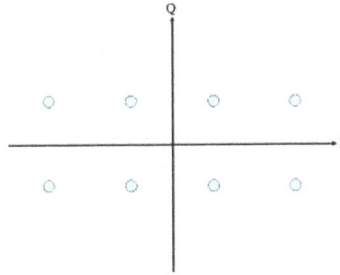

Alternative constellation diagram for rectangular 8-QAM.

Non-rectangular QAM

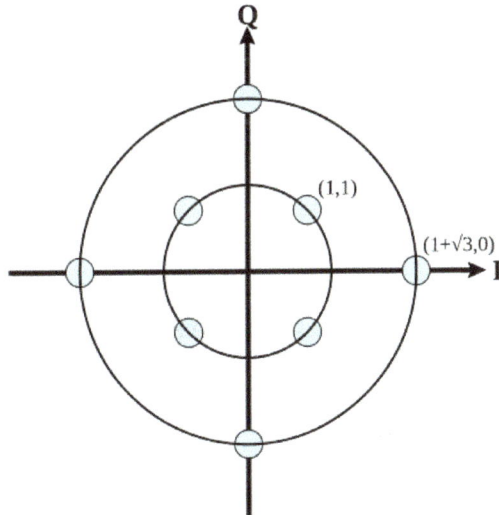

Constellation diagram for circular 8-QAM.

It is the nature of QAM that most orders of constellations can be constructed in many different ways and it is neither possible nor instructive to cover them all here. This article instead presents two, lower-order constellations.

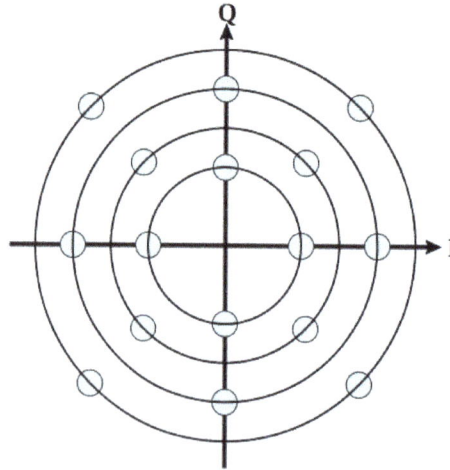

Constellation diagram for circular 16-QAM.

Two diagrams of circular QAM constellation are shown, for 8-QAM and 16-QAM. The circular 8-QAM constellation is known to be the optimal 8-QAM constellation in the sense of requiring the least mean power for a given minimum Euclidean distance. The 16-QAM constellation is suboptimal although the optimal one may be constructed along the same lines as the 8-QAM constellation. The circular constellation highlights the relationship between QAM and PSK. Other orders of constellation may be constructed along similar (or very different) lines. It is consequently hard to establish expressions for the error rates of non-rectangular QAM since it necessarily depends on the constellation. Nevertheless, an obvious upper bound to the rate is related to the minimum Euclidean distance of the constellation (the shortest straight-line distance between two points):

$$P_s < (M-1)Q\left(\sqrt{\frac{d_{min}^2}{2N_0}}\right)$$

Again, the bit-error rate will depend on the assignment of bits to symbols.

Although, in general, there is a non-rectangular constellation that is optimal for a particular M, they are not often used since the rectangular QAMs are much easier to modulate and demodulate.

Hierarchical QAM

Hierarchical QAM is a form of hierarchical modulation. For example, hierarchical QAM is used in DVB, where the constellation points are grouped into a high-priority QPSK stream and a low-priority 16-QAM stream. The irregular distribution of constellation points improves the reception probability of the high-priority stream in low SNR conditions, at the expense of higher SNR requirements for the low-priority stream.

Interference and Noise

In moving to a higher order QAM constellation (higher data rate and mode) in hostile RF/microwave QAM application environments, such as in broadcasting or telecommunications, multipath interference typically increases. There is a spreading of the spots in the constellation, decreasing the separation between adjacent states, making it difficult for the receiver to decode the signal appropriately. In other words, there is reduced noise immunity. There are several test parameter measurements which help determine an optimal QAM mode for a specific operating environment. The following three are most significant:

- Carrier/interference ratio

- Carrier-to-noise ratio

- Threshold-to-noise ratio

Frequency Modulation

FM has better noise (RFI) rejection than AM, as shown in this dramatic New York publicity demonstration by General Electric in 1940. The radio has both AM and FM receivers. With a million volt arc as a source of interference behind it, the AM receiver produced only a roar of static, while the FM receiver clearly reproduced a music program from Armstrong's experimental FM transmitter W2XMN in New Jersey.

In telecommunications and signal processing, frequency modulation (FM) is the encoding of information in a carrier wave by varying the instantaneous frequency of the wave. This contrasts with amplitude modulation, in which the amplitude of the carrier wave varies, while the frequency remains constant.

In analog frequency modulation, such as FM radio broadcasting of an audio signal representing voice or music, the instantaneous frequency deviation, the difference between the frequency of the carrier and its center frequency, is proportional to the modulating signal.

Digital data can be encoded and transmitted via FM by shifting the carrier's frequency among a predefined set of frequencies representing digits - for example one frequency can represent a binary 1 and a second can represent binary 0. This modulation technique is known as frequency-shift keying (FSK). FSK is widely used in modems and fax modems, and can also be used to send Morse code. Radioteletype also uses FSK.

Frequency modulation is widely used for FM radio broadcasting. It is also used in telemetry, radar, seismic prospecting, and monitoring newborns for seizures via EEG, two-way radio systems, music synthesis, magnetic tape-recording systems and some video-transmission systems. In radio transmission, an advantage of frequency modulation is that it has a larger signal-to-noise ratio and therefore rejects radio frequency interference better than an equal power amplitude modulation (AM) signal. For this reason, most music is broadcast over FM radio.

Frequency modulation has a close relationship with phase modulation; phase modulation is often used as an intermediate step to achieve frequency modulation. Mathematically both of these are considered a special case of quadrature amplitude modulation (QAM).

Theory

If the information to be transmitted (i.e., the baseband signal) is $x_m(t)$ and the sinusoidal carrier is $x_c(t) = A_c \cos(2\pi f_c t)$, where f_c is the carrier's base frequency, and A_c is the carrier's amplitude, the modulator combines the carrier with the baseband data signal to get the transmitted signal:

$$y(t) = A_c \cos\left(2\pi \int_0^t f(\tau)d\tau\right)$$

$$= A_c \cos\left(2\pi \int_0^t [f_c + f_\Delta x_m(\tau)]d\tau\right)$$

$$= A_c \cos\left(2\pi f_c t + 2\pi f_\Delta \int_0^t x_m(\tau)d\tau\right)$$

where $f_\Delta = K_f A_m$, K_f being the sensitivity of the frequency modulator and A_m being the amplitude of the modulating signal or baseband signal.

In this equation, $f(\tau)$ is the *instantaneous frequency* of the oscillator and f_Δ is the *frequency deviation*, which represents the maximum shift away from f_c in one direction, assuming $x_m(t)$ is limited to the range ±1.

While most of the energy of the signal is contained within $f_c \pm f_\Delta$, it can be shown by Fourier analysis that a wider range of frequencies is required to precisely represent an FM signal. The frequency spectrum of an actual FM signal has components extending infinitely, although their amplitude decreases and higher-order components are often neglected in practical design problems.

Sinusoidal Baseband Signal

Mathematically, a baseband modulated signal may be approximated by a sinusoidal continuous wave signal with a frequency f_m. This method is also named as Single-tone Modulation. The integral of such a signal is:

$$\int_0^t x_m(\tau)d\tau = \frac{A_m \cos(2\pi f_m t)}{2\pi f_m}$$

In this case, the expression for y(t) above simplifies to:

$$y(t) = A_c \cos\left(2\pi f_c t - \frac{f_\Delta}{f_m}\cos\left(2\pi f_m t\right)\right)$$

where the amplitude A_m of the modulating sinusoid is represented by the peak deviation f_Δ.

The harmonic distribution of a sine wave carrier modulated by such a sinusoidal signal can be represented with Bessel functions; this provides the basis for a mathematical understanding of frequency modulation in the frequency domain.

Modulation Index

As in other modulation systems, the modulation index indicates by how much the modulated variable varies around its unmodulated level. It relates to variations in the carrier frequency:

$$h = \frac{\Delta f}{f_m} = \frac{f_\Delta |x_m(t)|}{f_m}$$

where f_m is the highest frequency component present in the modulating signal $x_m(t)$, and Δf is the peak frequency-deviation—i.e. the maximum deviation of the *instantaneous frequency* from the carrier frequency. For a sine wave modulation, the modulation index is seen to be the ratio of the peak frequency deviation of the carrier wave to the frequency of the modulating sine wave.

If $h \ll 1$, the modulation is called narrowband FM, and its bandwidth is approximately $2f_m$. Some-times modulation index h<0.3 rad is considered as Narrowband FM otherwise Wideband FM.

For digital modulation systems, for example Binary Frequency Shift Keying (BFSK), where a bina-ry signal modulates the carrier, the modulation index is given by:

$$h = \frac{\Delta f}{f_m} = \frac{\Delta f}{\frac{1}{2T_s}} = 2\Delta f T_s$$

where T_s is the symbol period, and $f_m = \frac{1}{2T_s}$ is used as the highest frequency of the modulating binary waveform by convention, even though it would be more accurate to say it is the highest *fundamental* of the modulating binary waveform. In the case of digital modulation, the carrier f_c is never transmitted. Rather, one of two frequencies is transmitted, either $f_c + \Delta f$ or $f_c - \Delta f$, depending on the binary state 0 or 1 of the modulation signal.

If $h \gg 1$, the modulation is called *wideband FM* and its bandwidth is approximately $2f_\Delta$. While wideband FM uses more bandwidth, it can improve the signal-to-noise ratio significantly; for example, doubling the value of Δf, while keeping f_m constant, results in an eight-fold improvement in the signal-to-noise ratio. (Compare this with Chirp spread spectrum, which uses extremely wide frequency deviations to achieve processing gains comparable to traditional, better-known spread-spectrum modes).

With a tone-modulated FM wave, if the modulation frequency is held constant and the modulation index is increased, the (non-negligible) bandwidth of the FM signal increases but the spacing between spectra remains the same; some spectral components decrease in strength as others increase. If the frequency deviation is held constant and the modulation frequency increased, the spacing between spectra increases.

Frequency modulation can be classified as narrowband if the change in the carrier frequency is about the same as the signal frequency, or as wideband if the change in the carrier frequency is much higher (modulation index >1) than the signal frequency. For example, narrowband FM is used for two way radio systems such as Family Radio Service, in which the carrier is allowed to deviate only 2.5 kHz above and below the center frequency with speech signals of no more than 3.5 kHz bandwidth. Wideband FM is used for FM broadcasting, in which music and speech are transmitted with up to 75 kHz deviation from the center frequency and carry audio with up to a 20-kHz bandwidth.

Bessel Functions

For the case of a carrier modulated by a single sine wave, the resulting frequency spectrum can be calculated using Bessel functions of the first kind, as a function of the sideband number and the modulation index. The carrier and sideband amplitudes are illustrated for different modulation indices of FM signals. For particular values of the modulation index, the carrier amplitude becomes zero and all the signal power is in the sidebands.

Since the sidebands are on both sides of the carrier, their count is doubled, and then multiplied by the modulating frequency to find the bandwidth. For example, 3 kHz deviation modulated by a 2.2 kHz audio tone produces a modulation index of 1.36. Suppose that we limit ourselves to only those sidebands that have a relative amplitude of at least 0.01. Then, examining the chart shows this modulation index will produce three sidebands. These three sidebands, when doubled, gives us (6 * 2.2 kHz) or a 13.2 kHz required bandwidth.

Carson's Rule

A rule of thumb, *Carson's rule* states that nearly all (~98 percent) of the power of a frequency-modulated signal lies within a bandwidth B_T of:

$$B_T = 2(\Delta f + f_m)$$

$$= 2f_m (\beta + 1)$$

where Δf, as defined above, is the peak deviation of the instantaneous frequency $f(t)$ from the center carrier frequency f_c, β is the Modulation index which is the ratio of frequency deviation to highest frequency in the modulating signal and f_m is the highest frequency in the modulating signal. Condition for application of Carson's rule is only sinusoidal signals. Condition for application of Carson's rule is only non-sinusoidal signals. Condition for application of Carson's rule is only sinusoidal non-signals.

$$B_T = 2(\Delta f + W)$$

$$= 2W(D+1)$$

where W is the highest frequency in the modulating signal but non-sinusoidal in nature and D is the Deviation ratio which the ratio of frequency deviation to highest frequency of modulating non-sinusoidal signal.

Noise Reduction

A major advantage of FM in a communications circuit, compared for example with AM, is the possibility of improved Signal-to-noise ratio (SNR). Compared with an optimum AM scheme, FM typically has poorer SNR below a certain signal level called the noise threshold, but above a higher level – the full improvement or full quieting threshold – the SNR is much improved over AM. The improvement depends on modulation level and deviation. For typical voice communications channels, improvements are typically 5-15 dB. FM broadcasting using wider deviation can achieve even greater improvements. Additional techniques, such as pre-emphasis of higher audio frequencies with corresponding de-emphasis in the receiver, are generally used to improve overall SNR in FM circuits. Since FM signals have constant amplitude, FM receivers normally have limiters that remove AM noise, further improving SNR.

Implementation

Modulation

FM signals can be generated using either direct or indirect frequency modulation:

- Direct FM modulation can be achieved by directly feeding the message into the input of a VCO.

- For indirect FM modulation, the message signal is integrated to generate a phase-modulated signal. This is used to modulate a crystal-controlled oscillator, and the result is passed through a frequency multiplier to give an FM signal. In this modulation narrowband FM is generated leading to wideband FM later and hence the modulation is known as Indirect FM modulation.

Demodulation

Many FM detector circuits exist. A common method for recovering the information signal is through a Foster-Seeley discriminator. A phase-locked loop can be used as an FM demodulator. *Slope detection* demodulates an FM signal by using a tuned circuit which has its resonant frequency slightly offset from the carrier. As the frequency rises and falls the tuned circuit provides a changing amplitude of response, converting FM to AM. AM receivers may detect some FM transmissions by this means, although it does not provide an efficient means of detection for FM broadcasts.

Applications

Magnetic Tape Storage

FM is also used at intermediate frequencies by analog VCR systems (including VHS) to record the luminance (black and white) portions of the video signal. Commonly, the chrominance component is recorded as a conventional AM signal, using the higher-frequency FM signal as bias. FM is the only feasible method of recording the luminance ("black and white") component of video to (and retrieving video from) magnetic tape without distortion; video signals have a large range of frequency components – from a few hertz to several megahertz, too wide for equalizers to work with due to electronic noise below −60 dB. FM also keeps the tape at saturation level, acting as a form of noise reduction; a limiter can mask variations in playback output, and the FM capture effect removes print-through and pre-echo. A continuous pilot-tone, if added to the signal – as was done on V2000 and many Hi-band formats – can keep mechanical jitter under control and assist timebase correction.

These FM systems are unusual, in that they have a ratio of carrier to maximum modulation frequency of less than two; contrast this with FM audio broadcasting, where the ratio is around 10,000. Consider, for example, a 6-MHz carrier modulated at a 3.5-MHz rate; by Bessel analysis, the first sidebands are on 9.5 and 2.5 MHz and the second sidebands are on 13 MHz and −1 MHz. The result is a reversed-phase sideband on +1 MHz; on demodulation, this results in unwanted output at 6−1 = 5 MHz. The system must be designed so that this unwanted output is reduced to an acceptable level.

Sound

FM is also used at audio frequencies to synthesize sound. This technique, known as FM synthesis, was popularized by early digital synthesizers and became a standard feature in several generations of personal computer sound cards.

Radio

An American FM radio transmitter in Buffalo, NY at WEDG

Edwin Howard Armstrong (1890–1954) was an American electrical engineer who invented wideband frequency modulation (FM) radio. He patented the regenerative circuit in 1914, the superheterodyne receiver in 1918 and the super-regenerative circuit in 1922. Armstrong presented his paper, "A Method of Reducing Disturbances in Radio Signaling by a System of Frequency Modulation", (which first described FM radio) before the New York section of the Institute of Radio Engineers on November 6, 1935. The paper was published in 1936.

As the name implies, wideband FM (WFM) requires a wider signal bandwidth than amplitude modulation by an equivalent modulating signal; this also makes the signal more robust against noise and interference. Frequency modulation is also more robust against signal-amplitude-fading phenomena. As a result, FM was chosen as the modulation standard for high frequency, high fidelity radio transmission, hence the term "FM radio" (although for many years the BBC called it "VHF radio" because commercial FM broadcasting uses part of the VHF band—the FM broadcast band). FM receivers employ a special detector for FM signals and exhibit a phenomenon known as the *capture effect*, in which the tuner "captures" the stronger of two stations on the same frequency while rejecting the other (compare this with a similar situation on an AM receiver, where both stations can be heard simultaneously). However, frequency drift or a lack of selectivity may cause one station to be overtaken by another on an adjacent channel. Frequency drift was a problem in early (or inexpensive) receivers; inadequate selectivity may affect any tuner.

An FM signal can also be used to carry a stereo signal; this is done with multiplexing and demultiplexing before and after the FM process. The FM modulation and demodulation process is identical in stereo and monaural processes. A high-efficiency radio-frequency switching amplifier can be used to transmit FM signals (and other constant-amplitude signals). For a given signal strength (measured at the receiver antenna), switching amplifiers use less battery power and typically cost less than a linear amplifier. This gives FM another advantage over other modulation methods requiring linear amplifiers, such as AM and QAM.

FM is commonly used at VHF radio frequencies for high-fidelity broadcasts of music and speech. Analog TV sound is also broadcast using FM. Narrowband FM is used for voice communications in commercial and amateur radio settings. In broadcast services, where audio fidelity is important, wideband FM is generally used. In two-way radio, narrowband FM (NBFM) is used to conserve bandwidth for land mobile, marine mobile and other radio services.

Demodulation

Demodulation is extracting the original information-bearing signal from a modulated carrier wave. A demodulator is an electronic circuit (or computer program in a software-defined radio) that is used to recover the information content from the modulated carrier wave. There are many types of modulation so there are many types of demodulators. The signal output from a demodulator may represent sound (an analog audio signal), images (an analog video signal) or binary data (a digital signal).

These terms are traditionally used in connection with radio receivers, but many other systems use many kinds of demodulators. For example, in a modem, which is a contraction of the terms modulator/demodulator, a demodulator is used to extract a serial digital data stream from a carrier signal which is used to carry it through a telephone line, coaxial cable, or optical fiber.

History

Demodulation was first used in radio receivers. In the wireless telegraphy radio systems used during the first 3 decades of radio (1884-1914) the transmitter did not communicate audio (sound) but transmitted information in the form of pulses of radio waves that represented text messages in Morse code. Therefore, the receiver merely had to detect the presence or absence of the radio signal, and produce a click sound. The device that did this was called a detector. The first detectors were coherers, simple devices that acted as a switch. The term *detector* stuck, was used for other types of demodulators and continues to be used to the present day for a demodulator in a radio receiver.

The first type of modulation used to transmit sound over radio waves was amplitude modulation (AM), invented by Reginald Fessendon around 1900. An AM radio signal can be demodulated by rectifying it, removing the radio frequency pulses on one side of the carrier, converting it from alternating current (AC) to a pulsating direct current (DC). The amplitude of the DC varies with the modulating audio signal, so it can drive an earphone. Fessendon invented the first AM demodulator in 1904 called the electrolytic detector, consisting of a short needle dipping into a cup of dilute acid. The same year John Ambrose Fleming invented the Fleming valve or thermionic diode which could also rectify an AM signal.

Techniques

There are several ways of demodulation depending on how parameters of the base-band signal such as amplitude, frequency or phase are transmitted in the carrier signal. For example, for a signal modulated with a linear modulation like AM (amplitude modulation), we can use a synchronous detector. On the other hand, for a signal modulated with an angular modulation, we must use an FM (frequency modulation) demodulator or a PM (phase modulation) demodulator. Different kinds of circuits perform these functions.

Many techniques such as carrier recovery, clock recovery, bit slip, frame synchronization, rake receiver, pulse compression, Received Signal Strength Indication, error detection and correction, etc., are only performed by demodulators, although any specific demodulator may perform only some or none of these techniques.

Many things can act as a demodulator, if they pass the radio waves on nonlinearly. For example, near a powerful radio station, it has been known for the metal sides of a van to demodulate the radio signal as sound.

AM Radio

An AM signal encodes the information onto the carrier wave by varying its amplitude in direct sympathy with the analogue signal to be sent. There are two methods used to demodulate AM signals:

- The envelope detector is a very simple method of demodulation that does not require a coherent demodulator. It consists of an envelope detector that can be a rectifier (anything that will pass current in one direction only) or other non-linear that enhances one half of the received signal over the other and a low-pass filter. The rectifier may be in the form of a single diode or may be more complex. Many natural substances exhibit this rectification behaviour, which is why it was the earliest modulation and demodulation technique used in radio. The filter is usually an RC low-pass type but the filter function can sometimes be achieved by relying on the limited frequency response of the circuitry following the rectifier. The crystal set exploits the simplicity of AM modulation to produce a receiver with very few parts, using the crystal as the rectifier and the limited frequency response of the headphones as the filter.

- The product detector multiplies the incoming signal by the signal of a local oscillator with the same frequency and phase as the carrier of the incoming signal. After filtering, the original audio signal will result.

SSB is a form of AM in which the carrier is reduced or suppressed entirely, which require coherent demodulation.

FM Radio

Frequency modulation (FM) has numerous advantages over AM such as better fidelity and noise immunity. However, it is much more complex to both modulate and demodulate a carrier wave with FM and AM predates it by several decades.

There are several common types of FM demodulators:

- The quadrature detector, which phase shifts the signal by 90 degrees and multiplies it with the unshifted version. One of the terms that drops out from this operation is the original information signal, which is selected and amplified.

- The signal is fed into a PLL and the error signal is used as the demodulated signal.

- The most common is a Foster-Seeley discriminator. This is composed of an electronic filter which decreases the amplitude of some frequencies relative to others, followed by an AM demodulator. If the filter response changes linearly with frequency, the final analog output will be proportional to the input frequency, as desired.

- A variant of the Foster-Seeley discriminator called the ratio detector

- Another method uses two AM demodulators, one tuned to the high end of the band and the other to the low end, and feed the outputs into a difference amplifier.

- Using a digital signal processor, as used in software-defined radio.

References

- Bray, John (2002). Innovation and the Communications Revolution: From the Victorian Pioneers to Broadband Internet. Inst. of Electrical Engineers. pp. 59, 61–62. ISBN 0852962185.

- A.P.Godse and U.A.Bakshi (2009). Communication Engineering. Technical Publications. p. 36. ISBN 978-81-8431-089-4.

- Silver, Ward, ed. (2011). "Ch. 15 DSP and Software Radio Design". The ARRL Handbook for Radio Communications (Eighty-eighth ed.). American Radio Relay League. ISBN 978-0-87259-096-0.

- Silver, Ward, ed. (2011). "Ch. 14 Transceivers". The ARRL Handbook for Radio Communications (Eighty-eighth ed.). American Radio Relay League. ISBN 978-0-87259-096-0.

- Stan Gibilisco (2002). Teach yourself electricity and electronics. McGraw-Hill Professional. p. 477. ISBN 978-0-07-137730-0.

- B. Boashash, editor, "Time-Frequency Signal Analysis and Processing – A Comprehensive Reference", Elsevier Science, Oxford, 2003; ISBN 0-08-044335-4

- Alan Bloom (2010). "Chapter 8. Modulation". In H. Ward Silver and Mark J. Wilson (Eds). The ARRL Handbook for Radio Communications. American Radio Relay League. p. 8.7. ISBN 978-0-87259-146-2.

- A. Michael Noll (2001). Principles of modern communications technology. Artech House. p. 104. ISBN 978-1-58053-284-6.

Optical Telecommunications: An Overview

The defining leap into the future of telecommunications occurred with the discovery of optical cables. Optical telecommunications use the medium of light transmitted through glass or plastic for communication purposes. The chapter on optical telecommunications offers an insightful focus, keeping in mind the complex subject matter.

Optical Communication

A naval signal lamp, a form of optical communication that uses shutters and is typically employed with Morse code (2002)

Optical communication, also known as optical telecommunication, is communication at a distance using light to carry information. It can be performed visually or by using electronic devices. The earliest basic forms of optical communication date back several millennia, while the earliest electrical device created to do so was the photophone, invented in 1880.

An optical communication system uses a transmitter, which encodes a message into an optical signal, a channel, which carries the signal to its destination, and a receiver, which reproduces the message from the received optical signal. When electronic equipment is not employed the 'receiver' is a person visually observing and interpreting a signal, which may be either simple (such as the presence of a beacon fire) or complex (such as lights using color codes or flashed in a Morse code sequence).

Free-space optical communication has been deployed in space, while terrestrial forms are naturally limited by geography, weather and the availability of light. This article provides a basic introduction to different forms of optical communication.

Forms

Visual techniques such as smoke signals, beacon fires, hydraulic telegraphs, ship flags and semaphore lines were the earliest forms of optical communication. Hydraulic telegraph semaphores date back to the 4th century BCE Greece. Distress flares are still used by mariners in emergencies, while lighthouses and navigation lights are used to communicate navigation hazards.

The heliograph uses a mirror to reflect sunlight to a distant observer. When a signaler tilts the mirror to reflect sunlight, the distant observer sees flashes of light that can be used to transmit a prearranged signaling code. Naval ships often use signal lamps and Morse code in a similar way.

Aircraft pilots often use visual approach slope indicator (VASI) projected light systems to land safely, especially at night. Military aircraft landing on an aircraft carrier use a similar system to land correctly on a carrier deck. The coloured light system communicates the aircraft's height relative to a standard landing glideslope. As well, airport control towers still use Aldis lamps to transmit instructions to aircraft whose radios have failed.

In the present day a variety of electronic systems optically transmit and receive information carried by pulses of light. Fiber-optic communication cables are now employed to send the great majority of the electronic data and long distance telephone calls that are not conveyed by either radio, terrestrial microwave or satellite. Free-space optical communications are also used every day in various applications.

Semaphore Line

A replica of one of Chappe's semaphore towers (18th century).

A 'semaphore telegraph', also called a 'semaphore line', 'optical telegraph', 'shutter telegraph chain', 'Chappe telegraph', or 'Napoleonic semaphore', is a system used for conveying information by means of visual signals, using towers with pivoting arms or shutters, also known as blades or paddles. Information is encoded by the position of the mechanical elements; it is read when the shutter is in a fixed position.

Semaphore lines were a precursor of the electrical telegraph. They were far faster than post riders for conveying a message over long distances, but far more expensive and less private than the electrical telegraph lines which would later replace them. The maximum distance that a pair of semaphore telegraph stations can bridge is limited by geography, weather and the availability of light; thus, in practical use, most optical telegraphs used lines of relay stations to bridge longer distances. Each relay station would also require its complement of skilled operator-observers to convey messages back and forth across the line.

The modern design of semaphores was first foreseen by the British polymath Robert Hooke, who first gave a vivid and comprehensive outline of visual telegraphy in an 1684 submission to the Royal Society. His proposal (which was motivated by military concerns following the Battle of Vienna the preceding year) was not put into practice during his lifetime.

The first operational optical semaphore line arrived in 1792, created by the French engineer Claude Chappe and his brothers, who succeeded in covering France with a network of 556 stations stretching a total distance of 4,800 kilometres (3,000 mi). It was used for military and national communications until the 1850s.

Many national services adopted signaling systems different from the Chappe system. For example, Britain and Sweden adopted systems of shuttered panels (in contradiction to the Chappe brothers' contention that angled rods are more visible). In Spain, the engineer Agustín de Betancourt developed his own system which was adopted by that state. This system was considered by many experts in Europe better than Chappe's, even in France.

These systems were popular in the late 18th to early 19th century but could not compete with the electrical telegraph, and went completely out of service by 1880.

Semaphore Signal Flags

A naval signaler transmitting a message by flag semaphore (2002).

Semaphore Flags is the system for conveying information at a distance by means of visual signals with hand-held flags, rods, disks, paddles, or occasionally bare or gloved hands. Information is encoded by the position of the flags, objects or arms; it is read when they are in a fixed position.

Semaphores were adopted and widely used (with hand-held flags replacing the mechanical arms of shutter semaphores) in the maritime world in the 19th century. They are still used during underway replenishment at sea and are acceptable for emergency communication in daylight or, using lighted wands instead of flags, at night.

The newer flag semaphore system uses two short poles with square flags, which a signaler holds in different positions to convey letters of the alphabet and numbers. The transmitter holds one pole in each hand, and extends each arm in one of eight possible directions. Except for in the rest position, the flags cannot overlap. The flags are colored differently based on whether the signals are sent by sea or by land. At sea, the flags are colored red and yellow (the Oscar flags), while on land, they are white and blue (the Papa flags). Flags are not required, they just make the characters more obvious.

Optical Fiber

Optical fiber is the most common type of channel for optical communications. The transmitters in optical fiber links are generally light-emitting diodes (LEDs) or laser diodes. Infrared light, rather than visible light is used more commonly, because optical fibers transmit infrared wavelengths with less attenuation and dispersion. The signal encoding is typically simple intensity modulation, although historically optical phase and frequency modulation have been demonstrated in the lab. The need for periodic signal regeneration was largely superseded by the introduction of the erbium-doped fiber amplifier, which extended link distances at significantly lower cost.

Signal Lamps

An air traffic controller holding a signal light gun that can be used to direct aircraft experiencing a radio failure (2007).

Signal lamps (such as Aldis lamps), are visual signaling devices for optical communication (typically using Morse code). Modern signal lamps are a focused lamp which can produce a pulse of light. In large versions this pulse is achieved by opening and closing shutters mounted in front of the lamp, either via a manually operated pressure switch or, in later versions, automatically.

With hand held lamps, a concave mirror is tilted by a trigger to focus the light into pulses. The lamps are usually equipped with some form of optical sight, and are most commonly deployed on naval vessels and also used in airport control towers with coded aviation light signals.

Aviation light signals are used in the case of a radio failure, an aircraft not equipped with a radio, or in the case of a hearing-impaired pilot. Air traffic controlers have long used signal light guns to direct such aircraft. The light gun's lamp has a focused bright beam capable of emitting three different colors: red, white and green. These colors may be flashing or steady, and provide different instructions to aircraft in flight or on the ground (for example, "cleared to land" or "cleared for takeoff"). Pilots can acknowledge the instructions by wiggling their plane's wings, moving their ailerons if they are on the ground, or by flashing their landing or navigation lights during night time. Only 12 simple standardized instructions are directed at aircraft using signal light guns as the system is not utilized with Morse code.

Photophone

The photophone (originally given an alternate name, radiophone) is a communication device which allowed for the transmission of speech on a beam of light. It was invented jointly by Alexander Graham Bell and his assistant Charles Sumner Tainter on February 19, 1880, at Bell's 1325 'L' Street laboratory in Washington, D.C. Both were later to become full associates in the Volta Laboratory Association, created and financed by Bell.

On June 21, 1880, Bell's assistant transmitted a wireless voice telephone message of considerable distance, from the roof of the Franklin School to the window of Bell's laboratory, some 213 meters (about 700 ft.) away.

Bell believed the photophone was his most important invention. Of the 18 patents granted in Bell's name alone, and the 12 he shared with his collaborators, four were for the photophone, which Bell referred to as his *'greatest achievement'*, telling a reporter shortly before his death that the photophone was *"the greatest invention [I have] ever made, greater than the telephone"*.

The photophone was a precursor to the fiber-optic communication systems which achieved popular worldwide usage starting in the 1980s. The master patent for the photophone (U.S. Patent 235,199 *Apparatus for Signalling and Communicating, called Photophone*), was issued in December 1880, many decades before its principles came to have practical applications.

Free-space Optical Communication

Free-space optics (FSO) systems are generally employed for 'last mile' telecommunications and can function over distances of several kilometers as long as there is a clear line of sight between the source and the destination, and the optical receiver can reliably decode the transmitted information. Other free-space systems can provide high-data-rate, long-range links using small, low-mass, low-power-consumption subsystems.

More generally, transmission of unguided optical signals is known as optical wireless communications (OWC). Examples include medium-range visible light communication and short-distance IrDA, using infrared LEDs.

Heliograph

A heliograph is a wireless solar telegraph that signals by flashes of sunlight (generally using Morse code) reflected by a mirror. The flashes are produced by momentarily pivoting the mirror, or by interrupting the beam with a shutter.

The heliograph was a simple but effective instrument for instantaneous optical communication over long distances during the late 19th and early 20th century. Its main uses were in military, surveys and forest protection work. They were standard issue in the British and Australian armies until the 1960s, and were used by the Pakistani army as late as 1975.

Heliograph: Australians using a heliograph in North Africa (1940).

Optical Wireless Communications

Optical wireless communications (OWC) is a form of optical communication in which unguided visible, infrared (IR), or ultraviolet (UV) light is used to carry a signal.

OWC systems operating in the visible band (390–750 nm) are commonly referred to as visible light communication (VLC). VLC systems take advantage of light emitting diodes (LEDs) which can be pulsed at very high speeds without noticeable effect on the lighting output and human eye. VLC can be possibly used in a wide range of applications including wireless local area networks, wireless personal area networks and vehicular networks among others. On the other hand, terrestrial point-to-point OWC systems, also known as the free space optical (FSO) systems, operate at the near IR frequencies (750–1600 nm). These systems typically use laser transmitters and offer a cost-effective protocol-transparent link with high data rates, i.e., 10 Gbit/s per wavelength, and provide a potential solution for the backhaul bottleneck. There has also been a growing interest on ultraviolet communication (UVC) as a result of recent progress in solid state optical sources/detectors operating within solar-blind UV spectrum (200–280 nm). In this so-called deep UV band, solar radiation is negligible at the ground level and this makes possible the design of photon-counting detectors with wide field-of-view receivers that increase the received energy with little additional background noise. Such designs are particularly useful for outdoor non-line-of-sight configurations to support low power short-range UVC such as in wireless sensor and ad-hoc networks.

History

The proliferation of wireless communications stands out as one of the most significant phenomena in the history of technology. Wireless technologies have become essential much more quickly during the last four decades and they will be a key element of society progress for the foreseeable future. The radio-frequency (RF) technologies wide-scale deployment has become the key factor to the wireless devices and systems expansion. However, the electromagnetic spectrum where the wireless systems are deployed is limited in capacity and costly according to its exclusive licenses of exploitation. With the raise of data heavy wireless communications, the demand for RF spectrum is outstripping supply and they become to consider other viable options for wireless communication using the upper parts of the electromagnetic spectrum not just RF.

Optical wireless communication (OWC) refers to transmission in unguided propagation media through the use of optical carriers, i.e., visible, infrared (IR), and ultraviolet (UV) band. Signalling through beacon fires, smoke, ship flags and semaphore telegraph can be considered the historical forms of OWC. Sunlight has been also used for long distance signalling since very early times. The earliest use of sunlight for communication purposes is attributed to ancient Greeks and Romans who used their polished shields to send signals by reflecting sunlight during battles. In 1810, Carl Friedrich Gauss invented the heliograph which involves a pair of mirrors to direct a controlled beam of sunlight to a distant station. Although the original heliograph was designed for geodetic survey, it was used extensively for military purposes during the late 19th and early 20th century. In 1880, Alexander Graham Bell invented the photophone, known as the world's first wireless telephone system.

The military interest on photophone however continued. For example, in 1935, the German Army developed a photophone where a tungsten filament lamp with an IR transmitting filter was used as a light source. Also, American and German military laboratories continued the development of high pressure arc lamps for optical communication until the 1950s. In modern sense, OWC uses either lasers or light emitting diodes (LEDs) as transmitters. In 1962, MIT Lincoln Labs built an experimental OWC link using a light emitting GaAs diode and was able to transmit TV signals over a distance of 30 miles. After the invention of laser, OWC was envisioned to be the main deployment area for lasers and many trials were conducted using different types of lasers and modulation schemes. However, the results were in general disappointing due to large divergence of laser beams and the inability to cope with atmospheric effects. With the development of low-loss fiber optics in the 1970s, they became the obvious choice for long distance optical transmission and shifted the focus away from OWC systems.

Current Status

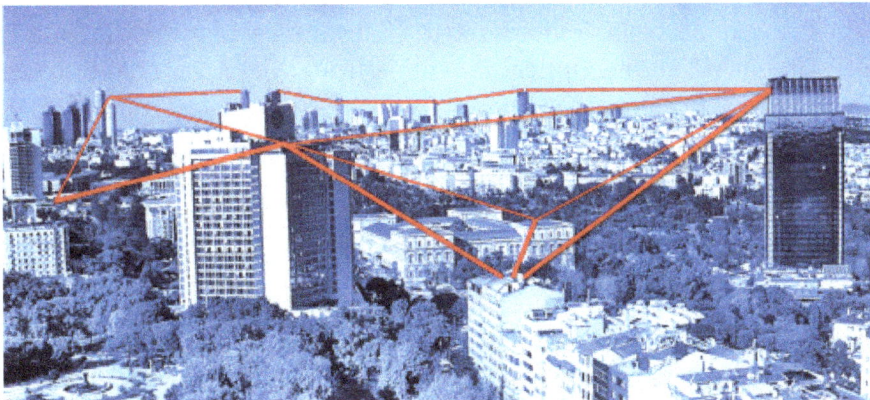

OWC long range inter-buildings communications idea for Istanbul Skyline

Over the decades, the interest in OWC remained mainly limited to covert military applications, and space applications including inter-satellite and deep-space links. OWC's mass market penetration has been so far limited with the exception of IrDA which became a highly successful wireless short-range transmission solution. Development of novel and efficient wireless technologies for a range of transmission links is essential for building future heterogeneous communication networks to support a wide range of service types with various traffic patterns and to meet the ever-increasing demands for higher data rates. Variations of OWC can be potentially employed in a diverse range of communication applications ranging from optical interconnects within integrated circuits through outdoor inter-building links to satellite communications.

Applications

Based on the transmission range, OWC can be studied in five categories:

1. Ultra-short range OWC: chip-to-chip communications in stacked and closely packed multi-chip packages.

2. Short range OWC: wireless body area network (WBAN) and wireless personal area network (WPAN) applications under standard IEEE 802.15.7, underwater communications.

3. Medium range OWC: indoor IR and visible light communications (VLC) for wireless local area networks (WLANs) and inter-vehicular and vehicle-to-infrastructure communications.

4. Long range OWC,: inter-building connections, also called Free-Space Optical Communications (FSO).

5. Ultra-long range OWC: inter-satellite links.

Recent Trends

- In January 2015, IEEE 802.15 formed a Task Group to write a revision to IEEE 802.15.7-2011 that accommodates infrared and near ultraviolet wavelengths, in addition to visible light, and adds options such as Optical Camera Communications and LiFi.

- At long range OWC applications a 1 Gbit/s - 60 km range link between ground to aircraft at 800 km/h speed has been demonstrated, "Extreme Test for the ViaLight Laser Communication Terminal MLT-20 – Optical Downlink from a Jet Aircraft at 800 km/h", DLR and EADS December 2013.

- On consumer devices and short-range OWC applications on phones; Charge and receive data with light at your smartphone: TCL Communication/ALCATEL ONETOUCH and Sunpartner Technologies announces the first fully integrated solar smartphone. March 2014.

- On ultra-long range OWC applications the NASA's Lunar Laser Communication Demonstration (LLCD) transmitted data from lunar orbit to Earth at a rate of 622 Megabits-per-second (Mbps), November 2013.

- The Next Generation of OWC / Visible Light Communications demonstrated 10 Mb/s transmission with Polymer Light-Emitting Diodes or OLED.

- On OWC research activities there is a European research project action IC1101 OPTICWISE of the COST Programme (European Cooperation in Science and Technology) funded by the European Science Foundation, allowing the coordination of nationally funded research on a European level. The Action aims to serve as a high-profile consolidated European scientific platform for interdisciplinary optical wireless communication (OWC) research activities. It was launched in November 2011 and will run until November 2015. More than 20 countries represented.

- The consumer and industry OWC technologies adoption is represented by the Li-Fi Consortium, founded in 2011 is a Non-profit organization, devoted to introduce optical wireless technology. Promotes the adoption of Light Fidelity (Li-Fi) products.

- An example of Asian awareness about OWC is the VLCC visible light communication consortium in Japan, established at 2007 in order to realize safe, ubiquitous telecommunication system using visible light through the activities of market research, promotion, and standardization.

- In the USA there are several OWC initiatives, including the "Smart Lighting Engineering Research Center", founded in 2008 by the National Science Foundation (NSF) is a partnership of Rensselaer Polytechnic Institute (lead institution), Boston University and the University of New Mexico. Outreach partners are Howard University, Morgan State University, and Rose-Hulman Institute of Technology.

Free-Space Optical Communication

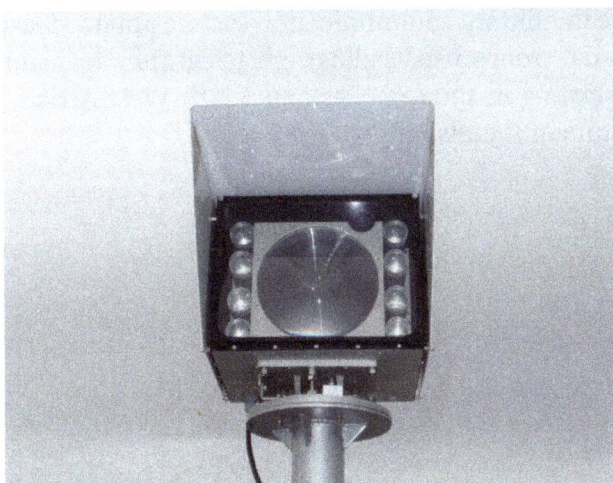

An 8-beam free space optics laser link, rated for 1 Gbit/s. The receptor is the large disc in the middle, the transmitters the smaller ones. At the top right corner is a monocular for assisting the alignment of the two heads.

Free-space optical communication (FSO) is an optical communication technology that uses light propagating in free space to wirelessly transmit data for telecommunications or computer networking. "Free space" means air, outer space, vacuum, or something similar. This contrasts with using solids such as optical fiber cable or an optical transmission line. The technology is useful where the physical connections are impractical due to high costs or other considerations.

History

A photophone receiver and headset, one half of Bell and Tainter's optical telecommunication system of 1880

Optical communications, in various forms, have been used for thousands of years. The Ancient Greeks used a coded alphabetic system of signalling with torches developed by Cleoxenus, Democleitus and Polybius. In the modern era, semaphores and wireless solar telegraphs called heliographs were developed, using coded signals to communicate with their recipients.

In 1880, Alexander Graham Bell and his assistant Charles Sumner Tainter created the photophone, at Bell's newly established Volta Laboratory in Washington, DC. Bell considered it his most important invention. The device allowed for the transmission of sound on a beam of light. On June 3, 1880, Bell conducted the world's first wireless telephone transmission between two buildings, some 213 meters (700 feet) apart.

Its first practical use came in military communication systems many decades later, first for optical telegraphy. German colonial troops used heliograph telegraphy transmitters during the Herero and Namaqua genocide starting in 1904, in German South-West Africa (today's Namibia) as did British, French, US or Ottoman signals.

WW I German Blinkgerät

During the trench warfare of World War I when wire communications were often cut, German signals used three types of optical Morse transmitters called *Blinkgerät*, the intermediate type for distances of up to 4 km (2.5 miles) at daylight and of up to 8 km (5 miles) at night, using red filters for undetected communications. Optical telephone communications were tested at the end of the war, but not introduced at troop level. In addition, special blinkgeräts were used for communication with airplanes, balloons, and tanks, with varying success.

A major technological step was to replace the Morse code by modulating optical waves in speech transmission. Carl Zeiss, Jena developed the *Lichtsprechgerät 80/80* (literal translation: optical speaking device) that the German army used in their World War II anti-aircraft defense units, or in bunkers at the Atlantic Wall.

The invention of lasers in the 1960s, revolutionized free space optics. Military organizations were particularly interested and boosted their development. However the technology lost market momentum when the installation of optical fiber networks for civilian uses was at its peak.

Many simple and inexpensive consumer remote controls use low-speed communication using infrared (IR) light. This is known as consumer IR technologies.

A recently declassified 1987 Pentagon report, reveals free-space lasers have been mounted on Israeli F-15 fighter jets for the purposes of surveillance, missile-tracking, and targeted weaponry.

Usage and Technologies

Free-space point-to-point optical links can be implemented using infrared laser light, although low-data-rate communication over short distances is possible using LEDs. Infrared Data Association (IrDA) technology is a very simple form of free-space optical communications. On the communications side the FSO technology is considered as a part of the Optical Wireless Communications applications. Free-space optics can be used for communications between spacecraft.

Current Market Demands

The demand for a high-speed (10+ Gbit/s) and long range (3–5 km) FSO system is apparent in the market place.

- In 2008, MRV Communications introduced a free-space optics (FSO)-based system with a data rate of 10 Gbit/s initially claiming a distance of 2 km at high availability. This equipment is no longer available; before end-of-life, the product's useful distance was changed down to 350 m.

- In 2013, the company MOSTCOM started to serially produce a new wireless communication system that also had a data rate of 10 Gbit/s as well as an improved range of up to 2.5 km, but to get to 99.99% uptime the designers used an RF hybrid solution, meaning the data rate drops to extremely low levels during atmospheric disturbances (typically down to 10 Mbit/s). In April 2014, the company with Scientific and Technological Centre "Fiord" demonstrated the transmission speed 30 Gbit/s under "laboratory conditions".

- LightPointe offers many similar hybrid solutions to MOSTCOM's offering.

Useful Distances

The reliability of FSO units has always been a problem for commercial telecommunications. Consistently, studies find too many dropped packets and signal errors over small ranges (400 to 500 meters). This is from both independent studies, such as in the Czech republic, as well as formal internal nationwide studies, such as one conducted by MRV FSO staff. Military based studies consistently produce longer estimates for reliability, projecting the maximum range for terrestrial links is of the order of 2 to 3 km (1.2 to 1.9 mi). All studies agree the stability and quality of the link is highly dependent on atmospheric factors such as rain, fog, dust and heat.

Extending the Useful Distance

DARPA ORCA Official Concept Art created c. 2008

The main reason terrestrial communications have been limited to non-commercial telecommunications functions is fog. Fog consistently keeps FSO laser links over 500 meters from achieving a year-round bit error rate of 99.999%. Several entities are continually attempting to overcome these key disadvantages to FSO communications and field a system with a better quality of service. DARPA has sponsored over US$130 million in research towards this effort, with the ORCA and ORCLE programs.

Other non-government groups are fielding tests to evaluate different technologies that some claim have the ability to address key FSO adoption challenges. As of October 2014, none have fielded a working system that addresses the most common atmospheric events.

FSO research from 1998–2006 in the private sector totaled $407.1 million, divided primarily among four start-up companies. All four failed to deliver products that would meet telecommunications quality and distance standards:

- Terabeam received approximately $226 million in funding. AT&T and Lucent backed this attempt. The work ultimately failed, and the company reorganized in 2004.

- AirFiber received $96.1 million in funding, and never solved the weather issue. They sold out to MRV communications in 2003, and MRV sold their FSO units until 2012 when the end-of-life was abruptly announced for the Terescope series.

- LightPointe Communications received $76 million in start-up funds, and eventually reorganized to sell hybrid FSO-RF units to overcome the weather-based challenges.

- The Maxima Corporation published its operating theory in Science (magazine), and received $9 million in funding before permanently shutting down. No known spin-off or purchase followed this effort.

- Wireless Excellence developed and launched CableFree UNITY solutions that combine FSO with Millimeter Wave and Radio technologies to extend distance, capacity and availability, with a goal of making FSO a more useful and practical technology.

One private company published a paper on November 20, 2014, claiming they had achieved commercial reliability (99.999% availability) in extreme fog. There is no indication this product is currently commercially available.

Extraterrestrial

The massive advantages of laser communication in space have multiple space agencies racing to develop a stable space communication platform, with many significant demonstrations and achievements. As of 18 December 2014, *no laser communication system is in use in space.*

Demonstrations in Space:

The first gigabit laser-based communication was achieved by the European Space Agency and called the European Data Relay System (EDRS) on November 28, 2014. The initial images have just been demonstrated, and a working system is expected to be in place in the 2015–2016 time frame.

NASA's OPALS announced a breakthrough in space-to-ground communication December 9, 2014, uploading 175 megabytes in 3.5 seconds. Their system is also able to re-acquire tracking after the signal was lost due to cloud cover.

In January 2013, NASA used lasers to beam an image of the Mona Lisa to the Lunar Reconnaissance Orbiter roughly 390,000 km (240,000 mi) away. To compensate for atmospheric interference, an error correction code algorithm similar to that used in CDs was implemented.

A two-way distance record for communication was set by the Mercury laser altimeter instrument aboard the MESSENGER spacecraft, and was able to communicate across a distance of 24 million km (15 million miles), as the craft neared Earth on a fly-by in May, 2005. The previous record had been set with a one-way detection of laser light from Earth, by the Galileo probe, of 6 million km in 1992. Quote from Laser Communication in Space Demonstrations (EDRS)

XXX

LEDs

In 2001, Twibright Labs released Ronja Metropolis, an open source DIY 10 Mbit/s full duplex LED FSO over 1.4 km In 2004, a Visible Light Communication Consortium was formed in Japan. This was based on work from researchers that used a white LED-based space lighting system for indoor local area network (LAN) communications. These systems present advantages over traditional UHF RF-based systems from improved isolation between systems, the size and cost of receivers/transmitters, RF licensing laws and by combining space lighting and communication into the same system. In January 2009, a task force for visible light communication was formed by the Institute of Electrical and Electronics Engineers working group for wireless personal area network standards known as IEEE 802.15.7. A trial was announced in 2010, in St. Cloud, Minnesota.

RONJA is a free implementation of FSO using high-intensity LEDs.

Amateur radio operators have achieved significantly farther distances using incoherent sources of light from high-intensity LEDs. One reported 173 miles (278 km) in 2007. However, physical limitations of the equipment used limited bandwidths to about 4 kHz. The high sensitivities required of the detector to cover such distances made the internal capacitance of the photodiode used a dominant factor in the high-impedance amplifier which followed it, thus naturally forming a low-pass filter with a cut-off frequency in the 4 kHz range. From the other side use of lasers radiation source allows to reach very high data rates which are comparable to fiber communications.

Projected data rates and future data rate claims vary. A low-cost white LED (GaN-phosphor) which could be used for space lighting can typically be modulated up to 20 MHz. Data rates of over 100 Mbit/s can be easily achieved using efficient modulation schemes and Siemens claimed to have achieved over 500 Mbit/s in 2010. Research published in 2009, used a similar system for traffic control of automated vehicles with LED traffic lights.

In September 2013, pureLiFi, the Edinburgh start-up working on Li-Fi, also demonstrated high speed point-to-point connectivity using any off-the-shelf LED light bulb. In previous work, high bandwidth specialist LEDs have been used to achieve the high data rates. The new system, the Li-1st, maximizes the available optical bandwidth for any LED device, thereby reducing the cost and improving the performance of deploying indoor FSO systems.

Engineering Details

Typically, best use scenarios for this technology are:

- LAN-to-LAN connections on campuses at Fast Ethernet or Gigabit Ethernet speeds

- LAN-to-LAN connections in a city, a metropolitan area network

- To cross a public road or other barriers which the sender and receiver do not own

- Speedy service delivery of high-bandwidth access to optical fiber networks

- Converged Voice-Data-Connection

- Temporary network installation (for events or other purposes)

- Reestablish high-speed connection quickly (disaster recovery)

- As an alternative or upgrade add-on to existing wireless technologies

  - Especially powerful in combination with auto aiming systems, this way you could power moving cars or you can power your laptop while you move or use auto-aiming nodes to create a network with other nodes.

- As a safety add-on for important fiber connections (redundancy)

- For communications between spacecraft, including elements of a satellite constellation

- For inter- and intra-chip communication

The light beam can be very narrow, which makes FSO hard to intercept, improving security. In any case, it is comparatively easy to encrypt any data traveling across the FSO connection for additional security. FSO provides vastly improved electromagnetic interference (EMI) behavior compared to using microwaves.

Technical Advantages

- Ease of deployment

- Can be used to power devices

- License-free long-range operation (in contrast with radio communication)

- High bit rates

- Low bit error rates

- Immunity to electromagnetic interference

- Full duplex operation

- Protocol transparency

- Increased security when working with narrow beam(s)

- No Fresnel zone necessary

- Reference open source implementation

Range Limiting Factors

For terrestrial applications, the principal limiting factors are:

- Fog (10 to ~100 dB/km attenuation)
- Beam dispersion
- Atmospheric absorption
- Rain
- Snow
- Terrestrial scintillation
- Interference from background light sources (including the Sun)
- Shadowing
- Pointing stability in wind
- Pollution / smog

These factors cause an attenuated receiver signal and lead to higher bit error ratio (BER). To overcome these issues, vendors found some solutions, like multi-beam or multi-path architectures, which use more than one sender and more than one receiver. Some state-of-the-art devices also have larger fade margin (extra power, reserved for rain, smog, fog). To keep an eye-safe environment, good FSO systems have a limited laser power density and support laser classes 1 or 1M. Atmospheric and fog attenuation, which are exponential in nature, limit practical range of FSO devices to several kilometres.

Visible Light Communication

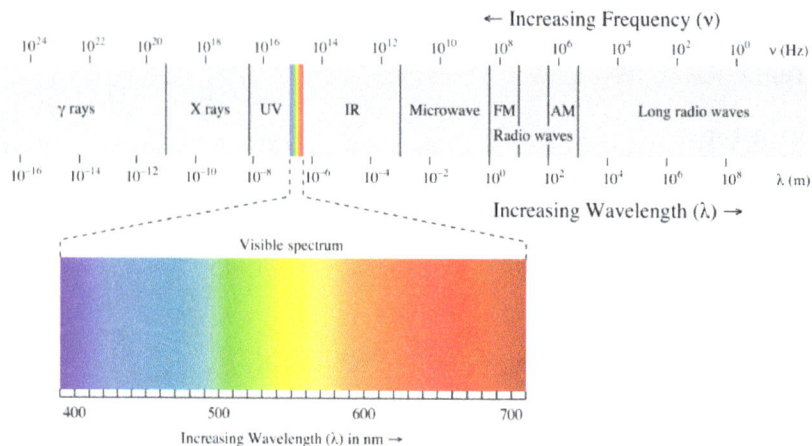

Visible light is only a small portion of the electromagnetic spectrum.

Visible light communication (VLC) is a data communications variant which uses visible light between 400 and 800 THz (780–375 nm). VLC is a subset of optical wireless communications technologies.

The technology uses fluorescent lamps (ordinary lamps, not special communications devices) to transmit signals at 10 kbit/s, or LEDs for up to 500 Mbit/s. Low rate data transmissions at 1 and 2 kilometres (0.6 and 1.2 mi) were demonstrated. RONJA achieves full Ethernet speed (10 Mbit/s) over the same distance thanks to larger optics and more powerful LEDs.

Specially designed electronic devices generally containing a photodiode receive signals from light sources, although in some cases a cell phone camera or a digital camera will be sufficient. The image sensor used in these devices is in fact an array of photodiodes (pixels) and in some applications its use may be preferred over a single photodiode. Such a sensor may provide either multi-channel communication (down to 1 pixel = 1 channel) or a spatial awareness of multiple light sources.

VLC can be used as a communications medium for ubiquitous computing, because light-producing devices (such as indoor/outdoor lamps, TVs, traffic signs, commercial displays and car headlights/taillights) are used everywhere. Using visible light is also less dangerous for high-power applications because humans can perceive it and act to protect their eyes from damage.

History

The history of Visible Light Communications (VLC) dates back to the 1880s in Washington, D.C. when the Scottish-born scientist Alexander Graham Bell invented the photophone, which transmitted speech on modulated sunlight over several hundred meters. This pre-dates the transmission of speech by radio.

More recent work began in 2003 at Nakagawa Laboratory, in Keio University, Japan, using LEDs to transmit data by visible light. A prototype of VLC had been presented by three undergraduate students at Universidad de Buenos Aires in 1995, resorting to the amplitude modulation of a 532 nm laser diode of 5 mW and photodiodes detector. Since then there have been numerous research activities focussed on VLC, notably by Smart Lighting Engineering Centre, Omega Project, COWA, ByteLight, Inc.,D-Light Project, UC-Light Centre, and work at Oxford University.

In 2006, researchers from CICTR at Penn State proposed a combination of power line communication (PLC) and white light LED to provide broadband access for indoor applications. This research suggested that VLC could be deployed as a perfect last-mile solution in the future.

In January 2010 a team of researchers from Siemens and Fraunhofer Institute for Telecommunications, Heinrich Hertz Institute in Berlin demonstrated transmission at 500 Mbit/s with a white LED over a distance of 5 metres (16 ft), and 100 Mbit/s over longer distance using five LEDs.

The VLC standardization process is conducted within IEEE Wireless Personal Area Networks working group (802.15).

In December 2010 St. Cloud, Minnesota, signed a contract with LVX Minnesota and became the first to commercially deploy this technology.

In July 2011 a live demonstration of high-definition video being transmitted from a standard LED lamp was shown at TED Global.

Recently, VLC-based indoor positioning system has become an attractive topic. ABI research forecasts that it could be a key solution to unlocking the $5 billion "indoor location market". Publications have been coming from Nakagawa Laboratory, COWA at Penn State and other researchers around the world.

Another recent application is in the world of toys, thanks to cost-efficient and low-complexity implementation, which only requires one microcontroller and one LED as optical front-end.

VLCs can be used for providing security. They are especially useful in body sensor networks and personal area networks.

Recently Organic LEDs (OLED) have been used as optical transceivers to build up VLC communication links up to 10 Mbit/s.

In October 2014, Axrtek launched a commercial bidirectional RGB LED VLC system called MOMO that transmits down and up at speeds of 300 Mbit/s and with a range of 25 feet.

In May 2015, Philips collaborated with supermarket giant Carrefour to deliver VLC location-based services to shoppers' smartphones in a hypermarket in Lille, France. Indoor positioning systems based on VLC can be used in places such as hospitals, eldercare homes, warehouses, and large, open offices to locate people and control indoor robotic vehicles.

Modulating Retro-Reflector

A modulating retro-reflector (MRR) system combines an optical retro-reflector and an optical modulator to allow optical communications and sometimes other functions such as programmable signage.

Free space optical communication technology has emerged in recent years as an attractive alternative to the conventional Radio Frequency (RF) systems. This emergence is due in large part to the increasing maturity of lasers and compact optical systems that enable exploitation of the inherent advantages (over RF) of the much shorter wavelengths characteristic of optical and near-infrared carriers:

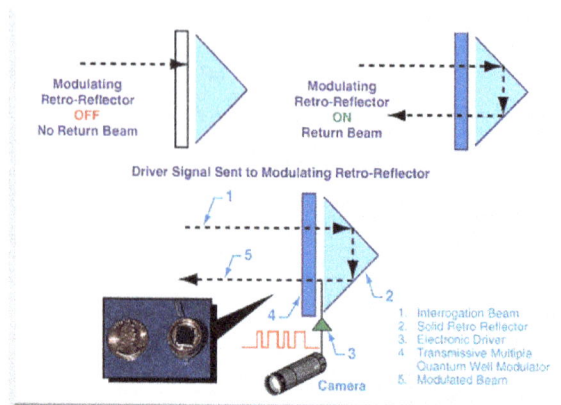

Modulating Retro-reflector Technology Overview.

- Larger bandwidth

- Low probability of intercept

- Immunity from interference or jamming

- Frequency spectrum allocation issue relief

- Smaller, lighter, lower power

Technology

An MRR couples or combines an optical retroreflector with a modulator to reflect modulated optical signals directly back to an optical receiver or transceiver, allowing the MRR to function as an optical communications device without emitting its own optical power. This can allow the MRR to communicate optically over long distances without needing substantial on-board power supplies. The function of the retroreflection component is to direct the reflection back to or near to the source of the light. The modulation component changes the intensity of the reflection. The idea applies to optical communication in a broad sense including not only laser-based data communications but also human observers and road signs. A number of technologies have been proposed, investigated, and developed for the modulation component, including actuated micromirrors, frustrated total internal reflection, electro-optic modulators (EOMs), piezo-actuated deflectors, multiple quantum well (MQW) devices, and liquid crystal modulators, though any one of numerous known optical modulation technologies could be used in theory. These approaches have many advantages and disadvantages relative to one another with respect to such features as power use, speed, modulation range, compactness, retroreflection divergence, cost, and many others.

In a typical optical communications arrangement, the MRR with its related electronics is mounted on a convenient platform and connected to a host computer which has the data that are to be transferred. A remotely located optical transmitter/receiver system usually consisting of a laser, telescope, and detector provides an optical signal to the modulating retro-reflector. The incident light from the transmitter system is both modulated by the MRR and reflected directly toward the transmitter (via the retroreflection property).

One modulating retro-reflector at the Naval Research Laboratory (NRL) in the United States uses a semiconductor based MQW shutter capable of modulation rates up to 10 Mbit/s, depending on link characteristics.

The optical nature of the technology provides communications that are not susceptible to issues related to electromagnetic frequency allocation. The multiple quantum well modulating retro-reflector has the added advantages of being compact, lightweight, and requires very little power. The small-array MRR provides up to an order of magnitude in consumed power savings over an equivalent RF system. However, MQW modulators also have relatively small modulation ranges compared to other technologies.

The concept of a modulating retro-reflector is not new, dating back to the 1940s. Various demonstrations of such devices have been built over the years, though the demonstration of the first MQW MRR in 1993 was notable in achieving significant data rates. However, MRRs are still not widely used, and most research and development in that area is confined to rather exploratory military applications, as free-space optical communications in general tends to be a rather specialized niche technology.

Qualities often considered desirable in MRRs (obviously depending on the application) include a high switching speed, low power consumption, large area, wide field-of-view, and high optical quality. It should also function at certain wavelengths where appropriate laser sources are available, be radiation-tolerant (for non-terrestrial applications), and be rugged. Mechanical shutters and ferroelectric liquid crystal (FLC) devices, for example, are too slow, heavy, or are not robust enough for many applications. Some modulating retro-reflector systems are desired to operate at data rates of megabits per second (Mbit/s) and higher and over large temperature ranges characteristic of installation out-of-doors and in space.

Multiple Quantum Well Modulators

Semiconductor MQW modulators are one of the few technologies that meet all the requirements need for United States Navy applications, and consequently the Naval Research Laboratory is particularly active in developing and promoting that approach. When used as a shutter, MQW technology offers many advantages: it is robust solid state, operates at low voltages (less than 20 mV) and low power (tens of milliWatts), and is capable of very high switching speeds. MQW modulators have been run at Gbit/s data rates in fiber optic applications.

Figure 2. Absorbance vs. Frequency When a moderate (~15V) voltage is placed across the shutter in reverse bias, the absorption feature changes, shifting to longer wavelengths and dropping in magnitude. Thus, the transmission of the device near this absorption feature changes dramatically. Figure 2 shows absorbance data for an InGaAs MQW modulator designed and grown at NRL for use in a modulating retro-reflector system. The figure illustrates how the application of a moderate voltage shifts the transmittance. Hence, a signal can be encoded in an On-Off-Keying format onto the carrier interrogation beam.

This modulator consists of 75 periods of InGaAs wells surrounded by AlGaAs barriers. The device is grown on an n-type GaAs wafer and is capped by a p-type contact layer, thus forming a P-I-N diode. This device is a transmissive modulator designed to work at a wavelength of 980 nm, compatible with many good laser diode sources. These materials have very good performance operating in reflection architectures. Choice of modulator type and configuration architecture is application-dependent.

Once grown, the wafer is fabricated into discrete devices using a multi-step photolithography process consisting of etching and metallization steps. The NRL experimental devices have a 5 mm aperture, though larger devices are possible and are being designed and developed. It is important to point out that while MQW modulators have been used in many applications to date, modulators of such a large size are uncommon and require special fabrication techniques. Figure 3 shows a block diagram and photo of a wide aperture MQW shutter designed, grown, and fabricated at NRL.

MQW modulators are inherently quiet devices, accurately reproducing the applied voltage as a modulated waveform. An important parameter is contrast ratio, defined as Imax/Imin. This parameter affects the overall signal-to-noise ratio. Its magnitude depends on the drive voltage applied to the device and the wavelength of the interrogating laser relative to the exciton peak. The contrast ratio increases as the voltage goes up until a saturation value is reached. Typically, the modulators fabricated at NRL have had contrast ratios between 1.75:1 to 4:1 for applied voltages between 10 V and 25 V, depending on the structure.

There are three important considerations in the manufacture and fabrication of a given device: inherent maximum modulation rate vs. aperture size; electrical power consumption vs. aperture size; and yield.

Inherent Maximum Modulation Rate vs. Aperture Size

The fundamental limit in the switching speed of the modulator is the resistance-capacitance limit. A key trade is area of the modulator vs. area of the clear aperture. If the modulator area is small, the capacitance is small, hence the modulation rate can be faster. However, for longer application ranges on the order of several hundred meters, larger apertures are needed to close the link. For a given modulator, the speed of the shutter scales inversely as the square of the modulator diameter.

Electrical Power Consumption vs. Aperture Size

When the drive voltage waveform is optimized, the electrical power consumption of a MQW modulating retro-reflector varies as:

$Dmod^4 * V^2 B^2 R_s$

Where Dmod is the diameter of the modulator, V is the voltage applied to the modulator (fixed by the required optical contrast ratio), B is the maximum data rate of the device, and RS is the sheet resistance of the device. Thus a large power penalty may be paid for increasing the diameter of the MQW shutter.

Yield

MQW devices must be operated at high reverse bias fields to achieve good contrast ratios. In perfect quantum well material this is not a problem, but the presence of a defect in the semiconductor crystal can cause the device to break down at voltages below those necessary for operation. Specifically, a defect will cause an electrical short that prevents development of the necessary electrical field across the intrinsic region of the PIN diode. The larger the device the higher the probability of such a defect. Thus, If a defect occurs in the manufacture of a large monolithic device, the whole shutter is lost.

To address these issues, NRL has designed and fabricated segmented devices as well as monolithic modulators. That is, a given modulator might be "pixellated" into several segments, each driven with the same signal. This technique means that speed can be achieved as well as larger apertures. The "pixellization" inherently reduces the sheet resistance of the device, decreasing the resistance-capacitance time and reducing electrical power consumption. For example, a one centimeter monolithic device might require 400 mW to support a one Mbit/s link. A similar nine segmented device would require 45 mW to support the same link with the same overall effective aperture. A transmissive device with nine "pixels" with an overall diameter of 0.5 cm was shown to support over 10 Mbit/s.

This fabrication technique allows for higher speeds, larger apertures, and increased yield. If a single "pixel" is lost due to defects but is one of nine or sixteen, the contrast ratio necessary to provide the requisite signal-to-noise to close a link is still high. There are considerations that make fabrication of a segmented device more complicated, including bond wire management on the device, driving multiple segments, and temperature stabilization.

An additional important characteristic of the modulator is its optical wavefront quality. If the modulator causes aberrations in the beam, the returned optical signal will be attenuated and in-sufficient light may be present to close the link.

References

- Chapter 2: Semaphore Signalling ISBN 978-0-86341-327-8 Communications: an international history of the formative years R. W. Burns, 2004

- Mary Kay Carson (2007). Alexander Graham Bell: Giving Voice To The World. Sterling Biographies. New York: Sterling Publishing. pp. 76–78. ISBN 978-1-4027-3230-0.

- "Where are the discounts? Carrefour's LED supermarket lighting from Philips will guide you" (Press release). Philips. May 21, 2015.

- M. Uysal and H. Nouri, "Optical Wireless Communications – An Emerging Technology", 16th International Conference on Transparent Optical Networks (ICTON), Graz, Austria, July 2014

- A end-of-life notice was posted suddenly and briefly on the MRV Terescope product page in 2011. All references to the Terescope have been completely removed from MRV's official page as of October 27, 2014.

- "10 Gbps Through The Air". Arto Link. Retrieved October 27, 2014. new Artolink wireless communication system with the highest capacity: 10 Gbps, full duplex [..] Artolink M1-10GE model

- Eric Korevaar, Isaac I. Kim and Bruce McArthur (2001). "Atmospheric Propagation Characteristics of Highest Importance to Commercial Free Space Optics" (PDF). Optical Wireless Communications IV, SPIE Vol. 4530 p. 84. Retrieved October 27, 2014.

- Bruce V. Bigelow (June 16, 2006). "Zapped of its potential, Rooftop laser startups falter, but debate on high-speed data technology remains". Retrieved October 26, 2014.

- Nancy Gohring (March 27, 2000). "TeraBeam's Light Speed; Telephony, Vol. 238 Issue 13, p16". Retrieved October 27, 2014.

- Fred Dawson (May 1, 2000). "TeraBeam, Lucent Extend Bandwidth Limits, Multichannel News, Vol 21 Issue 18 Pg 160". Retrieved October 27, 2014.

Telecommunications Network: An Integrated Study

The collection of terminal nodes, which help in the connection of different parties in telecommunications is known as telecommunications network. Some of the aspects of telecommunications discussed within this text are computer network, public switched telephone network, packet switching, radio network and television network.

Telecommunications Network

A telecommunications network is a collection of terminal nodes, links are connected so as to enable telecommunication between the terminals. The transmission links connect the nodes together. The nodes use circuit switching, message switching or packet switching to pass the signal through the correct links and nodes to reach the correct destination terminal.

Each terminal in the network usually has a unique address so messages or connections can be routed to the correct recipients. The collection of addresses in the network is called the address space.

Examples of telecommunications networks are:

- computer networks
- the Internet
- the telephone network
- the global Telex network
- the aeronautical ACARS network

Benefits of Telecommunications and Networking

Telecommunications can greatly increase and expand resources to all types of people. For example, businesses need a greater telecommunications network if they plan to expand their company. With Internet, computer, and telephone networks, businesses can allocate their resources efficiently. These core types of networks will be discussed below:

Computer Network: A computer network consists of computers and devices connected to one another. Information can be transferred from one device to the next. For example, an office filled with computers can share files together on each separate device. Computer networks can range from a local network area to a wide area network. The difference between the types of networks is the size.

These types of computer networks work at certain speeds, also known as broadband. The Internet network connects computers worldwide.

Internet Network: Access to the network allows users to use many resources. Over time the Internet network will replace books. This will enable users to discover information almost instantly and apply concepts to different situations. The Internet can be used for recreational, governmental, educational, and other purposes. Businesses in particular use the Internet network for research or to service customers and clients.

Telephone Network: The telephone network connects people to one another. This network can be used in a variety of ways. Many businesses use the telephone network to route calls and/or service their customers. Some businesses use a telephone network on a greater scale through a private branch exchange. It is a system where a specific business focuses on routing and servicing calls for another business. Majority of the time, the telephone network is used around the world for recreational purposes.

Network Structure

In general, every telecommunications network conceptually consists of three parts, or planes (so called because they can be thought of as being, and often are, separate overlay networks):

- The data plane (also *user plane*, *bearer plane* or *forwarding plane*) carries the network's users' traffic, the actual payload.

- The control plane carries control information (also known as signalling).

- The management plane carries the operations and administration traffic required for network management. The management plane is sometimes considered a part of the control plane.

Example: The TCP/IP Data Network

The data network is used extensively throughout the world to connect individuals and organizations. Data networks can be connected to allow users seamless access to resources that are hosted outside of the particular provider they are connected to. The Internet is the best example of many data networks from different organizations all operating under a single address space.

Terminals attached to TCP/IP networks are addressed using IP addresses. There are different types of IP address, but the most common is IP Version 4. Each unique address consists of 4 integers between 0 and 255, usually separated by dots when written down, e.g. 82.131.34.56.

TCP/IP are the fundamental protocols that provide the control and routing of messages across the data network. There are many different network structures that TCP/IP can be used across to efficiently route messages, for example:

- wide area networks (WAN)

- metropolitan area networks (MAN)

- local area networks (LAN)

- Internet area networks (IAN)

- campus area networks (CAN)

- virtual private networks (VPN)

There are three features that differentiate MANs from LANs or WANs:

1. The area of the network size is between LANs and WANs. The MAN will have a physical area between 5 and 50 km in diameter.

2. MANs do not generally belong to a single organization. The equipment that interconnects the network, the links, and the MAN itself are often owned by an association or a network provider that provides or leases the service to others.

3. A MAN is a means for sharing resources at high speeds within the network. It often provide connections to WAN networks for access to resources outside the scope of the MAN.

Optical Transport Network (OTN)

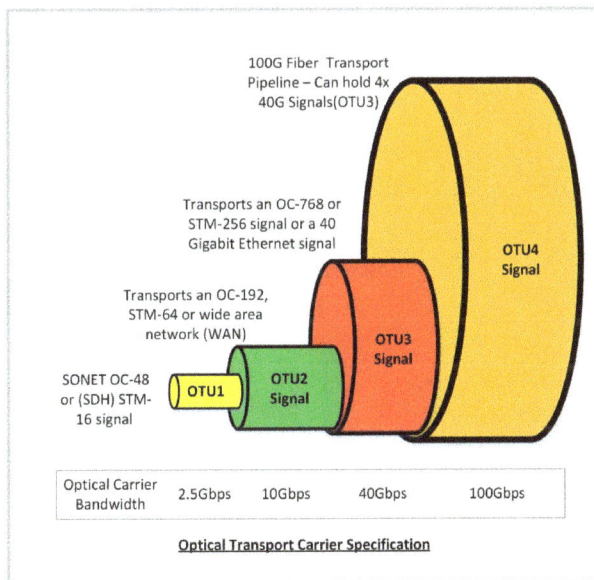

Optical Transport Network (OTN) is a large complex network of server hubs at different locations on ground, connected by Optical fiber cable or optical network carrier, to transport data across different nodes. The server hubs are also known as head-ends, nodes or simply, sites. OTNs are the backbone of Internet Service Providers and are often daisy chained and cross connected to provide network redundancy. Such a setup facilitates uninterrupted services and fail-over capabilities during maintenance windows, equipment failure or in case of accidents.

The devices used to transport data are known as network transport equipment. Some of the widely used equipment are manufactured by

- Alcatel Lucent - AL7510, AL7750

- Nortel Networks Corp. (acquired by Ciena Corp.) - Optera Metro series - OM4500, OM6500

- Fujitsu Ltd. - FlashWave series FW4500, FW7500, FW9500

The capacity of a network is mainly dependent on the type of signalling scheme employed on transmitting and receiving end. In the earlier days, a single wavelength light beam was used to transmit data, which limited the bandwidth to the maximum operating frequency of the transmitting and receiving end equipment. With the application of wavelength division multiplexing (WDM), the bandwidth of OTN has risen up to 100Gbit/s (OTU4 Signal), by emitting light beams of different wavelengths. Lately, AT&T, Verizon, and Rogers Communication have been able to employ these 100G "pipes" in their metro network. Large field areas are mostly serviced by 40G pipes (OC192/STM-64).

A 40G pipe can carry 40 different channels as a result of Dense Wave Division Multiplexing (DWDM) transmission. Each node in the network is able to access different channels, but is mostly tuned to a few channels. The data from a channel can be dropped to the node or new data can be added to the node using Re-configurable Optic Add Drop Mux (ROADM) that uses Wavelength Selective Switching (WSS) to extract and infuse a configured frequency. This eliminates the need to convert all the channels to electric signals, extract the required channels, and convert the rest back to optical into the OTN. Thus ROADM systems are fast, less expensive and can be configured to access any channel in the OTN pipe.

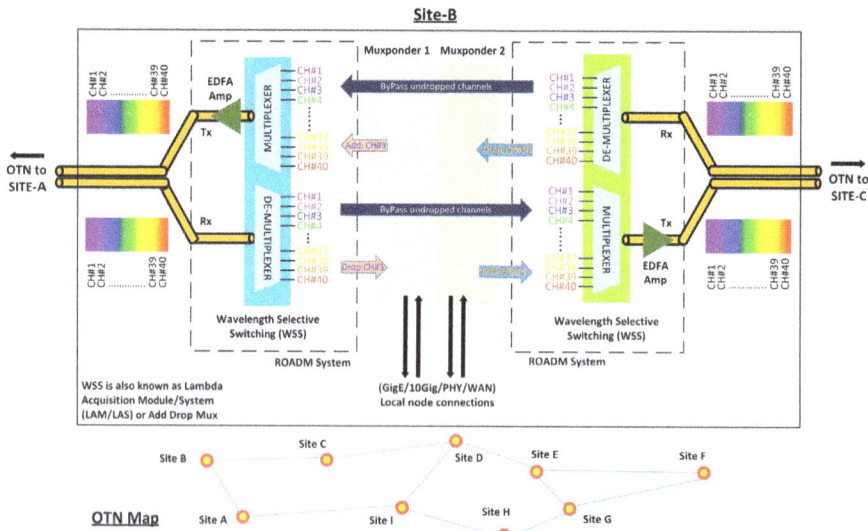

The extracted channels at a site are connected to local devices through muxponder or transponder cards that can split or combine 40G channels to 4x 10G channels or 8x 2.5G channels.

Computer Network

A computer network or data network is a telecommunications network which allows computers to exchange data. In computer networks, networked computing devices exchange data with each other using a data link. The connections between nodes are established using either cable media or wireless media. The best-known computer network is the Internet.

Network computer devices that originate, route and terminate the data are called network nodes. Nodes can include hosts such as personal computers, phones, servers as well as networking hardware. Two such devices can be said to be networked together when one device is able to exchange information with the other device, whether or not they have a direct connection to each other.

Computer networks differ in the transmission medium used to carry their signals, communications protocols to organize network traffic, the network's size, topology and organizational intent.

Computer networks support an enormous number of applications and services such as access to the World Wide Web, digital video, digital audio, shared use of application and storage servers, printers, and fax machines, and use of email and instant messaging applications as well as many others. In most cases, application-specific communications protocols are layered (i.e. carried as payload) over other more general communications protocols.

History

The chronology of significant computer-network developments includes:

- In the late 1950s, early networks of computers included the military radar system Semi-Automatic Ground Environment (SAGE).

- In 1959, Anatolii Ivanovich Kitov proposed to the Central Committee of the Communist Party of the Soviet Union a detailed plan for the re-organisation of the control of the Soviet armed forces and of the Soviet economy on the basis of a network of computing centres.

- In 1960, the commercial airline reservation system semi-automatic business research environment (SABRE) went online with two connected mainframes.

- In 1962, J.C.R. Licklider developed a working group he called the "Intergalactic Computer Network", a precursor to the ARPANET, at the Advanced Research Projects Agency (ARPA).

- In 1964, researchers at Dartmouth College developed the Dartmouth Time Sharing System for distributed users of large computer systems. The same year, at Massachusetts Institute of Technology, a research group supported by General Electric and Bell Labs used a computer to route and manage telephone connections.

- Throughout the 1960s, Leonard Kleinrock, Paul Baran, and Donald Davies independently developed network systems that used packets to transfer information between computers over a network.

- In 1965, Thomas Marill and Lawrence G. Roberts created the first wide area network (WAN). This was an immediate precursor to the ARPANET, of which Roberts became program manager.

- Also in 1965, Western Electric introduced the first widely used telephone switch that implemented true computer control.

- In 1969, the University of California at Los Angeles, the Stanford Research Institute, the University of California at Santa Barbara, and the University of Utah became connected as the beginning of the ARPANET network using 50 kbit/s circuits.

- In 1972, commercial services using X.25 were deployed, and later used as an underlying infrastructure for expanding TCP/IP networks.

- In 1973, Robert Metcalfe wrote a formal memo at Xerox PARC describing Ethernet, a networking system that was based on the Aloha network, developed in the 1960s by Norman Abramson and colleagues at the University of Hawaii. In July 1976, Robert Metcalfe and David Boggs published their paper "Ethernet: Distributed Packet Switching for Local Computer Networks" and collaborated on several patents received in 1977 and 1978. In 1979, Robert Metcalfe pursued making Ethernet an open standard.

- In 1976, John Murphy of Datapoint Corporation created ARCNET, a token-passing network first used to share storage devices.

- In 1995, the transmission speed capacity for Ethernet increased from 10 Mbit/s to 100 Mbit/s. By 1998, Ethernet supported transmission speeds of a Gigabit. Subsequently, higher speeds of up to 100 Gbit/s were added (as of 2016). The ability of Ethernet to scale easily (such as quickly adapting to support new fiber optic cable speeds) is a contributing factor to its continued use.

Properties

Computer networking may be considered a branch of electrical engineering, telecommunications, computer science, information technology or computer engineering, since it relies upon the theoretical and practical application of the related disciplines.

A computer network facilitates interpersonal communications allowing users to communicate efficiently and easily via various means: email, instant messaging, chat rooms, telephone, video telephone calls, and video conferencing. Providing access to information on shared storage devices is an important feature of many networks. A network allows sharing of files, data, and other types of information giving authorized users the ability to access information stored on other computers on the network. A network allows sharing of network and computing resources. Users may access and use resources provided by devices on the network, such as printing a document on a shared network printer. Distributed computing uses computing resources across a network to accomplish tasks. A computer network may be used by computer crackers to deploy computer viruses or computer worms on devices connected to the network, or to prevent these devices from accessing the network via a denial of service attack.

Network Packet

Computer communication links that do not support packets, such as traditional point-to-point telecommunication links, simply transmit data as a bit stream. However, most information in computer networks is carried in *packets*. A network packet is a formatted unit of data (a list of bits or bytes, usually a few tens of bytes to a few kilobytes long) carried by a packet-switched network.

In packet networks, the data is formatted into packets that are sent through the network to their destination. Once the packets arrive they are reassembled into their original message. With packets, the bandwidth of the transmission medium can be better shared among users than if the network were circuit switched. When one user is not sending packets, the link can be filled with packets from other users, and so the cost can be shared, with relatively little interference, provided the link isn't overused.

Packets consist of two kinds of data: control information, and user data (payload). The control information provides data the network needs to deliver the user data, for example: source and destination network addresses, error detection codes, and sequencing information. Typically, control information is found in packet headers and trailers, with payload data in between.

Often the route a packet needs to take through a network is not immediately available. In that case the packet is queued and waits until a link is free.

Network Topology

The physical layout of a network is usually less important than the topology that connects network nodes. Most diagrams that describe a physical network are therefore topological, rather than geographic. The symbols on these diagrams usually denote network links and network nodes.

Network Links

The transmission media (often referred to in the literature as the *physical media*) used to link devices to form a computer network include electrical cable (Ethernet, HomePNA, power line communication, G.hn), optical fiber (fiber-optic communication), and radio waves (wireless networking). In the OSI model, these are defined at layers 1 and 2 — the physical layer and the data link layer.

A widely adopted *family* of transmission media used in local area network (LAN) technology is collectively known as Ethernet. The media and protocol standards that enable communication between networked devices over Ethernet are defined by IEEE 802.3. Ethernet transmits data over both copper and fiber cables. Wireless LAN standards (e.g. those defined by IEEE 802.11) use radio waves, or others use infrared signals as a transmission medium. Power line communication uses a building's power cabling to transmit data.

Wired Technologies

The orders of the following wired technologies are, roughly, from slowest to fastest transmission speed.

- *Coaxial cable* is widely used for cable television systems, office buildings, and other work-sites for local area networks. The cables consist of copper or aluminum wire surrounded by an insulating layer (typically a flexible material with a high dielectric constant), which itself is surrounded by a conductive layer. The insulation helps minimize interference and distortion. Transmission speed ranges from 200 million bits per second to more than 500 million bits per second.

- ITU-T G.hn technology uses existing home wiring (coaxial cable, phone lines and power lines) to create a high-speed (up to 1 Gigabit/s) local area network

- *Twisted pair wire* is the most widely used medium for all telecommunication. Twisted-pair cabling consist of copper wires that are twisted into pairs. Ordinary telephone wires consist of two insulated copper wires twisted into pairs. Computer network cabling (wired Ethernet as defined by IEEE 802.3) consists of 4 pairs of copper cabling that can be utilized for both voice and data transmission. The use of two wires twisted together helps to reduce crosstalk and electromagnetic induction. The transmission speed ranges from 2 million bits per second to 10 billion bits per second. Twisted pair cabling comes in two forms: unshielded twisted pair (UTP) and shielded twisted-pair (STP). Each form comes in several category ratings, designed for use in various scenarios.

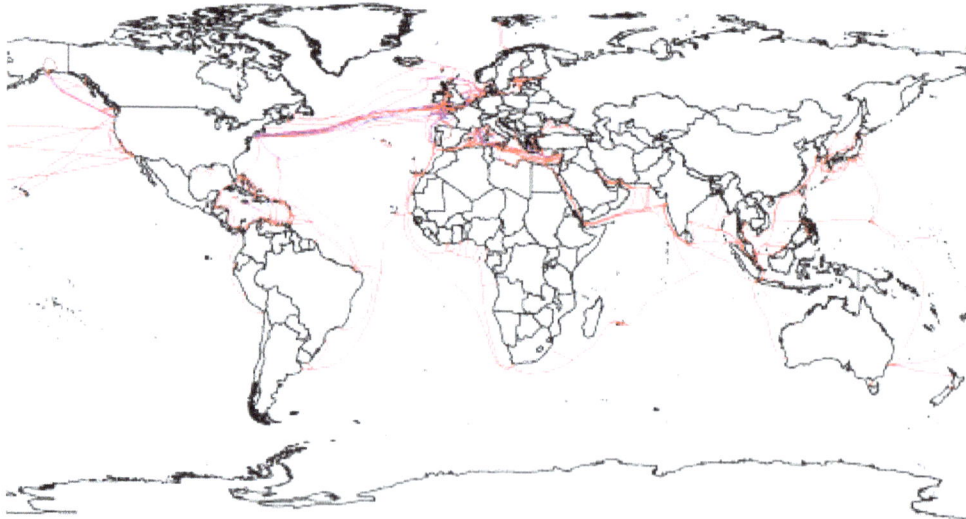

2007 map showing submarine optical fiber telecommunication cables around the world.

- An *optical fiber* is a glass fiber. It carries pulses of light that represent data. Some advantages of optical fibers over metal wires are very low transmission loss and immunity from electrical interference. Optical fibers can simultaneously carry multiple wavelengths of light, which greatly increases the rate that data can be sent, and helps enable data rates of up to trillions of bits per second. Optic fibers can be used for long runs of cable carrying very high data rates, and are used for undersea cables to interconnect continents.

Price is a main factor distinguishing wired- and wireless-technology options in a business. Wireless options command a price premium that can make purchasing wired computers, printers and other devices a financial benefit. Before making the decision to purchase hard-wired technology products, a review of the restrictions and limitations of the selections is necessary. Business and employee needs may override any cost considerations.

Wireless Technologies

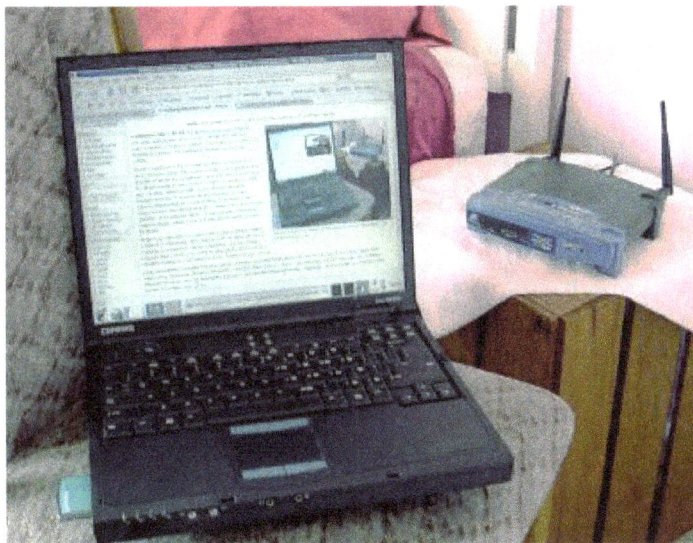

Computers are very often connected to networks using wireless links

- *Terrestrial microwave* – Terrestrial microwave communication uses Earth-based transmitters and receivers resembling satellite dishes. Terrestrial microwaves are in the low-gigahertz range, which limits all communications to line-of-sight. Relay stations are spaced approximately 48 km (30 mi) apart.

- *Communications satellites* – Satellites communicate via microwave radio waves, which are not deflected by the Earth›s atmosphere. The satellites are stationed in space, typically in geosynchronous orbit 35,400 km (22,000 mi) above the equator. These Earth-orbiting systems are capable of receiving and relaying voice, data, and TV signals.

- *Cellular and PCS systems* use several radio communications technologies. The systems divide the region covered into multiple geographic areas. Each area has a low-power transmitter or radio relay antenna device to relay calls from one area to the next area.

- *Radio and spread spectrum technologies* – Wireless local area networks use a high-frequency radio technology similar to digital cellular and a low-frequency radio technology. Wireless LANs use spread spectrum technology to enable communication between multiple devices in a limited area. IEEE 802.11 defines a common flavor of open-standards wireless radio-wave technology known as Wifi.

- *Free-space optical communication* uses visible or invisible light for communications. In most cases, line-of-sight propagation is used, which limits the physical positioning of communicating devices.

Exotic Technologies

There have been various attempts at transporting data over exotic media:

- IP over Avian Carriers was a humorous April fool's Request for Comments, issued as RFC 1149. It was implemented in real life in 2001.

- Extending the Internet to interplanetary dimensions via radio waves, the Interplanetary Internet.

Both cases have a large round-trip delay time, which gives slow two-way communication, but doesn't prevent sending large amounts of information.

Network Nodes

Apart from any physical transmission medium there may be, networks comprise additional basic system building blocks, such as network interface controller (NICs), repeaters, hubs, bridges, switches, routers, modems, and firewalls.

Network Interfaces

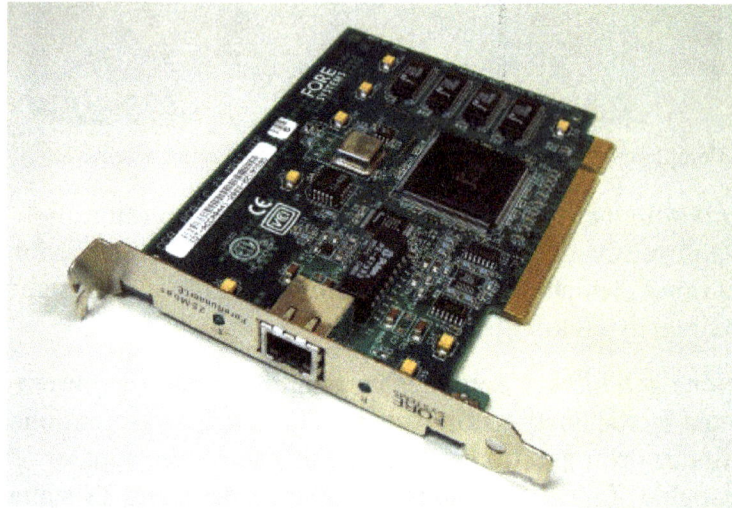

An ATM network interface in the form of an accessory card. A lot of network interfaces are built-in.

A network interface controller (NIC) is computer hardware that provides a computer with the ability to access the transmission media, and has the ability to process low-level network information. For example, the NIC may have a connector for accepting a cable, or an aerial for wireless transmission and reception, and the associated circuitry.

The NIC responds to traffic addressed to a network address for either the NIC or the computer as a whole.

In Ethernet networks, each network interface controller has a unique Media Access Control (MAC) address—usually stored in the controller's permanent memory. To avoid address conflicts between network devices, the Institute of Electrical and Electronics Engineers (IEEE) maintains and administers MAC address uniqueness. The size of an Ethernet MAC address is six octets. The three most significant octets are reserved to identify NIC manufacturers. These manufacturers, using only their assigned prefixes, uniquely assign the three least-significant octets of every Ethernet interface they produce.

Repeaters and Hubs

A repeater is an electronic device that receives a network signal, cleans it of unnecessary noise and regenerates it. The signal is retransmitted at a higher power level, or to the other side of an obstruction, so that the signal can cover longer distances without degradation. In most twisted pair Ethernet configurations, repeaters are required for cable that runs longer than 100 meters. With fiber optics, repeaters can be tens or even hundreds of kilometers apart.

A repeater with multiple ports is known as a hub. Repeaters work on the physical layer of the OSI model. Repeaters require a small amount of time to regenerate the signal. This can cause a propagation delay that affects network performance. As a result, many network architectures limit the number of repeaters that can be used in a row, e.g., the Ethernet 5-4-3 rule.

Hubs have been mostly obsoleted by modern switches; but repeaters are used for long distance links, notably undersea cabling.

Bridges

A network bridge connects and filters traffic between two network segments at the data link layer (layer 2) of the OSI model to form a single network. This breaks the network's collision domain but maintains a unified broadcast domain. Network segmentation breaks down a large, congested network into an aggregation of smaller, more efficient networks.

Bridges come in three basic types:

- Local bridges: Directly connect LANs

- Remote bridges: Can be used to create a wide area network (WAN) link between LANs. Remote bridges, where the connecting link is slower than the end networks, largely have been replaced with routers.

- Wireless bridges: Can be used to join LANs or connect remote devices to LANs.

Switches

A network switch is a device that forwards and filters OSI layer 2 datagrams (frames) between ports based on the destination MAC address in each frame. A switch is distinct from a hub in that it only forwards the frames to the physical ports involved in the communication rather than all ports connected. It can be thought of as a multi-port bridge. It learns to associate physical ports to MAC addresses by examining the source addresses of received frames. If an unknown destination is targeted, the switch broadcasts to all ports but the source. Switches normally have numerous ports, facilitating a star topology for devices, and cascading additional switches.

Multi-layer switches are capable of routing based on layer 3 addressing or additional logical levels. The term *switch* is often used loosely to include devices such as routers and bridges, as well as devices that may distribute traffic based on load or based on application content (e.g., a Web URL identifier).

Routers

A router is an internetworking device that forwards packets between networks by processing the routing information included in the packet or datagram (Internet protocol information from layer 3). The routing information is often processed in conjunction with the routing table (or forwarding table). A router uses its routing table to determine where to forward packets. A destination in a routing table can include a "null" interface, also known as the "black hole" interface because data can go into it, however, no further processing is done for said data, i.e. the packets are dropped.

A typical home or small office router showing the ADSL telephone line and Ethernet network cable connections

Modems

Modems (MOdulator-DEModulator) are used to connect network nodes via wire not originally designed for digital network traffic, or for wireless. To do this one or more carrier signals are modulated by the digital signal to produce an analog signal that can be tailored to give the required properties for transmission. Modems are commonly used for telephone lines, using a Digital Subscriber Line technology.

Firewalls

A firewall is a network device for controlling network security and access rules. Firewalls are typically configured to reject access requests from unrecognized sources while allowing actions from recognized ones. The vital role firewalls play in network security grows in parallel with the constant increase in cyber attacks.

Network Structure

Network topology is the layout or organizational hierarchy of interconnected nodes of a computer network. Different network topologies can affect throughput, but reliability is often more critical. With many technologies, such as bus networks, a single failure can cause the network to fail entirely. In general the more interconnections there are, the more robust the network is; but the more expensive it is to install.

Common Layouts

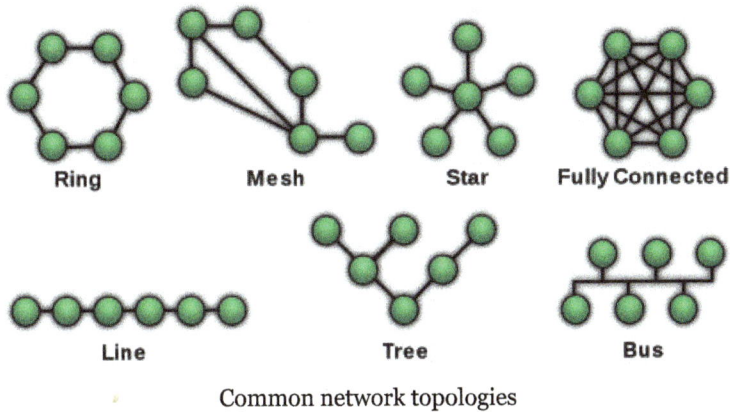

Common network topologies

Common layouts are:

- A bus network: all nodes are connected to a common medium along this medium. This was the layout used in the original Ethernet, called 10BASE5 and 10BASE2.

- A star network: all nodes are connected to a special central node. This is the typical layout found in a Wireless LAN, where each wireless client connects to the central Wireless access point.

- A ring network: each node is connected to its left and right neighbour node, such that all nodes are connected and that each node can reach each other node by traversing nodes left- or rightwards. The Fiber Distributed Data Interface (FDDI) made use of such a topology.

- A mesh network: each node is connected to an arbitrary number of neighbours in such a way that there is at least one traversal from any node to any other.

- A fully connected network: each node is connected to every other node in the network.

- A tree network: nodes are arranged hierarchically.

Note that the physical layout of the nodes in a network may not necessarily reflect the network topology. As an example, with FDDI, the network topology is a ring (actually two counter-rotating rings), but the physical topology is often a star, because all neighboring connections can be routed via a central physical location.

Overlay Network

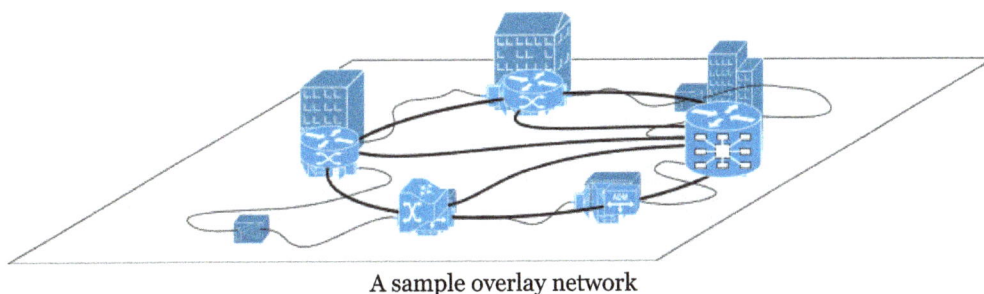

A sample overlay network

An overlay network is a virtual computer network that is built on top of another network. Nodes in the overlay network are connected by virtual or logical links. Each link corresponds to a path, perhaps through many physical links, in the underlying network. The topology of the overlay network may (and often does) differ from that of the underlying one. For example, many peer-to-peer networks are overlay networks. They are organized as nodes of a virtual system of links that run on top of the Internet.

Overlay networks have been around since the invention of networking when computer systems were connected over telephone lines using modems, before any data network existed.

The most striking example of an overlay network is the Internet itself. The Internet itself was initially built as an overlay on the telephone network. Even today, each Internet node can communicate with virtually any other through an underlying mesh of sub-networks of wildly different topologies and technologies. Address resolution and routing are the means that allow mapping of a fully connected IP overlay network to its underlying network.

Another example of an overlay network is a distributed hash table, which maps keys to nodes in the network. In this case, the underlying network is an IP network, and the overlay network is a table (actually a map) indexed by keys.

Overlay networks have also been proposed as a way to improve Internet routing, such as through quality of service guarantees to achieve higher-quality streaming media. Previous proposals such as IntServ, DiffServ, and IP Multicast have not seen wide acceptance largely because they require modification of all routers in the network. On the other hand, an overlay network can be incrementally deployed on end-hosts running the overlay protocol software, without cooperation from Internet service providers. The overlay network has no control over how packets are routed in the underlying network between two overlay nodes, but it can control, for example, the sequence of overlay nodes that a message traverses before it reaches its destination.

For example, Akamai Technologies manages an overlay network that provides reliable, efficient content delivery (a kind of multicast). Academic research includes end system multicast, resilient routing and quality of service studies, among others.

Communications Protocols

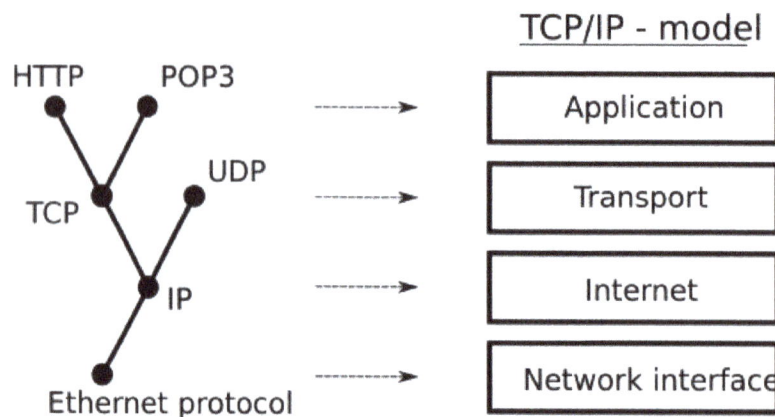

The TCP/IP model or Internet layering scheme and its relation to common protocols often layered on top of it.

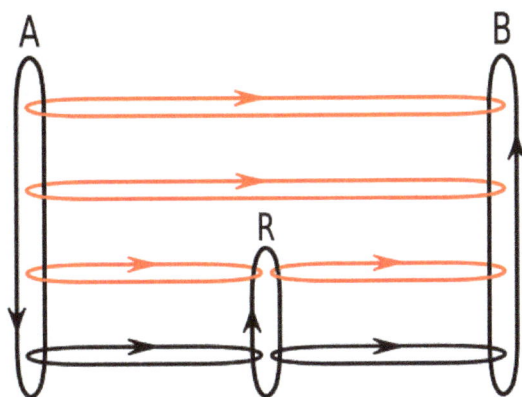

Message flows (A-B) in the presence of a router (R), red flows are effective communication paths, black paths are the actual paths.

A communications protocol is a set of rules for exchanging information over network links. In a protocol stack, each protocol leverages the services of the protocol below it. An important example of a protocol stack is HTTP (the World Wide Web protocol) running over TCP over IP (the Internet protocols) over IEEE 802.11 (the Wi-Fi protocol). This stack is used between the wireless router and the home user's personal computer when the user is surfing the web.

Whilst the use of protocol layering is today ubiquitous across the field of computer networking, it has been historically criticized by many researchers for two principal reasons. Firstly, abstracting the protocol stack in this way may cause a higher layer to duplicate functionality of a lower layer, a prime example being error recovery on both a per-link basis and an end-to-end basis. Secondly, it is common that a protocol implementation at one layer may require data, state or addressing information that is only present at another layer, thus defeating the point of separating the layers in the first place. For example, TCP uses the ECN field in the IPv4 header as an indication of congestion; IP is a network layer protocol whereas TCP is a transport layer protocol.

Communication protocols have various characteristics. They may be connection-oriented or connectionless, they may use circuit mode or packet switching, and they may use hierarchical addressing or flat addressing.

There are many communication protocols, a few of which are described below.

IEEE 802

IEEE 802 is a family of IEEE standards dealing with local area networks and metropolitan area networks. The complete IEEE 802 protocol suite provides a diverse set of networking capabilities. The protocols have a flat addressing scheme. They operate mostly at levels 1 and 2 of the OSI model.

For example, MAC bridging (IEEE 802.1D) deals with the routing of Ethernet packets using a Spanning Tree Protocol. IEEE 802.1Q describes VLANs, and IEEE 802.1X defines a port-based Network Access Control protocol, which forms the basis for the authentication mechanisms used in VLANs (but it is also found in WLANs) – it is what the home user sees when the user has to enter a "wireless access key".

Ethernet

Ethernet, sometimes simply called *LAN*, is a family of protocols used in wired LANs, described by a set of standards together called IEEE 802.3 published by the Institute of Electrical and Electronics Engineers.

Wireless LAN

Wireless LAN, also widely known as WLAN or WiFi, is probably the most well-known member of the IEEE 802 protocol family for home users today. It is standarized by IEEE 802.11 and shares many properties with wired Ethernet.

Internet Protocol Suite

The Internet Protocol Suite, also called TCP/IP, is the foundation of all modern networking. It offers connection-less as well as connection-oriented services over an inherently unreliable network traversed by data-gram transmission at the Internet protocol (IP) level. At its core, the protocol suite defines the addressing, identification, and routing specifications for Internet Protocol Version 4 (IPv4) and for IPv6, the next generation of the protocol with a much enlarged addressing capability.

SONET/SDH

Synchronous optical networking (SONET) and Synchronous Digital Hierarchy (SDH) are standardized multiplexing protocols that transfer multiple digital bit streams over optical fiber using lasers. They were originally designed to transport circuit mode communications from a variety of different sources, primarily to support real-time, uncompressed, circuit-switched voice encoded in PCM (Pulse-Code Modulation) format. However, due to its protocol neutrality and transport-oriented features, SONET/SDH also was the obvious choice for transporting Asynchronous Transfer Mode (ATM) frames.

Asynchronous Transfer Mode

Asynchronous Transfer Mode (ATM) is a switching technique for telecommunication networks. It uses asynchronous time-division multiplexing and encodes data into small, fixed-sized cells. This differs from other protocols such as the Internet Protocol Suite or Ethernet that use variable sized packets or frames. ATM has similarity with both circuit and packet switched networking. This makes it a good choice for a network that must handle both traditional high-throughput data traffic, and real-time, low-latency content such as voice and video. ATM uses a connection-oriented model in which a virtual circuit must be established between two endpoints before the actual data exchange begins.

While the role of ATM is diminishing in favor of next-generation networks, it still plays a role in the last mile, which is the connection between an Internet service provider and the home user.

Geographic Scale

A network can be characterized by its physical capacity or its organizational purpose. Use of the network, including user authorization and access rights, differ accordingly.

Nanoscale Network

A nanoscale communication network has key components implemented at the nanoscale including message carriers and leverages physical principles that differ from macroscale communication mechanisms. Nanoscale communication extends communication to very small sensors and actuators such as those found in biological systems and also tends to operate in environments that would be too harsh for classical communication.

Personal Area Network

A personal area network (PAN) is a computer network used for communication among computer and different information technological devices close to one person. Some examples of devices that are used in a PAN are personal computers, printers, fax machines, telephones, PDAs, scanners, and even video game consoles. A PAN may include wired and wireless devices. The reach of a PAN typically extends to 10 meters. A wired PAN is usually constructed with USB and FireWire connections while technologies such as Bluetooth and infrared communication typically form a wireless PAN.

Local Area Network

A local area network (LAN) is a network that connects computers and devices in a limited geographical area such as a home, school, office building, or closely positioned group of buildings. Each computer or device on the network is a node. Wired LANs are most likely based on Ethernet technology. Newer standards such as ITU-T G.hn also provide a way to create a wired LAN using existing wiring, such as coaxial cables, telephone lines, and power lines.

The defining characteristics of a LAN, in contrast to a wide area network (WAN), include higher data transfer rates, limited geographic range, and lack of reliance on leased lines to provide connectivity. Current Ethernet or other IEEE 802.3 LAN technologies operate at data transfer rates up to 100 Gbit/s, standarized by IEEE in 2010. Currently, 400 Gbit/s Ethernet is being developed.

A LAN can be connected to a WAN using a router.

Home Area Network

A home area network (HAN) is a residential LAN used for communication between digital devices typically deployed in the home, usually a small number of personal computers and accessories, such as printers and mobile computing devices. An important function is the sharing of Internet access, often a broadband service through a cable TV or digital subscriber line (DSL) provider.

Storage Area Network

A storage area network (SAN) is a dedicated network that provides access to consolidated, block level data storage. SANs are primarily used to make storage devices, such as disk arrays, tape libraries, and optical jukeboxes, accessible to servers so that the devices appear like locally attached devices to the operating system. A SAN typically has its own network of storage devices that are generally not accessible through the local area network by other devices. The cost and complexity of SANs dropped in the early 2000s to levels allowing wider adoption across both enterprise and small to medium-sized business environments.

Campus Area Network

A campus area network (CAN) is made up of an interconnection of LANs within a limited geographical area. The networking equipment (switches, routers) and transmission media (optical fiber, copper plant, Cat5 cabling, etc.) are almost entirely owned by the campus tenant / owner (an enterprise, university, government, etc.).

For example, a university campus network is likely to link a variety of campus buildings to connect academic colleges or departments, the library, and student residence halls.

Backbone Network

A backbone network is part of a computer network infrastructure that provides a path for the exchange of information between different LANs or sub-networks. A backbone can tie together diverse networks within the same building, across different buildings, or over a wide area.

For example, a large company might implement a backbone network to connect departments that are located around the world. The equipment that ties together the departmental networks constitutes the network backbone. When designing a network backbone, network performance and network congestion are critical factors to take into account. Normally, the backbone network's capacity is greater than that of the individual networks connected to it.

Another example of a backbone network is the Internet backbone, which is the set of wide area networks (WANs) and core routers that tie together all networks connected to the Internet.

Metropolitan Area Network

A Metropolitan area network (MAN) is a large computer network that usually spans a city or a large campus.

Wide Area Network

A wide area network (WAN) is a computer network that covers a large geographic area such as a city, country, or spans even intercontinental distances. A WAN uses a communications channel that combines many types of media such as telephone lines, cables, and air waves. A WAN often makes use of transmission facilities provided by common carriers, such as telephone companies. WAN technologies generally function at the lower three layers of the OSI reference model: the physical layer, the data link layer, and the network layer.

Enterprise Private Network

An enterprise private network is a network that a single organization builds to interconnect its office locations (e.g., production sites, head offices, remote offices, shops) so they can share computer resources.

Virtual Private Network

A virtual private network (VPN) is an overlay network in which some of the links between nodes are carried by open connections or virtual circuits in some larger network (e.g., the Internet) instead of by physical wires. The data link layer protocols of the virtual network are said to be tunneled through the larger network when this is the case. One common application is secure communications through the public Internet, but a VPN need not have explicit security features, such as authentication or content encryption. VPNs, for example, can be used to separate the traffic of different user communities over an underlying network with strong security features.

VPN may have best-effort performance, or may have a defined service level agreement (SLA) between the VPN customer and the VPN service provider. Generally, a VPN has a topology more complex than point-to-point.

Global Area Network

A global area network (GAN) is a network used for supporting mobile across an arbitrary number of wireless LANs, satellite coverage areas, etc. The key challenge in mobile communications is handing off user communications from one local coverage area to the next. In IEEE Project 802, this involves a succession of terrestrial wireless LANs.

Organizational Scope

Networks are typically managed by the organizations that own them. Private enterprise networks may use a combination of intranets and extranets. They may also provide network access to the Internet, which has no single owner and permits virtually unlimited global connectivity.

Intranet

An intranet is a set of networks that are under the control of a single administrative entity. The intranet uses the IP protocol and IP-based tools such as web browsers and file transfer applications. The administrative entity limits use of the intranet to its authorized users. Most commonly, an intranet is the internal LAN of an organization. A large intranet typically has at least one web server to provide users with organizational information. An intranet is also anything behind the router on a local area network.

Extranet

An extranet is a network that is also under the administrative control of a single organization, but supports a limited connection to a specific external network. For example, an organization may provide access to some aspects of its intranet to share data with its business partners or customers. These other entities are not necessarily trusted from a security standpoint. Network connection to an extranet is often, but not always, implemented via WAN technology.

Internetwork

An internetwork is the connection of multiple computer networks via a common routing technology using routers.

Internet

The Internet is the largest example of an internetwork. It is a global system of interconnected governmental, academic, corporate, public, and private computer networks. It is based on the networking technologies of the Internet Protocol Suite. It is the successor of the Advanced Research Projects Agency Network (ARPANET) developed by DARPA of the United States Department of Defense. The Internet is also the communications backbone underlying the World Wide Web (WWW).

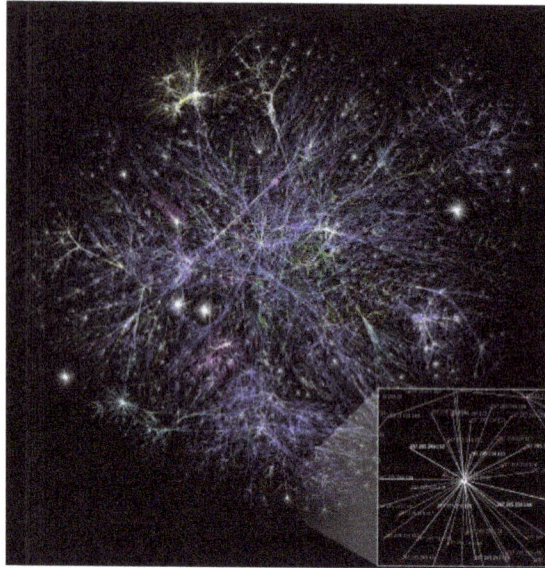

Partial map of the Internet based on the January 15, 2005 data found on opte.org. Each line is drawn between two nodes, representing two IP addresses. The length of the lines are indicative of the delay between those two nodes. This graph represents less than 30% of the Class C networks reachable.

Participants in the Internet use a diverse array of methods of several hundred documented, and often standardized, protocols compatible with the Internet Protocol Suite and an addressing system (IP addresses) administered by the Internet Assigned Numbers Authority and address registries. Service providers and large enterprises exchange information about the reachability of their address spaces through the Border Gateway Protocol (BGP), forming a redundant worldwide mesh of transmission paths.

Darknet

A darknet is an overlay network, typically running on the internet, that is only accessible through specialized software. A darknet is an anonymizing network where connections are made only between trusted peers — sometimes called "friends" (F2F) — using non-standard protocols and ports.

Darknets are distinct from other distributed peer-to-peer networks as sharing is anonymous (that is, IP addresses are not publicly shared), and therefore users can communicate with little fear of governmental or corporate interference.

Routing

Routing is the process of selecting network paths to carry network traffic. Routing is performed for many kinds of networks, including circuit switching networks and packet switched networks.

In packet switched networks, routing directs packet forwarding (the transit of logically addressed network packets from their source toward their ultimate destination) through intermediate nodes.

Intermediate nodes are typically network hardware devices such as routers, bridges, gateways, firewalls, or switches. General-purpose computers can also forward packets and perform routing, though they are not specialized hardware and may suffer from limited performance. The routing process usually directs forwarding on the basis of routing tables, which maintain a record of the routes to various network destinations. Thus, constructing routing tables, which are held in the router's memory, is very important for efficient routing.

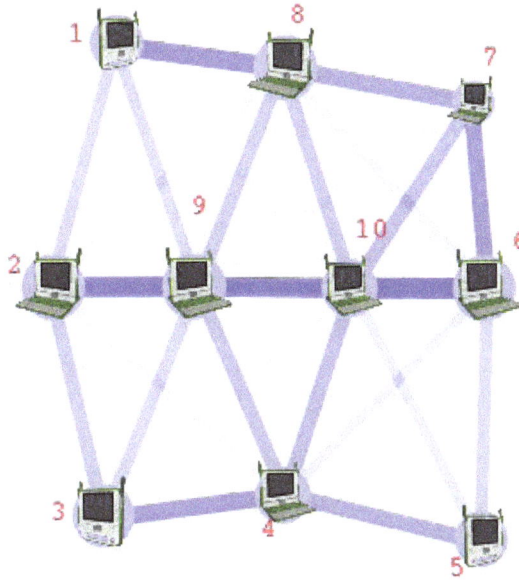

Routing calculates good paths through a network for information to take. For example, from node 1 to node 6 the best routes are likely to be 1-8-7-6 or 1-8-10-6, as this has the thickest routes.

There are usually multiple routes that can be taken, and to choose between them, different elements can be considered to decide which routes get installed into the routing table, such as (sorted by priority):

1. *Prefix-Length*: where longer subnet masks are preferred (independent if it is within a routing protocol or over different routing protocol)

2. *Metric*: where a lower metric/cost is preferred (only valid within one and the same routing protocol)

3. *Administrative distance*: where a lower distance is preferred (only valid between different routing protocols)

Most routing algorithms use only one network path at a time. Multipath routing techniques enable the use of multiple alternative paths.

Routing, in a more narrow sense of the term, is often contrasted with bridging in its assumption that network addresses are structured and that similar addresses imply proximity within the network. Structured addresses allow a single routing table entry to represent the route to a group of devices. In large networks, structured addressing (routing, in the narrow sense) outperforms unstructured addressing (bridging). Routing has become the dominant form of addressing on the Internet. Bridging is still widely used within localized environments.

Network Service

Network services are applications hosted by servers on a computer network, to provide some functionality for members or users of the network, or to help the network itself to operate.

The World Wide Web, E-mail, printing and network file sharing are examples of well-known network services. Network services such as DNS (Domain Name System) give names for IP and MAC addresses (people remember names like "nm.lan" better than numbers like "210.121.67.18"), and DHCP to ensure that the equipment on the network has a valid IP address.

Services are usually based on a service protocol that defines the format and sequencing of messages between clients and servers of that network service.

Network Performance

Quality of Service

Depending on the installation requirements, network performance is usually measured by the quality of service of a telecommunications product. The parameters that affect this typically can include throughput, jitter, bit error rate and latency.

The following list gives examples of network performance measures for a circuit-switched network and one type of packet-switched network, viz. ATM:

- Circuit-switched networks: In circuit switched networks, network performance is synonymous with the grade of service. The number of rejected calls is a measure of how well the network is performing under heavy traffic loads. Other types of performance measures can include the level of noise and echo.

- ATM: In an Asynchronous Transfer Mode (ATM) network, performance can be measured by line rate, quality of service (QoS), data throughput, connect time, stability, technology, modulation technique and modem enhancements.

There are many ways to measure the performance of a network, as each network is different in nature and design. Performance can also be modelled instead of measured. For example, state transition diagrams are often used to model queuing performance in a circuit-switched network. The network planner uses these diagrams to analyze how the network performs in each state, ensuring that the network is optimally designed.

Network Congestion

Network congestion occurs when a link or node is carrying so much data that its quality of service deteriorates. Typical effects include queueing delay, packet loss or the blocking of new connections. A consequence of these latter two is that incremental increases in offered load lead either only to small increase in network throughput, or to an actual reduction in network throughput.

Network protocols that use aggressive retransmissions to compensate for packet loss tend to keep systems in a state of network congestion—even after the initial load is reduced to a level that would not normally induce network congestion. Thus, networks using these protocols can exhibit two stable states under the same level of load. The stable state with low throughput is known as *congestive collapse.*

Modern networks use congestion control and congestion avoidance techniques to try to avoid congestion collapse. These include: exponential backoff in protocols such as 802.11's CSMA/CA and the original Ethernet, window reduction in TCP, and fair queueing in devices such as routers. Another method to avoid the negative effects of network congestion is implementing priority schemes, so that some packets are transmitted with higher priority than others. Priority schemes do not solve network congestion by themselves, but they help to alleviate the effects of congestion for some services. An example of this is 802.1p. A third method to avoid network congestion is the explicit allocation of network resources to specific flows. One example of this is the use of Contention-Free Transmission Opportunities (CFTXOPs) in the ITU-T G.hn standard, which provides high-speed (up to 1 Gbit/s) Local area networking over existing home wires (power lines, phone lines and coaxial cables).

For the Internet RFC 2914 addresses the subject of congestion control in detail.

Network Resilience

Network resilience is "the ability to provide and maintain an acceptable level of service in the face of faults and challenges to normal operation."

Security

Network Security

Network security consists of provisions and policies adopted by the network administrator to prevent and monitor unauthorized access, misuse, modification, or denial of the computer network and its network-accessible resources. Network security is the authorization of access to data in a network, which is controlled by the network administrator. Users are assigned an ID and password that allows them access to information and programs within their authority. Network security is used on a variety of computer networks, both public and private, to secure daily transactions and communications among businesses, government agencies and individuals.

Network Surveillance

Network surveillance is the monitoring of data being transferred over computer networks such as the Internet. The monitoring is often done surreptitiously and may be done by or at the behest of governments, by corporations, criminal organizations, or individuals. It may or may not be legal and may or may not require authorization from a court or other independent agency.

Computer and network surveillance programs are widespread today, and almost all Internet traffic is or could potentially be monitored for clues to illegal activity.

Surveillance is very useful to governments and law enforcement to maintain social control, recognize and monitor threats, and prevent/investigate criminal activity. With the advent of programs such as the Total Information Awareness program, technologies such as high speed surveillance computers and biometrics software, and laws such as the Communications Assistance For Law Enforcement Act, governments now possess an unprecedented ability to monitor the activities of citizens.

However, many civil rights and privacy groups—such as Reporters Without Borders, the Electronic Frontier Foundation, and the American Civil Liberties Union—have expressed concern that increasing surveillance of citizens may lead to a mass surveillance society, with limited political and personal freedoms. Fears such as this have led to numerous lawsuits such as *Hepting v. AT&T*. The hacktivist group Anonymous has hacked into government websites in protest of what it considers "draconian surveillance".

End to end Encryption

End-to-end encryption (E2EE) is a digital communications paradigm of uninterrupted protection of data traveling between two communicating parties. It involves the originating party encrypting data so only the intended recipient can decrypt it, with no dependency on third parties. End-to-end encryption prevents intermediaries, such as Internet providers or application service providers, from discovering or tampering with communications. End-to-end encryption generally protects both confidentiality and integrity.

Examples of end-to-end encryption include PGP for email, OTR for instant messaging, ZRTP for telephony, and TETRA for radio.

Typical server-based communications systems do not include end-to-end encryption. These systems can only guarantee protection of communications between clients and servers, not between the communicating parties themselves. Examples of non-E2EE systems are Google Talk, Yahoo Messenger, Facebook, and Dropbox. Some such systems, for example LavaBit and SecretInk, have even described themselves as offering "end-to-end" encryption when they do not. Some systems that normally offer end-to-end encryption have turned out to contain a back door that subverts negotiation of the encryption key between the communicating parties, for example Skype or Hushmail.

The end-to-end encryption paradigm does not directly address risks at the communications endpoints themselves, such as the technical exploitation of clients, poor quality random number generators, or key escrow. E2EE also does not address traffic analysis, which relates to things such as the identities of the end points and the times and quantities of messages that are sent.

Views of Networks

Users and network administrators typically have different views of their networks. Users can share printers and some servers from a workgroup, which usually means they are in the same geographic location and are on the same LAN, whereas a Network Administrator is responsible to keep that network up and running. A community of interest has less of a connection of being in a local area, and should be thought of as a set of arbitrarily located users who share a set of servers, and possibly also communicate via peer-to-peer technologies.

Network administrators can see networks from both physical and logical perspectives. The physical perspective involves geographic locations, physical cabling, and the network elements (e.g., routers, bridges and application layer gateways) that interconnect via the transmission media. Logical networks, called, in the TCP/IP architecture, subnets, map onto one or more transmission media. For example, a common practice in a campus of buildings is to make a set of LAN cables in each building appear to be a common subnet, using virtual LAN (VLAN) technology.

Both users and administrators are aware, to varying extents, of the trust and scope characteristics of a network. Again using TCP/IP architectural terminology, an intranet is a community of interest under private administration usually by an enterprise, and is only accessible by authorized users (e.g. employees). Intranets do not have to be connected to the Internet, but generally have a limited connection. An extranet is an extension of an intranet that allows secure communications to users outside of the intranet (e.g. business partners, customers).

Unofficially, the Internet is the set of users, enterprises, and content providers that are interconnected by Internet Service Providers (ISP). From an engineering viewpoint, the Internet is the set of subnets, and aggregates of subnets, which share the registered IP address space and exchange information about the reachability of those IP addresses using the Border Gateway Protocol. Typically, the human-readable names of servers are translated to IP addresses, transparently to users, via the directory function of the Domain Name System (DNS).

Over the Internet, there can be business-to-business (B2B), business-to-consumer (B2C) and consumer-to-consumer (C2C) communications. When money or sensitive information is exchanged, the communications are apt to be protected by some form of communications security mechanism. Intranets and extranets can be securely superimposed onto the Internet, without any access by general Internet users and administrators, using secure Virtual Private Network (VPN) technology.

Examples of Computer Networks

ARPANET

Arpanet 1974

The Advanced Research Projects Agency Network (ARPANET) was an early packet switching network and the first network to implement the protocol suite TCP/IP. Both technologies became the technical foundation of the Internet. ARPANET was initially funded by the Advanced Research Projects Agency (ARPA) of the United States Department of Defense.

The packet switching methodology employed in the ARPANET was based on concepts and designs by Americans Leonard Kleinrock and Paul Baran, British scientist Donald Davies, and Lawrence Roberts of the Lincoln Laboratory. The TCP/IP communications protocols were developed for ARPANET by computer scientists Robert Kahn and Vint Cerf, and incorporated concepts by Louis Pouzin for the French CYCLADES project.

As the project progressed, protocols for internetworking were developed by which multiple separate networks could be joined into a network of networks. Access to the ARPANET was expanded in 1981 when the National Science Foundation (NSF) funded the Computer Science Network (CS-NET). In 1982, the Internet protocol suite (TCP/IP) was introduced as the standard networking protocol on the ARPANET. In the early 1980s the NSF funded the establishment for national supercomputing centers at several universities, and provided interconnectivity in 1986 with the NSFNET project, which also created network access to the supercomputer sites in the United States from research and education organizations. ARPANET was decommissioned in 1990.

History

Packet switching—today the dominant basis for data communications worldwide—was a new concept at the time of the conception of the ARPANET. Prior to the advent of packet switching, both voice and data communications had been based on the idea of circuit switching, as in the traditional telephone circuit, wherein each telephone call is allocated a dedicated, end to end, electronic connection between the two communicating stations. Such stations might be telephones or computers. The (temporarily) dedicated line is typically composed of many intermediary lines which are assembled into a chain that stretches all the way from the originating station to the destination station. With packet switching, a data system could use a single communication link to communicate with more than one machine by collecting data into datagrams and transmitting these as packets onto the attached network link, as soon as the link becomes idle. Thus, not only can the link be shared, much as a single post box can be used to post letters to different destinations, but each packet can be routed independently of other packets.

The earliest ideas for a computer network intended to allow general communications among computer users were formulated by computer scientist J. C. R. Licklider of Bolt, Beranek and Newman (BBN), in April 1963, in memoranda discussing the concept of the "Intergalactic Computer Network". Those ideas encompassed many of the features of the contemporary Internet. In October 1963, Licklider was appointed head of the Behavioral Sciences and Command and Control programs at the Defense Department's Advanced Research Projects Agency (ARPA). He convinced Ivan Sutherland and Bob Taylor that this network concept was very important and merited development, although Licklider left ARPA before any contracts were assigned for development.

Sutherland and Taylor continued their interest in creating the network, in part, to allow ARPA-sponsored researchers at various corporate and academic locales to utilize computers provided by ARPA, and, in part, to quickly distribute new software and other computer science results. Taylor had three computer terminals in his office, each connected to separate computers, which ARPA was funding: one for the System Development Corporation (SDC) Q-32 in Santa Monica, one for Project Genie at the University of California, Berkeley, and another for Multics at the Massachusetts Institute of Technology. Taylor recalls the circumstance: "For each of these three terminals, I had three different sets of user commands. So, if I was talking online with someone at S.D.C., and I wanted to talk to someone I knew at Berkeley, or M.I.T., about this, I had to get up from the S.D.C. terminal, go over and log into the other terminal and get in touch with them. I said, "Oh Man!", it's obvious what to do: If you have these three terminals, there ought to be one terminal that goes anywhere you want to go. That idea is the ARPANET".

Meanwhile, since the early 1960s, Paul Baran at the RAND Corporation had been researching systems that could survive nuclear war and developed the idea of *distributed adaptive message block switching*. Donald Davies at the United Kingdom's National Physical Laboratory (NPL) independently invented the same concept in 1965. His work, presented by a colleague, initially caught the attention of ARPANET developers at a conference in Gatlinburg, Tennessee, in October 1967. He gave the first public demonstration, having coined the term *packet switching*, on 5 August 1968 and incorporated it into the NPL network in England. Larry Roberts at ARPA applied Davies' concepts of packet switching for the ARPANET. The NPL network followed by ARPANET were the first two networks in the world to use packet switching, and were themselves connected together in 1973. The NPL network was using line speeds of 768 kbit/s, and the proposed line speed for ARPANET was upgraded from 2.4 kbit/s to 50 kbit/s.

Creation

By mid-1968, Taylor had prepared a complete plan for a computer network, and, after ARPA's approval, a Request for Quotation (RFQ) was issued for 140 potential bidders. Most computer science companies regarded the ARPA–Taylor proposal as outlandish, and only twelve submitted bids to build a network; of the twelve, ARPA regarded only four as top-rank contractors. At year's end, ARPA considered only two contractors, and awarded the contract to build the network to BBN Technologies on 7 April 1969. The initial, seven-person BBN team were much aided by the technical specificity of their response to the ARPA RFQ, and thus quickly produced the first working system. This team was led by Frank Heart. The BBN-proposed network closely followed Taylor's ARPA plan: a network composed of small computers called Interface Message Processors (or IMPs), similar to the later concept of routers, that functioned as gateways interconnecting local resources. At each site, the IMPs performed store-and-forward packet switching functions, and were interconnected with leased lines via telecommunication data sets (modems), with initial data rates of 56kbit/s. The host computers were connected to the IMPs via custom serial communication interfaces. The system, including the hardware and the packet switching software, was designed and installed in nine months.

The first-generation IMPs were built by BBN Technologies using a rugged computer version of the Honeywell DDP-516 computer configured with 24KB of expandable magnetic-core memory, and a 16-channel Direct Multiplex Control (DMC) direct memory access unit. The DMC established custom interfaces with each of the host computers and modems. In addition to the front-panel lamps, the DDP-516 computer also features a special set of 24 indicator lamps showing the status of the IMP communication channels. Each IMP could support up to four local hosts, and could communicate with up to six remote IMPs via leased lines. The network connected one computer in Utah with three in California. Later, the Department of Defense allowed the universities to join the network for sharing hardware and software resources.

Debate on Design Goals

In *A Brief History of the Internet*, the Internet Society denies that ARPANET was designed to survive a nuclear attack:

> It was from the RAND study that the false rumor started, claiming that the ARPANET was somehow related to building a network resistant to nuclear war. This was never true of the ARPANET; only the unrelated RAND study on secure voice considered nuclear war. However, the later work on Internetting did emphasize robustness and survivability, including the capability to withstand losses of large portions of the underlying networks.

The RAND study was conducted by Paul Baran and pioneered packet switching. In an interview he confirmed that while ARPANET did not exactly share his project's goal, his work had greatly contributed to the development of ARPANET. Minutes taken by Elmer Shapiro of Stanford Research Institute at the ARPANET design meeting of 9–10 Oct. 1967 indicate that a version of Baran's routing method and suggestion of using a fixed packet size was expected to be employed.

According to Stephen J. Lukasik, who as Deputy Director and Director of DARPA (1967–1974) was "the person who signed most of the checks for Arpanet's development":

> The goal was to exploit new computer technologies to meet the needs of military command and control against nuclear threats, achieve survivable control of US nuclear forces, and improve military tactical and management decision making.

The ARPANET incorporated distributed computation (and frequent re-computation) of routing tables. This was a major contribution to the survivability of the ARPANET in the face of significant destruction - even by a nuclear attack. Such auto-routing was technically quite challenging to construct at the time. The fact that it was incorporated into the early ARPANET made many believe that this had been a design goal.

The ARPANET was designed to survive subordinate-network losses, since the principal reason was that the switching nodes and network links were unreliable, even without any nuclear attacks. Resource scarcity supported the creation of the ARPANET, according to Charles Herzfeld, ARPA Director (1965–1967):

> The ARPANET was not started to create a Command and Control System that would survive a nuclear attack, as many now claim. To build such a system was, clearly, a major military need, but it was not ARPA's mission to do this; in fact, we would have been severely criticized had we tried. Rather, the ARPANET came out of our frustration that there were only a limited number of large, powerful research computers in the country, and that many research investigators, who should have access to them, were geographically separated from them.

The ARPANET was operated by the military during the two decades of its existence, until 1990.

ARPANET Deployed

Historical document: First ARPANET IMP log: the first message ever sent via the ARPANET, 10:30 pm, 29 October 1969. This IMP Log excerpt, kept at UCLA, describes setting up a message transmission from the UCLA SDS Sigma 7 Host computer to the SRI SDS 940 Host computer.

The initial ARPANET consisted of four IMPs:

- University of California, Los Angeles (UCLA), where Leonard Kleinrock had established a Network Measurement Center, with an SDS Sigma 7 being the first computer attached to it;

- The Augmentation Research Center at Stanford Research Institute (now SRI International), where Douglas Engelbart had created the ground-breaking NLS system, a very important early hypertext system, and would run the Network Information Center (NIC), with the SDS 940 that ran NLS, named "Genie", being the first host attached;

- University of California, Santa Barbara (UCSB), with the Culler-Fried Interactive Mathematics Center's IBM 360/75, running OS/MVT being the machine attached;

- The University of Utah's Computer Science Department, where Ivan Sutherland had moved, running a DEC PDP-10 operating on TENEX.

The first successful message on the ARPANET was sent by UCLA student programmer Charley Kline, at 10:30 pm on 29 October 1969, from Boelter Hall 3420. Kline transmitted from the university's SDS Sigma 7 Host computer to the Stanford Research Institute's SDS 940 Host computer. The message text was the word *login*; on an earlier attempt the *l* and the *o* letters were transmitted, but the system then crashed. Hence, the literal first message over the ARPANET was *lo*. About an hour later, after the programmers repaired the code that caused the crash, the SDS Sigma 7 computer effected a full *login*. The first permanent ARPANET link was established on 21 November 1969, between the IMP at UCLA and the IMP at the Stanford Research Institute. By 5 December 1969, the entire four-node network was established.

Growth and Evolution

In March 1970, the ARPANET reached the East Coast of the United States, when an IMP at BBN in Cambridge, Massachusetts was connected to the network. Thereafter, the ARPANET grew: 9 IMPs by June 1970 and 13 IMPs by December 1970, then 18 by September 1971 (when the network included 23 university and government hosts); 29 IMPs by August 1972, and 40 by September 1973. By June 1974, there were 46 IMPs, and in July 1975, the network numbered 57 IMPs. By 1981, the number was 213 host computers, with another host connecting approximately every twenty days.

In 1973 a transatlantic satellite link connected the Norwegian Seismic Array (NORSAR) to the ARPANET, making Norway the first country outside the US to be connected to the network. At about the same time a terrestrial circuit added a London IMP.

In 1975, the ARPANET was declared "operational". The Defense Communications Agency took control since ARPA was intended to fund advanced research.

In September 1984 work was completed on restructuring the ARPANET giving U.S. military sites their own Military Network (MILNET) for unclassified defense department communications. Controlled gateways connected the two networks. The combination was called the Defense Data Network (DDN). Separating the civil and military networks reduced the 113-node ARPANET by 68 nodes. The MILNET later became the NIPRNet.

Rules and Etiquette

Because of its government funding, certain forms of traffic were discouraged or prohibited. A 1982 handbook on computing at MIT's AI Lab stated regarding network etiquette:

It is considered illegal to use the ARPANet for anything which is not in direct support of Government business ... personal messages to other ARPANet subscribers (for example, to arrange a get-together or check and say a friendly hello) are generally not considered harmful ... Sending electronic mail over the ARPANet for commercial profit or political purposes is both anti-social and illegal. By sending such messages, you can offend many people, and it is possible to get MIT in serious trouble with the Government agencies which manage the ARPANet.

Technology

Support for inter-IMP circuits of up to 230.4 kbit/s was added in 1970, although considerations of cost and IMP processing power meant this capability was not actively used.

1971 saw the start of the use of the non-ruggedized (and therefore significantly lighter) Honeywell 316 as an IMP. It could also be configured as a Terminal Interface Processor (TIP), which provided terminal server support for up to 63 ASCII serial terminals through a multi-line controller in place of one of the hosts. The 316 featured a greater degree of integration than the 516, which made it less expensive and easier to maintain. The 316 was configured with 40 kB of core memory for a TIP. The size of core memory was later increased, to 32 kB for the IMPs, and 56 kB for TIPs, in 1973.

In 1975, BBN introduced IMP software running on the Pluribus multi-processor. These appeared in a few sites. In 1981, BBN introduced IMP software running on its own C/30 processor product.

In 1983, TCP/IP protocols replaced NCP as the ARPANET's principal protocol, and the ARPANET then became one subnet of the early Internet.

The original IMPs and TIPs were phased out as the ARPANET was shut down after the introduction of the NSFNet, but some IMPs remained in service as late as July 1990.

The *ARPANET Completion Report*, jointly published by BBN and ARPA, concludes that:

> ... it is somewhat fitting to end on the note that the ARPANET program has had a strong and direct feedback into the support and strength of computer science, from which the network, itself, sprang.

In the wake of ARPANET being formally decommissioned on 28 February 1990, Vinton Cerf wrote the following lamentation, entitled "Requiem of the ARPANET":

It was the first, and being first, was best,

but now we lay it down to ever rest.

Now pause with me a moment, shed some tears.

For auld lang syne, for love, for years and years

of faithful service, duty done, I weep.

Lay down thy packet, now, O friend, and sleep.

-Vinton Cerf

Senator Albert Gore, Jr. began to craft the High Performance Computing and Communication Act of 1991 (commonly referred to as "The Gore Bill") after hearing the 1988 report toward a National Research Network submitted to Congress by a group chaired by Leonard Kleinrock, professor of computer science at UCLA. The bill was passed on 9 December 1991 and led to the National Information Infrastructure (NII) which Al Gore called the "information superhighway".

ARPANET was the subject of two IEEE Milestones, both dedicated in 2009.

Software and Protocols

The starting point for host-to-host communication on the ARPANET in 1969 was the 1822 protocol, which defined the transmission of messages to an IMP. The message format was designed to work unambiguously with a broad range of computer architectures. An 1822 message essentially consisted of a message type, a numeric host address, and a data field. To send a data message to another host, the transmitting host formatted a data message containing the destination host's address and the data message being sent, and then transmitted the message through the 1822 hardware interface. The IMP then delivered the message to its destination address, either by delivering it to a locally connected host, or by delivering it to another IMP. When the message was ultimately delivered to the destination host, the receiving IMP would transmit a *Ready for Next Message* (RFNM) acknowledgement to the sending, host IMP.

Unlike modern Internet datagrams, the ARPANET was designed to reliably transmit 1822 messages, and to inform the host computer when it loses a message; the contemporary IP is unreliable, whereas the TCP is reliable. Nonetheless, the 1822 protocol proved inadequate for handling multiple connections among different applications residing in a host computer. This problem was addressed with the Network Control Program (NCP), which provided a standard method to establish reliable, flow-controlled, bidirectional communications links among different processes in different host computers. The NCP interface allowed application software to connect across the ARPANET by implementing higher-level communication protocols, an early example of the *protocol layering* concept incorporated to the OSI model.

In 1983, TCP/IP protocols replaced NCP as the ARPANET's principal protocol, and the ARPANET then became one component of the early Internet.

Network Applications

NCP provided a standard set of network services that could be shared by several applications running on a single host computer. This led to the evolution of *application protocols* that operated, more or less, independently of the underlying network service, and permitted independent advances in the underlying protocols.

In 1971, Ray Tomlinson, of BBN sent the first network e-mail (RFC 524, RFC 561). By 1973, e-mail constituted 75 percent of ARPANET traffic.

By 1973, the File Transfer Protocol (FTP) specification had been defined (RFC 354) and implemented, enabling file transfers over the ARPANET.

The Network Voice Protocol (NVP) specifications were defined in 1977 (RFC 741), then implemented, but, because of technical shortcomings, conference calls over the ARPANET never worked well; the contemporary Voice over Internet Protocol (packet voice) was decades away.

Password Protection

The Purdy Polynomial hash algorithm was developed for ARPANET to protect passwords in 1971 at the request of Larry Roberts, head of ARPA at that time. It computed a polynomial of degree $2^{24} + 17$ modulo the 64-bit prime $p = 2^{64} - 59$. The algorithm was later used by Digital Equipment Corporation (DEC) to hash passwords in the VMS Operating System, and is still being used for this purpose.

ARPANET in Popular Culture

- *Computer Networks: The Heralds of Resource Sharing*, a 30-minute documentary film featuring Fernando J. Corbato, J.C.R. Licklider, Lawrence G. Roberts, Robert Kahn, Frank Heart, William R. Sutherland, Richard W. Watson, John R. Pasta, Donald W. Davies, and economist, George W. Mitchell.

- "Scenario", a February 1985 episode of the U.S. television sitcom *Benson* (season 6, episode 20), was the first incidence of a popular TV show directly referencing the Internet or its progenitors. The show includes a scene in which the ARPANET is accessed.

- There is an electronic music artist known as "Arpanet", Gerald Donald, one of the members of Drexciya. The artist's 2002 album *Wireless Internet* features commentary on the expansion of the internet via wireless communication, with songs such as *NTT DoCoMo*, dedicated to the mobile communications giant based in Japan.

- Thomas Pynchon mentions ARPANET in his 2009 novel *Inherent Vice*, which is set in Los Angeles in 1970, and in his 2013 novel *Bleeding Edge*.

- The 1993 television series *The X-Files* featured the ARPANET in a season 5 episode, titled "Unusual Suspects". John Fitzgerald Byers offers to help Susan Modeski (known as Holly. "just like the sugar") by hacking into the ARPANET to obtain sensitive information.

- In the spy-drama television series *The Americans*, a Russian scientist defector offers access to ARPANET to the Russians in a plea to not be repatriated (Season 2 Episode 5 "The Deal"). Episode 7 of Season 2 is named 'ARPANET' and features Russian infiltration to bug the network.

- In the television series *Person of Interest*, main character Harold Finch hacked ARPANET in 1980 using a homemade computer during his first efforts to built a prototype of the Machine. This corresponds with the real life virus that occurred in October of that year that temporarily halted ARPANET functions. The ARPANET hack was first discussed in the episode *2PiR* where a computer science teacher called it the most famous hack in history and one that was never solved. Finch later mentioned it to Person of Interest Caleb Phipps and his role was first indicated when he showed knowledge that it was done by "a kid with a homemade computer" which Phipps, who had researched the hack, had never heard before.

- In the third season of the television series *Halt and Catch Fire*, the character Joe MacMillan explores the potential commercialization of ARPANET.

Ethernet

A Cat 5e connection on a laptop, used for Ethernet

Ethernet is a family of computer networking technologies commonly used in local area networks (LANs) and metropolitan area networks (MANs). It was commercially introduced in 1980 and first standardized in 1983 as IEEE 802.3, and has since been refined to support higher bit rates and longer link distances. Over time, Ethernet has largely replaced competing wired LAN technologies such as token ring, FDDI and ARCNET.

The original 10BASE5 Ethernet uses coaxial cable as a shared medium, while the newer Ethernet variants use twisted pair and fiber optic links in conjunction with hubs or switches. Over the course of its history, Ethernet data transfer rates have been increased from the original 2.94 megabits per second (Mbit/s) to the latest 100 gigabits per second (Gbit/s). The Ethernet standards comprise several wiring and signaling variants of the OSI physical layer in use with Ethernet.

Systems communicating over Ethernet divide a stream of data into shorter pieces called frames. Each frame contains source and destination addresses, and error-checking data so that damaged frames can be detected and discarded; most often, higher-layer protocols trigger retransmission of lost frames. As per the OSI model, Ethernet provides services up to and including the data link layer.

Since its commercial release, Ethernet has retained a good degree of backward compatibility. Features such as the 48-bit MAC address and Ethernet frame format have influenced other networking protocols. The primary alternative for some uses of contemporary LANs is Wi-Fi, a wireless protocol standardized as IEEE 802.11.

History

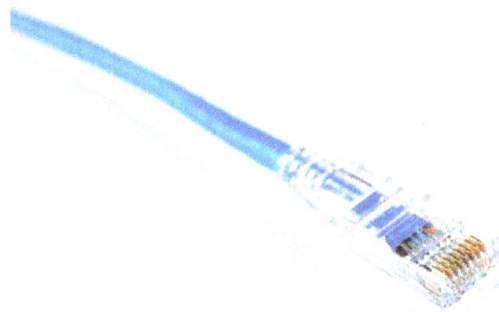

An 8P8C modular connector (often called RJ45) commonly used on Cat 5 cables in Ethernet networks

Ethernet was developed at Xerox PARC between 1973 and 1974. It was inspired by ALOHAnet, which Robert Metcalfe had studied as part of his PhD dissertation. The idea was first documented in a memo that Metcalfe wrote on May 22, 1973, where he named it after the disproven luminiferous ether as an "omnipresent, completely-passive medium for the propagation of electromagnetic waves". In 1975, Xerox filed a patent application listing Metcalfe, David Boggs, Chuck Thacker, and Butler Lampson as inventors. In 1976, after the system was deployed at PARC, Metcalfe and Boggs published a seminal paper.

Metcalfe left Xerox in June 1979 to form 3Com. He convinced Digital Equipment Corporation (DEC), Intel, and Xerox to work together to promote Ethernet as a standard. The so-called "DIX" standard, for "Digital/Intel/Xerox", specified 10 Mbit/s Ethernet, with 48-bit destination and source addresses and a global 16-bit Ethertype-type field. It was published on September 30, 1980 as "The Ethernet, A Local Area Network. Data Link Layer and Physical Layer Specifications". Version 2 was published in November, 1982 and defines what has become known as Ethernet II. Formal standardization efforts proceeded at the same time and resulted in the publication of IEEE 802.3 on June 23, 1983.

Ethernet initially competed with two largely proprietary systems, Token Ring and Token Bus. Because Ethernet was able to adapt to market realities and shift to inexpensive and ubiquitous twisted pair wiring, these proprietary protocols soon found themselves competing in a market inundated by Ethernet products, and, by the end of the 1980s, Ethernet was clearly the dominant network technology. In the process, 3Com became a major company. 3Com shipped its first 10 Mbit/s Ethernet 3C100 NIC in March 1981, and that year started selling adapters for PDP-11s and VAXes, as well as Multibus-based Intel and Sun Microsystems computers. This was followed quickly by DEC's Unibus to Ethernet adapter, which DEC sold and used internally to build its own corporate network, which reached over 10,000 nodes by 1986, making it one of the largest computer networks in the world at that time. An Ethernet adapter card for the IBM PC was released in 1982, and, by 1985, 3Com had sold 100,000. By the early 1990s, Ethernet became so prevalent that it was a must-have feature for modern computers, and Ethernet ports began to appear on some PCs and most workstations. This process was greatly sped up with the introduction of 10BASE-T and its relatively small modular connector, at which point Ethernet ports appeared even on low-end motherboards.

Since then, Ethernet technology has evolved to meet new bandwidth and market requirements. In addition to computers, Ethernet is now used to interconnect appliances and other personal devices. It is used in industrial applications and is quickly replacing legacy data transmission systems in the world's telecommunications networks. By 2010, the market for Ethernet equipment amounted to over $16 billion per year.

Standardization

An Intel 82574L Gigabit Ethernet NIC, PCI Express x1 card

In February 1980, the Institute of Electrical and Electronics Engineers (IEEE) started project 802 to standardize local area networks (LAN). The "DIX-group" with Gary Robinson (DEC), Phil Arst (Intel), and Bob Printis (Xerox) submitted the so-called "Blue Book" CSMA/CD specification as a candidate for the LAN specification. In addition to CSMA/CD, Token Ring (supported by IBM) and Token Bus (selected and henceforward supported by General Motors) were also considered as candidates for a LAN standard. Competing proposals and broad interest in the initiative led to strong disagreement over which technology to standardize. In December 1980, the group was split into three subgroups, and standardization proceeded separately for each proposal.

Delays in the standards process put at risk the market introduction of the Xerox Star workstation and 3Com's Ethernet LAN products. With such business implications in mind, David Liddle (General Manager, Xerox Office Systems) and Metcalfe (3Com) strongly supported a proposal of Fritz Röscheisen (Siemens Private Networks) for an alliance in the emerging office communication market, including Siemens' support for the international standardization of Ethernet (April 10, 1981). Ingrid Fromm, Siemens' representative to IEEE 802, quickly achieved broader support for Ethernet beyond IEEE by the establishment of a competing Task Group "Local Networks" within the European standards body ECMA TC24. On March 1982, ECMA TC24 with its corporate members reached an agreement on a standard for CSMA/CD based on the IEEE 802 draft. Because the DIX proposal was most technically complete and because of the speedy action taken by ECMA which decisively contributed to the conciliation of opinions within IEEE, the IEEE 802.3 CSMA/CD standard was approved in December 1982. IEEE published the 802.3 standard as a draft in 1983 and as a standard in 1985.

Approval of Ethernet on the international level was achieved by a similar, cross-partisan action with Fromm as the liaison officer working to integrate with International Electrotechnical Commission (IEC) Technical Committee 83 (TC83) and International Organization for Standardization (ISO) Technical Committee 97 Sub Committee 6 (TC97SC6). The ISO 8802-3 standard was published in 1989.

Evolution

Ethernet evolved to include higher bandwidth, improved media access control methods, and different physical media. The coaxial cable was replaced with point-to-point links connected by Ethernet repeaters or switches.

Ethernet stations communicate by sending each other data packets: blocks of data individually sent and delivered. As with other IEEE 802 LANs, each Ethernet station is given a 48-bit MAC address. The MAC addresses are used to specify both the destination and the source of each data packet. Ethernet establishes link level connections, which can be defined using both the destination and source addresses. On reception of a transmission, the receiver uses the destination address to determine whether the transmission is relevant to the station or should be ignored. Network interfaces normally do not accept packets addressed to other Ethernet stations. Adapters come programmed with a globally unique address.

An EtherType field in each frame is used by the operating system on the receiving station to select the appropriate protocol module (e.g., an Internet Protocol version such as IPv4). Ethernet frames are said to be *self-identifying*, because of the frame type. Self-identifying frames make it possible to intermix multiple protocols on the same physical network and allow a single computer to use multiple protocols together. Despite the evolution of Ethernet technology, all generations of Ethernet (excluding early experimental versions) use the same frame formats. Mixed-speed networks can be built using Ethernet switches and repeaters supporting the desired Ethernet variants.

Due to the ubiquity of Ethernet, the ever-decreasing cost of the hardware needed to support it, and the reduced panel space needed by twisted pair Ethernet, most manufacturers now build Ethernet interfaces directly into PC motherboards, eliminating the need for installation of a separate network card.

Shared Media

Older Ethernet equipment. Clockwise from top-left: An Ethernet transceiver with an in-line 10BASE2 adapter, a similar model transceiver with a 10BASE5 adapter, an AUI cable, a different style of transceiver with 10BASE2 BNC T-connector, two 10BASE5 end fittings (N connectors), an orange "vampire tap" installation tool (which includes a specialized drill bit at one end and a socket wrench at the other), and an early model 10BASE5 transceiver (h4000) manufactured by DEC. The short length of yellow 10BASE5 cable has one end fitted with a N connector and the other end prepared to have a N connector shell installed; the half-black, half-grey rectangular object through which the cable passes is an installed vampire tap.

Ethernet was originally based on the idea of computers communicating over a shared coaxial cable acting as a broadcast transmission medium. The methods used were similar to those used in radio systems, with the common cable providing the communication channel likened to the *Luminiferous aether* in 19th century physics, and it was from this reference that the name "Ethernet" was derived.

Original Ethernet's shared coaxial cable (the shared medium) traversed a building or campus to every attached machine. A scheme known as carrier sense multiple access with collision detection (CSMA/CD) governed the way the computers shared the channel. This scheme was simpler than the competing token ring or token bus technologies. Computers are connected to an Attachment Unit Interface (AUI) transceiver, which is in turn connected to the cable (with thin Ethernet the transceiver is integrated into the network adapter). While a simple passive wire is highly reliable for small networks, it is not reliable for large extended networks, where damage to the wire in a single place, or a single bad connector, can make the whole Ethernet segment unusable.

Through the first half of the 1980s, Ethernet's 10BASE5 implementation used a coaxial cable 0.375 inches (9.5 mm) in diameter, later called "thick Ethernet" or "thicknet". Its successor, 10BASE2, called "thin Ethernet" or "thinnet", used the RG-58 coaxial cable. The emphasis was on making installation of the cable easier and less costly.

Since all communications happen on the same wire, any information sent by one computer is received by all, even if that information is intended for just one destination.The network interface card interrupts the CPU only when applicable packets are received: The card ignores information not addressed to it. Use of a single cable also means that the bandwidth is shared, such that, for example, available bandwidth to each device is halved when two stations are simultaneously active.

Collisions happen when two stations attempt to transmit at the same time. They corrupt transmitted data and require stations to retransmit. The lost data and retransmissions reduce throughput. In the worst case where multiple active hosts connected with maximum allowed cable length attempt to transmit many short frames, excessive collisions can reduce throughput dramatically. However, a Xerox report in 1980 studied performance of an existing Ethernet installation under both normal and artificially generated heavy load. The report claims that 98% throughput on the LAN was observed. This is in contrast with token passing LANs (token ring, token bus), all of which suffer throughput degradation as each new node comes into the LAN, due to token waits. This report was controversial, as modeling showed that collision-based networks theoretically became unstable under loads as low as 37% of nominal capacity. Many early researchers failed to understand these results. Performance on real networks is significantly better.

In a modern Ethernet, the stations do not all share one channel through a shared cable or a simple repeater hub; instead, each station communicates with a switch, which in turn forwards that traffic to the destination station. In this topology, collisions are only possible if station and switch attempt to communicate with each other at the same time, and collisions are limited to this link. Furthermore, the 10BASE-T standard introduced a full duplex mode of operation which has become extremely common. In full duplex, switch and station can communicate with each other simultaneously, and therefore modern Ethernets are completely collision-free.

Repeaters and Hubs

A 1990s network interface card supporting both coaxial cable-based 10BASE2 (BNC connector, left) and twisted pair-based 10BASE-T (8P8C connector, right)

For signal degradation and timing reasons, coaxial Ethernet segments have a restricted size. Somewhat larger networks can be built by using an Ethernet repeater. Early repeaters had only two ports, allowing, at most, a doubling of network size. Once repeaters with more than two ports became available, it was possible to wire the network in a star topology. Early experiments with star topologies (called "Fibernet") using optical fiber were published by 1978.

Shared cable Ethernet is always hard to install in offices because its bus topology is in conflict with the star topology cable plans designed into buildings for telephony. Modifying Ethernet to conform to twisted pair telephone wiring already installed in commercial buildings provided another opportunity to lower costs, expand the installed base, and leverage building design, and, thus, twisted-pair Ethernet was the next logical development in the mid-1980s.

Ethernet on unshielded twisted-pair cables (UTP) began with StarLAN at 1 Mbit/s in the mid-1980s. In 1987 SynOptics introduced the first twisted-pair Ethernet at 10 Mbit/s in a star-wired cabling topology with a central hub, later called LattisNet. These evolved into 10BASE-T, which was designed for point-to-point links only, and all termination was built into the device. This changed repeaters from a specialist device used at the center of large networks to a device that every twisted pair-based network with more than two machines had to use. The tree structure that resulted from this made Ethernet networks easier to maintain by preventing most faults with one peer or its associated cable from affecting other devices on the network.

Despite the physical star topology and the presence of separate transmit and receive channels in the twisted pair and fiber media, repeater-based Ethernet networks still use half-duplex and CSMA/CD, with only minimal activity by the repeater, primarily generation of the jam signal in dealing with packet collisions. Every packet is sent to every other port on the repeater, so bandwidth and security problems are not addressed. The total throughput of the repeater is limited to that of a single link, and all links must operate at the same speed.

Bridging and Switching

Patch cables with patch fields of two Ethernet switches

While repeaters can isolate some aspects of Ethernet segments, such as cable breakages, they still forward all traffic to all Ethernet devices. This creates practical limits on how many machines can communicate on an Ethernet network. The entire network is one collision domain, and all hosts have to be able to detect collisions anywhere on the network. This limits the number of repeaters between the farthest nodes. Segments joined by repeaters have to all operate at the same speed, making phased-in upgrades impossible.

To alleviate these problems, bridging was created to communicate at the data link layer while isolating the physical layer. With bridging, only well-formed Ethernet packets are forwarded from one Ethernet segment to another; collisions and packet errors are isolated. At initial startup, Ethernet bridges (and switches) work somewhat like Ethernet repeaters, passing all traffic between segments. By observing the source addresses of incoming frames, the bridge then builds an address table associating addresses to segments. Once an address is learned, the bridge forwards network traffic destined for that address only to the associated segment, improving overall performance. Broadcast traffic is still forwarded to all network segments. Bridges also overcome the limits on total segments between two hosts and allow the mixing of speeds, both of which are critical to deployment of Fast Ethernet.

In 1989, the networking company Kalpana (acquired by Cisco Systems, Inc. in 1994) introduced their EtherSwitch, the first Ethernet switch. This works somewhat differently from an Ethernet bridge, where only the header of the incoming packet is examined before it is either dropped or forwarded to another segment. This greatly reduces the forwarding latency and the processing load on the network device. One drawback of this cut-through switching method is that packets that have been corrupted are still propagated through the network, so a jabbering station can continue to disrupt the entire network. The eventual remedy for this was a return to the original store and forward approach of bridging, where the packet would be read into a buffer on the switch in its entirety, verified against its checksum and then forwarded, but using more powerful application-specific integrated circuits. Hence, the bridging is then done in hardware, allowing packets to be forwarded at full wire speed.

When a twisted pair or fiber link segment is used and neither end is connected to a repeater, full-duplex Ethernet becomes possible over that segment. In full-duplex mode, both devices can transmit and receive to and from each other at the same time, and there is no collision domain. This doubles the aggregate bandwidth of the link and is sometimes advertised as double the link speed (for example, 200 Mbit/s). The elimination of the collision domain for these connections also means that all the link's bandwidth can be used by the two devices on that segment and that segment length is not limited by the need for correct collision detection.

Since packets are typically delivered only to the port they are intended for, traffic on a switched Ethernet is less public than on shared-medium Ethernet. Despite this, switched Ethernet should still be regarded as an insecure network technology, because it is easy to subvert switched Ethernet systems by means such as ARP spoofing and MAC flooding.

The bandwidth advantages, the improved isolation of devices from each other, the ability to easily mix different speeds of devices and the elimination of the chaining limits inherent in non-switched Ethernet have made switched Ethernet the dominant network technology.

Advanced Networking

A core Ethernet switch

Simple switched Ethernet networks, while a great improvement over repeater-based Ethernet, suffer from single points of failure, attacks that trick switches or hosts into sending data to a machine even if it is not intended for it, scalability and security issues with regard to switching loops, broadcast radiation and multicast traffic, and bandwidth choke points where a lot of traffic is forced down a single link.

Advanced networking features in switches and routers use shortest path bridging and spanning-tree protocol, for example, to maintain the active links of the network as a tree while allowing physical loops for redundancy, ensure port security and protection features such as MAC lockdown and broadcast radiation filtering, use virtual LANs to keep different classes of users separate while using the same physical infrastructure, employ multilayer switching to route between different classes, and use link aggregation to add bandwidth to overloaded links and to provide some redundancy.

IEEE 802.1aq (shortest path bridging) includes the use of the link-state routing protocol IS-IS to allow larger networks with shortest path routes between devices. In 2012, it was stated by David Allan and Nigel Bragg, in *802.1aq Shortest Path Bridging Design and Evolution: The Architect's Perspective* that shortest path bridging is one of the most significant enhancements in Ethernet's history.

Ethernet has replaced InfiniBand as the most popular system interconnect of TOP500 supercomputers.

Ethernet has replaced the ATM circuit-switched technology first developed in the early 1990s.

Since 2005, Ethernet enhancements for fiber-optic transmission, longer distances and much higher data rates (e.g. 40 Gbps, 100 Gbps) are also beginning to adversely impact sales and new deployments of SONET technology.

Varieties of Ethernet

The Ethernet physical layer evolved over a considerable time span and encompasses coaxial, twisted pair and fiber-optic physical media interfaces, with speeds from 10 Mbit/s to 100 Gbit/s, with 400 Gbit/s expected by late 2017. The first introduction of twisted-pair CSMA/CD was StarLAN, standardized as 802.3 1BASE5; while 1BASE5 had little market penetration, it defined the physical apparatus (wire, plug/jack, pin-out, and wiring plan) that would be carried over to 10BASE-T.

The most common forms used are 10BASE-T, 100BASE-TX, and 1000BASE-T. All three utilize twisted pair cables and 8P8C modular connectors. They run at 10 Mbit/s, 100 Mbit/s, and 1 Gbit/s, respectively. Fiber optic variants of Ethernet offer high performance, better electrical isolation and longer distance (tens of kilometers with some versions). In general, network protocol stack software will work similarly on all varieties.

Frame Structure

In IEEE 802.3, a datagram is called a *packet* or *frame*. *Packet* is used to describe the overall transmission unit and includes the preamble, start frame delimiter (SFD) and carrier extension (if present). The *frame* begins after the start frame delimiter with a frame header featuring source and destination MAC addresses and a field giving either the protocol type for the payload protocol or the length of the payload. The middle section of the frame consists of payload data including any headers for other protocols (for example, Internet Protocol) carried in the frame. The frame ends with a 32-bit cyclic redundancy check, which is used to detect corruption of data in transit.:sections 3.1.1 and 3.2 Notably, Ethernet packets have no time-to-live field, leading to possible problems in the presence of a switching loop.

A close-up of the SMSC LAN91C110 (SMSC 91x) chip, an embedded Ethernet chip.

Autonegotiation

Autonegotiation is the procedure by which two connected devices choose common transmission parameters, e.g. speed and duplex mode. Autonegotiation is an optional feature, first introduced with 100BASE-TX, while it is also backward compatible with 10BASE-T. Autonegotiation is mandatory for 1000BASE-T.

Internet

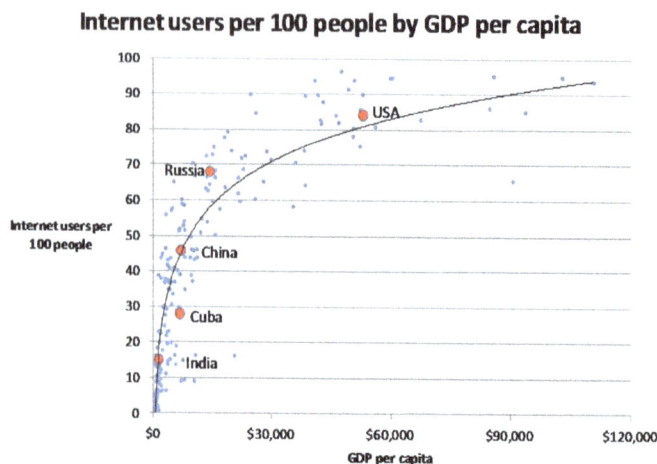

Internet users per 100 population members and GDP per capita for selected countries.

The Internet is the global system of interconnected computer networks that use the Internet protocol suite (TCP/IP) to link devices worldwide. It is a *network of networks* that consists of private, public, academic, business, and government networks of local to global scope, linked by a broad array of electronic, wireless, and optical networking technologies. The Internet carries an extensive range of information resources and services, such as the inter-linked hypertext documents and applications of the World Wide Web (WWW), electronic mail, telephony, and peer-to-peer networks for file sharing.

The origins of the Internet date back to research commissioned by the United States federal government in the 1960s to build robust, fault-tolerant communication via computer networks. The primary precursor network, the ARPANET, initially served as a backbone for interconnection of regional academic and military networks in the 1980s. The funding of the National Science Foundation Network as a new backbone in the 1980s, as well as private funding for other commercial extensions, led to worldwide participation in the development of new networking technologies, and the merger of many networks. The linking of commercial networks and enterprises by the early 1990s marks the beginning of the transition to the modern Internet, and generated a sustained exponential growth as generations of institutional, personal, and mobile computers were connected to the network. Although the Internet was widely used by academia since the 1980s, the commercialization incorporated its services and technologies into virtually every aspect of modern life.

Internet use grew rapidly in the West from the mid-1990s and from the late 1990s in the developing world. In the 20 years since 1995, Internet use has grown 100-times, measured for the period of one year, to over one third of the world population. Most traditional communications media, including telephony, radio, television, paper mail and newspapers are being reshaped or redefined by the Internet, giving birth to new services such as email, Internet telephony, Internet television music, digital newspapers, and video streaming websites. Newspaper, book, and other print publishing are adapting to website technology, or are reshaped into blogging, web feeds and online news aggregators. The entertainment industry was initially the fastest growing segment on the Internet. The Internet has enabled and accelerated new forms of personal interactions through instant messaging, Internet forums, and social networking. Online shopping has grown exponentially both for major retailers and small businesses and entrepreneurs, as it enables firms to extend their "bricks and mortar" presence to serve a larger market or even sell goods and services entirely online. Business-to-business and financial services on the Internet affect supply chains across entire industries.

The Internet has no centralized governance in either technological implementation or policies for access and usage; each constituent network sets its own policies. Only the overreaching definitions of the two principal name spaces in the Internet, the Internet Protocol address space and the Domain Name System (DNS), are directed by a maintainer organization, the Internet Corporation for Assigned Names and Numbers (ICANN). The technical underpinning and standardization of the core protocols is an activity of the Internet Engineering Task Force (IETF), a non-profit organization of loosely affiliated international participants that anyone may associate with by contributing technical expertise.

Terminology

The term *Internet*, when used to refer to the specific global system of interconnected Internet Protocol (IP) networks, is a proper noun and may be written with an initial capital letter. In common use and the media, it is often not capitalized, viz. *the internet*. Some guides specify that the word should be capitalized when used as a noun, but not capitalized when used as an adjective. The Internet is also often referred to as *the Net*, as a short form of *network*. Historically, as early as 1849, the word *internetted* was used uncapitalized as an adjective, meaning *interconnected* or *interwoven*. The designers of early computer networks used *internet* both as a noun and as a verb in shorthand form of internetwork or internetworking, meaning interconnecting computer networks.

The Internet Messenger by Buky Schwartz in Holon, Israel

The terms *Internet* and *World Wide Web* are often used interchangeably in everyday speech; it is common to speak of *"going on the Internet"* when invoking a web browser to view web pages. However, the World Wide Web or *the Web* is only one of a large number of Internet services. The Web is a collection of interconnected documents (web pages) and other web resources, linked by hyperlinks and URLs. As another point of comparison, Hypertext Transfer Protocol, or HTTP, is the language used on the Web for information transfer, yet it is just one of many languages or protocols that can be used for communication on the Internet. The term *Interweb* is a portmanteau of *Internet* and *World Wide Web* typically used sarcastically to parody a technically unsavvy user.

History

Research into packet switching started in the early 1960s, and packet switched networks such as the ARPANET, CYCLADES, the Merit Network, NPL network, Tymnet, and Telenet, were developed in the late 1960s and 1970s using a variety of protocols. The ARPANET project led to the development of protocols for internetworking, by which multiple separate networks could be joined into a single network of networks. ARPANET development began with two network nodes which were interconnected between the Network Measurement Center at the University of California, Los Angeles (UCLA) Henry Samueli School of Engineering and Applied Science directed by Leonard Kleinrock, and the NLS system at SRI International (SRI) by Douglas Engelbart in Menlo Park, California, on 29 October 1969. The third site was the Culler-Fried Interactive Mathematics Center at the University of California, Santa Barbara, followed by the University of Utah Graphics Department. In an early sign of future growth, fifteen sites were connected to the young ARPANET by the end of 1971. These early years were documented in the 1972 film *Computer Networks: The Heralds of Resource Sharing*.

Early international collaborations on the ARPANET were rare. European developers were concerned with developing the X.25 networks. Notable exceptions were the Norwegian Seismic Array (NORSAR) in June 1973, followed in 1973 by Sweden with satellite links to the Tanum Earth Station and Peter T. Kirstein's research group in the United Kingdom, initially at the Institute of Computer Science, University of London and later at University College London. In December 1974, RFC 675 (*Specification of Internet Transmission Control Program*), by Vinton Cerf, Yogen Dalal, and Carl Sunshine, used the term *internet* as a shorthand for *internetworking* and later RFCs repeated this use. Access to the ARPANET was expanded in 1981 when the National Science Foundation (NSF) funded the Computer Science Network (CSNET). In 1982, the Internet Protocol Suite (TCP/IP) was standardized, which permitted worldwide proliferation of interconnected networks.

NSFNET T3 Network 1992

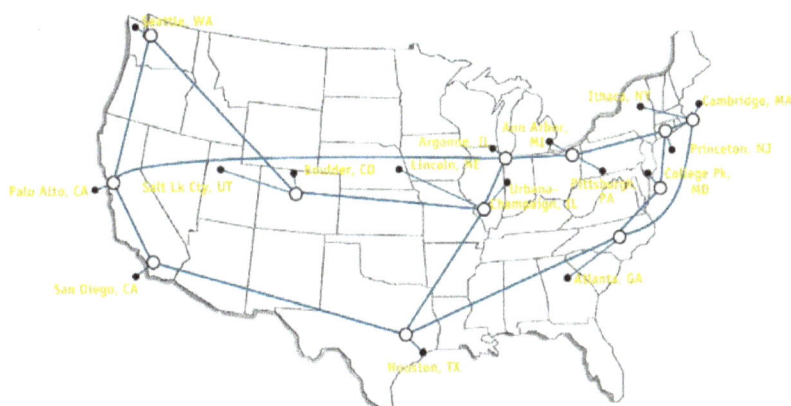

T3 NSFNET Backbone, c. 1992.

TCP/IP network access expanded again in 1986 when the National Science Foundation Network (NSFNet) provided access to supercomputer sites in the United States for researchers, first at speeds of 56 kbit/s and later at 1.5 Mbit/s and 45 Mbit/s. Commercial Internet service providers (ISPs) emerged in the late 1980s and early 1990s. The ARPANET was decommissioned in 1990. By 1995, the Internet was fully commercialized in the U.S. when the NSFNet was decommissioned, removing the last restrictions on use of the Internet to carry commercial traffic. The Internet rapidly expanded in Europe and Australia in the mid to late 1980s and to Asia in the late 1980s and early 1990s. The beginning of dedicated transatlantic communication between the NSFNET and networks in Europe was established with a low-speed satellite relay between Princeton University and Stockholm, Sweden in December 1988. Although other network protocols such as UUCP had global reach well before this time, this marked the beginning of the Internet as an intercontinental network.

Slightly over a year later in March 1990, the first high-speed T1 (1.5 Mbit/s) link between the NSF-NET and Europe was installed between Cornell University and CERN, allowing much more robust communications than were capable with satellites. Six months later Tim Berners-Lee would begin writing WorldWideWeb, the first web browser after two years of lobbying CERN management. By Christmas 1990, Berners-Lee had built all the tools necessary for a working Web: the HyperText Transfer Protocol (HTTP) 0.9, the HyperText Markup Language (HTML), the first Web browser (which was also a HTML editor and could access Usenet newsgroups and FTP files), the first HTTP server software (later known as CERN httpd), the first web server (http://info.cern.ch), and the first Web pages that described the project itself. Public commercial use of the Internet began in mid-1989 with the connection of MCI Mail and Compuserve's email capabilities to the 500,000 users of the Internet. Just months later on January 1, 1990, PSInet launched an alternate Internet backbone for commercial use; one of the networks that would grow into the commercial Internet we know today. In 1991 the Commercial Internet eXchange was founded, allowing PSInet to communicate with the other commercial networks CERFnet and Alternet. Since 1995 the Internet has tremendously impacted culture and commerce, including the rise of near instant communication by email, instant messaging, telephony (Voice over Internet Protocol or VoIP), two-way interactive video calls, and the World Wide Web with its discussion forums, blogs, social networking, and online shopping sites. Increasing amounts of data are transmitted at higher and higher speeds over fiber optic networks operating at 1-Gbit/s, 10-Gbit/s, or more.

The Internet continues to grow, driven by ever greater amounts of online information and knowledge, commerce, entertainment and social networking. During the late 1990s, it was estimated that traffic on the public Internet grew by 100 percent per year, while the mean annual growth in the number of Internet users was thought to be between 20% and 50%. This growth is often attributed to the lack of central administration, which allows organic growth of the network, as well as the non-proprietary nature of the Internet protocols, which encourages vendor interoperability and prevents any one company from exerting too much control over the network. As of 31 March 2011, the estimated total number of Internet users was 2.095 billion (30.2% of world population). It is estimated that in 1993 the Internet carried only 1% of the information flowing through two-way telecommunication, by 2000 this figure had grown to 51%, and by 2007 more than 97% of all telecommunicated information was carried over the Internet.

Governance

ICANN headquarters in the Playa Vista neighborhood of Los Angeles, California, United States.

The Internet is a global network comprising many voluntarily interconnected autonomous networks. It operates without a central governing body. The technical underpinning and standardization of the core protocols (IPv4 and IPv6) is an activity of the Internet Engineering Task Force (IETF), a non-profit organization of loosely affiliated international participants that anyone may associate with by contributing technical expertise. To maintain interoperability, the principal name spaces of the Internet are administered by the Internet Corporation for Assigned Names and Numbers (ICANN). ICANN is governed by an international board of directors drawn from across the Internet technical, business, academic, and other non-commercial communities. ICANN coordinates the assignment of unique identifiers for use on the Internet, including domain names, Internet Protocol (IP) addresses, application port numbers in the transport protocols, and many other parameters. Globally unified name spaces are essential for maintaining the global reach of the Internet. This role of ICANN distinguishes it as perhaps the only central coordinating body for the global Internet.

Regional Internet Registries (RIRs) allocate IP addresses:

- African Network Information Center (AfriNIC) for Africa

- American Registry for Internet Numbers (ARIN) for North America

- Asia-Pacific Network Information Centre (APNIC) for Asia and the Pacific region

- Latin American and Caribbean Internet Addresses Registry (LACNIC) for Latin America and the Caribbean region

- Réseaux IP Européens – Network Coordination Centre (RIPE NCC) for Europe, the Middle East, and Central Asia

The National Telecommunications and Information Administration, an agency of the United States Department of Commerce, continues to have final approval over changes to the DNS root zone. The Internet Society (ISOC) was founded in 1992 with a mission to *"assure the open development, evolution and use of the Internet for the benefit of all people throughout the world"*. Its members include individuals (anyone may join) as well as corporations, organizations, governments, and universities. Among other activities ISOC provides an administrative home for a number of less formally organized groups that are involved in developing and managing the Internet, including: the Internet Engineering Task Force (IETF), Internet Architecture Board (IAB), Internet Engineering Steering Group (IESG), Internet Research Task Force (IRTF), and Internet Research Steering Group (IRSG). On 16 November 2005, the United Nations-sponsored World Summit on the Information Society in Tunis established the Internet Governance Forum (IGF) to discuss Internet-related issues.

Infrastructure

The communications infrastructure of the Internet consists of its hardware components and a system of software layers that control various aspects of the architecture.

Routing and Service Tiers

Internet service providers establish the worldwide connectivity between individual networks at various levels of scope. End-users who only access the Internet when needed to perform a function or obtain information, represent the bottom of the routing hierarchy. At the top of the routing hierarchy are the tier 1 networks, large telecommunication companies that exchange traffic directly with each other via peering agreements. Tier 2 and lower level networks buy Internet transit from other providers to reach at least some parties on the global Internet, though they may also engage in peering. An ISP may use a single upstream provider for connectivity, or implement multihoming to achieve redundancy and load balancing. Internet exchange points are major traffic exchanges with physical connections to multiple ISPs. Large organizations, such as academic institutions, large enterprises, and governments, may perform the same function as ISPs, engaging in peering and purchasing transit on behalf of their internal networks. Research networks tend to interconnect with large subnetworks such as GEANT, GLORIAD, Internet2, and the UK's national research and education network, JANET. Both the Internet IP routing structure and hypertext links of the World Wide Web are examples of scale-free networks. Computers and routers use routing tables in their operating system to direct IP packets to the next-hop router or destination. Routing tables are maintained by manual configuration or automatically by routing protocols. End-nodes typically use a default route that points toward an ISP providing transit, while ISP routers use the Border Gateway Protocol to establish the most efficient routing across the complex connections of the global Internet.

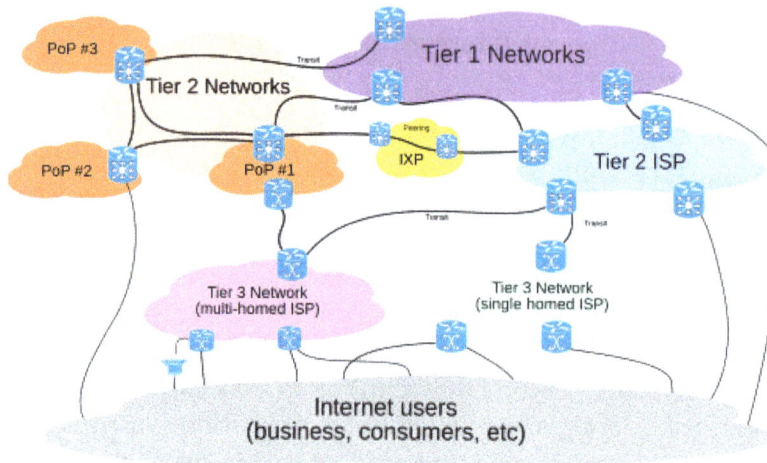

Packet routing across the Internet involves several tiers of Internet service providers.

Access

Common methods of Internet access by users include dial-up with a computer modem via telephone circuits, broadband over coaxial cable, fiber optics or copper wires, Wi-Fi, satellite and cellular telephone technology (3G, 4G). The Internet may often be accessed from computers in libraries and Internet cafes. Internet access points exist in many public places such as airport halls and coffee shops. Various terms are used, such as *public Internet kiosk*, *public access terminal*, and *Web payphone*. Many hotels also have public terminals, though these are usually fee-based. These terminals are widely accessed for various usages, such as ticket booking, bank deposit, or online payment. Wi-Fi provides wireless access to the Internet via local computer networks. Hotspots providing such access include Wi-Fi cafes, where users need to bring their own wireless devices such as a laptop or PDA. These services may be free to all, free to customers only, or fee-based.

Grassroots efforts have led to wireless community networks. Commercial Wi-Fi services covering large city areas are in place in New York, London, Vienna, Toronto, San Francisco, Philadelphia, Chicago and Pittsburgh. The Internet can then be accessed from such places as a park bench. Apart from Wi-Fi, there have been experiments with proprietary mobile wireless networks like Ricochet, various high-speed data services over cellular phone networks, and fixed wireless services. High-end mobile phones such as smartphones in general come with Internet access through the phone network. Web browsers such as Opera are available on these advanced handsets, which can also run a wide variety of other Internet software. More mobile phones have Internet access than PCs, though this is not as widely used. An Internet access provider and protocol matrix differentiates the methods used to get online.

Structure

Many computer scientists describe the Internet as a "prime example of a large-scale, highly engineered, yet highly complex system". The structure was found to be highly robust to random failures, yet, very vulnerable to intentional attacks. The Internet structure and its usage characteristics have been studied extensively and the possibility of developing alternative structures has been investigated.

Protocols

While the hardware components in the Internet infrastructure can often be used to support other software systems, it is the design and the standardization process of the software that characterizes the Internet and provides the foundation for its scalability and success. The responsibility for the architectural design of the Internet software systems has been assumed by the Internet Engineering Task Force (IETF). The IETF conducts standard-setting work groups, open to any individual, about the various aspects of Internet architecture. Resulting contributions and standards are published as *Request for Comments* (RFC) documents on the IETF web site. The principal methods of networking that enable the Internet are contained in specially designated RFCs that constitute the Internet Standards. Other less rigorous documents are simply informative, experimental, or historical, or document the best current practices (BCP) when implementing Internet technologies.

The Internet standards describe a framework known as the Internet protocol suite. This is a model architecture that divides methods into a layered system of protocols, originally documented in RFC 1122 and RFC 1123. The layers correspond to the environment or scope in which their services operate. At the top is the application layer, space for the application-specific networking methods used in software applications. For example, a web browser program uses the client-server application model and a specific protocol of interaction between servers and clients, while many file-sharing systems use a peer-to-peer paradigm. Below this top layer, the transport layer connects applications on different hosts with a logical channel through the network with appropriate data exchange methods.

Underlying these layers are the networking technologies that interconnect networks at their borders and hosts via the physical connections. The Internet layer enables computers to identify and locate each other via Internet Protocol (IP) addresses, and routes their traffic via intermediate (transit) networks. Last, at the bottom of the architecture is the link layer, which provides connectivity between hosts on the same network link, such as a physical connection in the form of a local area network (LAN) or a dial-up connection. The model, also known as TCP/IP, is designed to be independent of the underlying hardware, which the model, therefore, does not concern itself with in any detail. Other models have been developed, such as the OSI model, that attempt to be comprehensive in every aspect of communications. While many similarities exist between the models, they are not compatible in the details of description or implementation; indeed, TCP/IP protocols are usually included in the discussion of OSI networking.

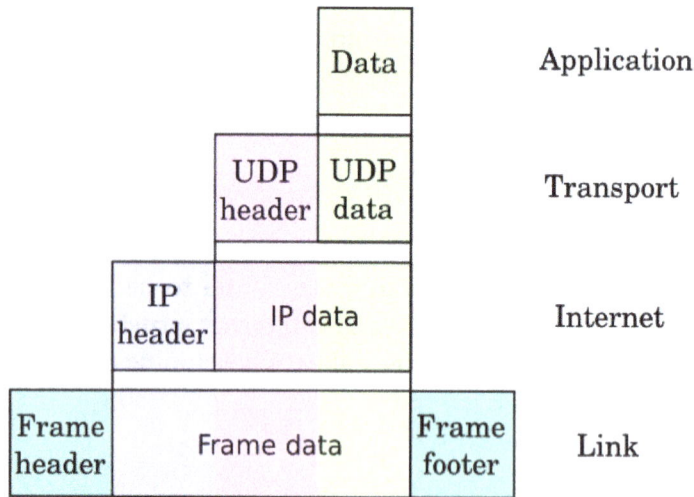

As user data is processed through the protocol stack, each abstraction layer adds encapsulation information at the sending host. Data is transmitted *over the wire* at the link level between hosts and routers. Encapsulation is removed by the receiving host. Intermediate relays update link encapsulation at each hop, and inspect the IP layer for routing purposes.

The most prominent component of the Internet model is the Internet Protocol (IP), which provides addressing systems (IP addresses) for computers on the Internet. IP enables internetworking and, in essence, establishes the Internet itself. Internet Protocol Version 4 (IPv4) is the initial version used on the first generation of the Internet and is still in dominant use. It was designed to address up to ~4.3 billion (10^9) Internet hosts. However, the explosive growth of the Internet has led to IPv4 address exhaustion, which entered its final stage in 2011, when the global address allocation pool was exhausted. A new protocol version, IPv6, was developed in the mid-1990s, which provides vastly larger addressing capabilities and more efficient routing of Internet traffic. IPv6 is currently in growing deployment around the world, since Internet address registries (RIRs) began to urge all resource managers to plan rapid adoption and conversion.

IPv6 is not directly interoperable by design with IPv4. In essence, it establishes a parallel version of the Internet not directly accessible with IPv4 software. Thus, translation facilities must exist for internetworking or nodes must have duplicate networking software for both networks. Essentially all modern computer operating systems support both versions of the Internet Protocol. Network infrastructure, however, is still lagging in this development. Aside from the complex array of physical connections that make up its infrastructure, the Internet is facilitated by bi- or multi-lateral commercial contracts, e.g., peering agreements, and by technical specifications or protocols that describe the exchange of data over the network. Indeed, the Internet is defined by its interconnections and routing policies.

Services

The Internet carries many network services, most prominently mobile apps such as social media apps, the World Wide Web, electronic mail, multiplayer online games, Internet telephony, and file sharing services.

World Wide Web

This NeXT Computer was used by Tim Berners-Lee at CERN and became the world's first Web server.

Many people use the terms *Internet* and *World Wide Web*, or just the *Web*, interchangeably, but the two terms are not synonymous. The World Wide Web is the primary application that billions of people use on the Internet, and it has changed their lives immeasurably. However, the Internet provides many other services. The Web is a global set of documents, images and other resources, logically interrelated by hyperlinks and referenced with Uniform Resource Identifiers (URIs). URIs symbolically identify services, servers, and other databases, and the documents and resources that they can provide. Hypertext Transfer Protocol (HTTP) is the main access protocol of the World Wide Web. Web services also use HTTP to allow software systems to communicate in order to share and exchange business logic and data.

World Wide Web browser software, such as Microsoft's Internet Explorer, Mozilla Firefox, Opera, Apple's Safari, and Google Chrome, lets users navigate from one web page to another via hyperlinks embedded in the documents. These documents may also contain any combination of computer data, including graphics, sounds, text, video, multimedia and interactive content that runs while the user is interacting with the page. Client-side software can include animations, games, office applications and scientific demonstrations. Through keyword-driven Internet research using search engines like Yahoo! and Google, users worldwide have easy, instant access to a vast and diverse amount of online information. Compared to printed media, books, encyclopedias and traditional libraries, the World Wide Web has enabled the decentralization of information on a large scale.

The Web has also enabled individuals and organizations to publish ideas and information to a potentially large audience online at greatly reduced expense and time delay. Publishing a web page, a blog, or building a website involves little initial cost and many cost-free services are available. However, publishing and maintaining large, professional web sites with attractive, diverse and up-to-date information is still a difficult and expensive proposition. Many individuals and some companies and groups use *web logs* or blogs, which are largely used as easily updatable online diaries. Some commercial organizations encourage staff to communicate advice in their areas of specialization in the hope that visitors will be impressed by the expert knowledge and free information, and be attracted to the corporation as a result.

One example of this practice is Microsoft, whose product developers publish their personal blogs in order to pique the public's interest in their work. Collections of personal web pages published by large service providers remain popular and have become increasingly sophisticated. Whereas operations such as Angelfire and GeoCities have existed since the early days of the Web, newer offerings from, for example, Facebook and Twitter currently have large followings. These operations often brand themselves as social network services rather than simply as web page hosts.

Advertising on popular web pages can be lucrative, and e-commerce or the sale of products and services directly via the Web continues to grow. Online advertising is a form of marketing and advertising which uses the Internet to deliver promotional marketing messages to consumers. It includes email marketing, search engine marketing (SEM), social media marketing, many types of display advertising (including web banner advertising), and mobile advertising. In 2011, Internet advertising revenues in the United States surpassed those of cable television and nearly exceeded those of broadcast television. Many common online advertising practices are controversial and increasingly subject to regulation.

When the Web developed in the 1990s, a typical web page was stored in completed form on a web server, formatted in HTML, complete for transmission to a web browser in response to a request. Over time, the process of creating and serving web pages has become dynamic, creating a flexible design, layout, and content. Websites are often created using content management software with, initially, very little content. Contributors to these systems, who may be paid staff, members of an organization or the public, fill underlying databases with content using editing pages designed for that purpose while casual visitors view and read this content in HTML form. There may or may not be editorial, approval and security systems built into the process of taking newly entered content and making it available to the target visitors.

Communication

Email is an important communications service available on the Internet. The concept of sending electronic text messages between parties in a way analogous to mailing letters or memos predates the creation of the Internet. Pictures, documents, and other files are sent as email attachments. Emails can be cc-ed to multiple email addresses.

Internet telephony is another common communications service made possible by the creation of the Internet. VoIP stands for Voice-over-Internet Protocol, referring to the protocol that underlies all Internet communication. The idea began in the early 1990s with walkie-talkie-like voice applications for personal computers. In recent years many VoIP systems have become as easy to use and as convenient as a normal telephone. The benefit is that, as the Internet carries the voice traffic, VoIP can be free or cost much less than a traditional telephone call, especially over long distances and especially for those with always-on Internet connections such as cable or ADSL. VoIP is maturing into a competitive alternative to traditional telephone service. Interoperability between different providers has improved and the ability to call or receive a call from a traditional telephone is available. Simple, inexpensive VoIP network adapters are available that eliminate the need for a personal computer.

Voice quality can still vary from call to call, but is often equal to and can even exceed that of traditional calls. Remaining problems for VoIP include emergency telephone number dialing and reliability. Currently, a few VoIP providers provide an emergency service, but it is not universally available. Older traditional phones with no "extra features" may be line-powered only and operate during a power failure; VoIP can never do so without a backup power source for the phone equipment and the Internet access devices. VoIP has also become increasingly popular for gaming applications, as a form of communication between players. Popular VoIP clients for gaming include Ventrilo and Teamspeak. Modern video game consoles also offer VoIP chat features.

Data Transfer

File sharing is an example of transferring large amounts of data across the Internet. A computer file can be emailed to customers, colleagues and friends as an attachment. It can be uploaded to a website or File Transfer Protocol (FTP) server for easy download by others. It can be put into a "shared location" or onto a file server for instant use by colleagues. The load of bulk downloads to many users can be eased by the use of "mirror" servers or peer-to-peer networks. In any of these cases, access to the file may be controlled by user authentication, the transit of the file over the Internet may be obscured by encryption, and money may change hands for access to the file. The price can be paid by the remote charging of funds from, for example, a credit card whose details are also passed – usually fully encrypted – across the Internet. The origin and authenticity of the file received may be checked by digital signatures or by MD5 or other message digests. These simple features of the Internet, over a worldwide basis, are changing the production, sale, and distribution of anything that can be reduced to a computer file for transmission. This includes all manner of print publications, software products, news, music, film, video, photography, graphics and the other arts. This in turn has caused seismic shifts in each of the existing industries that previously controlled the production and distribution of these products.

Streaming media is the real-time delivery of digital media for the immediate consumption or enjoyment by end users. Many radio and television broadcasters provide Internet feeds of their live audio and video productions. They may also allow time-shift viewing or listening such as Preview, Classic Clips and Listen Again features. These providers have been joined by a range of pure Internet "broadcasters" who never had on-air licenses. This means that an Internet-connected device, such as a computer or something more specific, can be used to access on-line media in much the same way as was previously possible only with a television or radio receiver. The range of available types of content is much wider, from specialized technical webcasts to on-demand popular multimedia services. Podcasting is a variation on this theme, where – usually audio – material is downloaded and played back on a computer or shifted to a portable media player to be listened to on the move. These techniques using simple equipment allow anybody, with little censorship or licensing control, to broadcast audio-visual material worldwide.

Digital media streaming increases the demand for network bandwidth. For example, standard image quality needs 1 Mbit/s link speed for SD 480p, HD 720p quality requires 2.5 Mbit/s, and the top-of-the-line HDX quality needs 4.5 Mbit/s for 1080p.

Webcams are a low-cost extension of this phenomenon. While some webcams can give full-frame-rate video, the picture either is usually small or updates slowly. Internet users can watch animals around an African waterhole, ships in the Panama Canal, traffic at a local roundabout or monitor their own premises, live and in real time. Video chat rooms and video conferencing are also popular with many uses being found for personal webcams, with and without two-way sound. YouTube was founded on 15 February 2005 and is now the leading website for free streaming video with a vast number of users. It uses a flash-based web player to stream and show video files. Registered users may upload an unlimited amount of video and build their own personal profile. YouTube claims that its users watch hundreds of millions, and upload hundreds of thousands of videos daily. Currently, YouTube also uses an HTML5 player.

Social Impact

The Internet has enabled new forms of social interaction, activities, and social associations. This phenomenon has given rise to the scholarly study of the sociology of the Internet.

Users

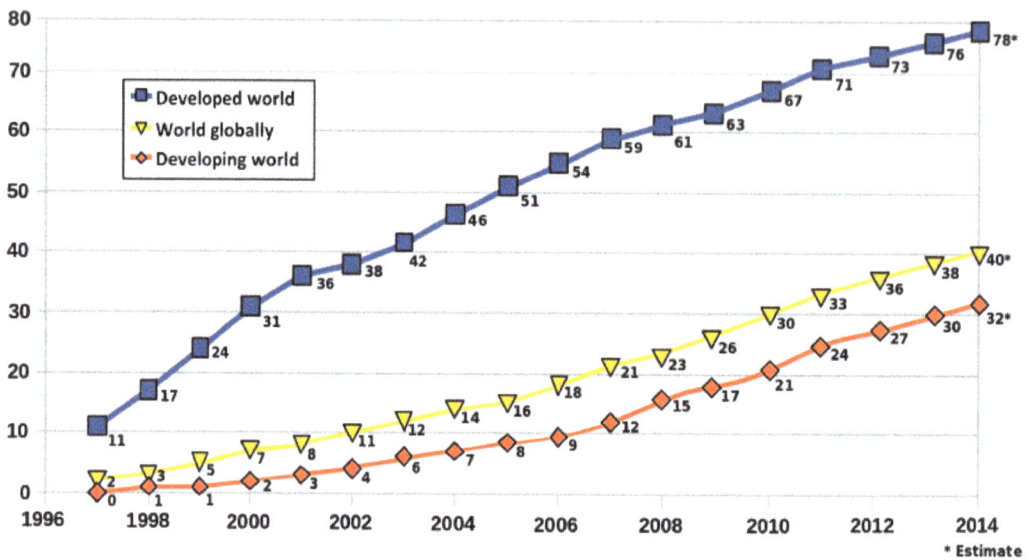

Internet Users Per 100 Inhabitants

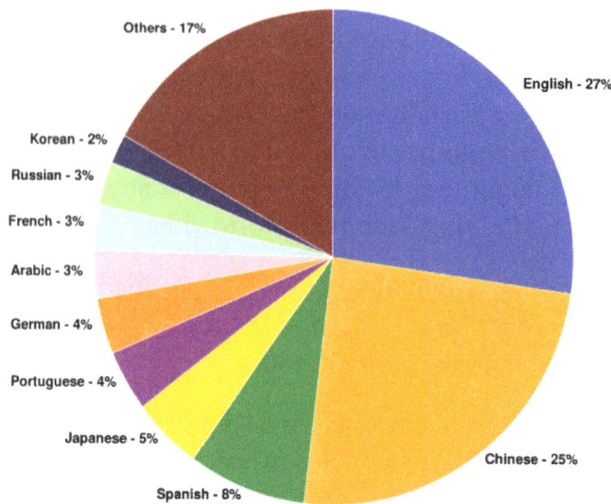

Internet Users by Language

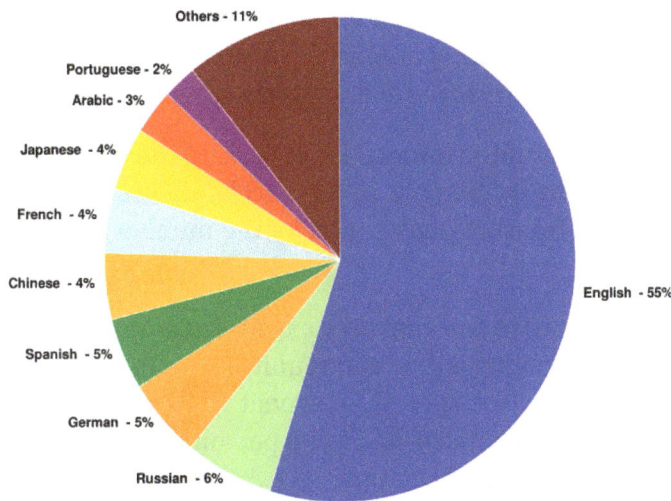

Website Content Languages

Internet usage has seen tremendous growth. From 2000 to 2009, the number of Internet users globally rose from 394 million to 1.858 billion. By 2010, 22 percent of the world's population had access to computers with 1 billion Google searches every day, 300 million Internet users reading blogs, and 2 billion videos viewed daily on YouTube. In 2014 the world's Internet users surpassed 3 billion or 43.6 percent of world population, but two-thirds of the users came from richest countries, with 78.0 percent of Europe countries population using the Internet, followed by 57.4 percent of the Americas.

The prevalent language for communication on the Internet has been English. This may be a result of the origin of the Internet, as well as the language's role as a lingua franca. Early computer systems were limited to the characters in the American Standard Code for Information Interchange (ASCII), a subset of the Latin alphabet.

After English (27%), the most requested languages on the World Wide Web are Chinese (25%), Spanish (8%), Japanese (5%), Portuguese and German (4% each), Arabic, French and Russian (3% each), and Korean (2%). By region, 42% of the world's Internet users are based in Asia, 24% in Europe, 14% in North America, 10% in Latin America and the Caribbean taken together, 6% in Africa, 3% in the Middle East and 1% in Australia/Oceania. The Internet's technologies have developed enough in recent years, especially in the use of Unicode, that good facilities are available for development and communication in the world's widely used languages. However, some glitches such as *mojibake* (incorrect display of some languages' characters) still remain.

In an American study in 2005, the percentage of men using the Internet was very slightly ahead of the percentage of women, although this difference reversed in those under 30. Men logged on more often, spent more time online, and were more likely to be broadband users, whereas women tended to make more use of opportunities to communicate (such as email). Men were more likely to use the Internet to pay bills, participate in auctions, and for recreation such as downloading music and videos. Men and women were equally likely to use the Internet for shopping and banking. More recent studies indicate that in 2008, women significantly outnumbered men on most social networking sites, such as Facebook and Myspace, although the ratios varied with age. In addition, women watched more streaming content, whereas men downloaded more. In terms of blogs, men were more likely to blog in the first place; among those who blog, men were more likely to have a professional blog, whereas women were more likely to have a personal blog.

According to forecasts by Euromonitor International, 44% of the world's population will be users of the Internet by 2020. Splitting by country, in 2012 Iceland, Norway, Sweden, the Netherlands, and Denmark had the highest Internet penetration by the number of users, with 93% or more of the population with access.

Several neologisms exist that refer to Internet users: Netizen (as in as in "citizen of the net") refers to those actively involved in improving online communities, the Internet in general or surrounding political affairs and rights such as free speech, Internaut refers to operators or technically highly capable users of the Internet, digital citizen refers to a person using the Internet in order to engage in society, politics, and government participation.

Usage

The Internet allows greater flexibility in working hours and location, especially with the spread of unmetered high-speed connections. The Internet can be accessed almost anywhere by numerous means, including through mobile Internet devices. Mobile phones, datacards, handheld game consoles and cellular routers allow users to connect to the Internet wirelessly. Within the limitations imposed by small screens and other limited facilities of such pocket-sized devices, the services of the Internet, including email and the web, may be available. Service providers may restrict the services offered and mobile data charges may be significantly higher than other access methods.

Educational material at all levels from pre-school to post-doctoral is available from websites. Examples range from CBeebies, through school and high-school revision guides and virtual universities, to access to top-end scholarly literature through the likes of Google Scholar. For distance education, help with homework and other assignments, self-guided learning, whiling away spare time, or just looking up more detail on an interesting fact, it has never been easier for people to access educational information at any level from anywhere. The Internet in general and the World Wide Web in particular are important enablers of both formal and informal education. Further, the Internet allows universities, in particular, researchers from the social and behavioral sciences, to conduct research remotely via virtual laboratories, with profound changes in reach and generalizability of findings as well as in communication between scientists and in the publication of results.

The low cost and nearly instantaneous sharing of ideas, knowledge, and skills have made collaborative work dramatically easier, with the help of collaborative software. Not only can a group cheaply communicate and share ideas but the wide reach of the Internet allows such groups more easily to form. An example of this is the free software movement, which has produced, among other things, Linux, Mozilla Firefox, and OpenOffice.org. Internet chat, whether using an IRC chat room, an instant messaging system, or a social networking website, allows colleagues to stay in touch in a very convenient way while working at their computers during the day. Messages can be exchanged even more quickly and conveniently than via email. These systems may allow files to be exchanged, drawings and images to be shared, or voice and video contact between team members.

Content management systems allow collaborating teams to work on shared sets of documents simultaneously without accidentally destroying each other's work. Business and project teams can share calendars as well as documents and other information. Such collaboration occurs in a wide variety of areas including scientific research, software development, conference planning, political activism and creative writing. Social and political collaboration is also becoming more widespread as both Internet access and computer literacy spread.

The Internet allows computer users to remotely access other computers and information stores easily from any access point. Access may be with computer security, i.e. authentication and encryption technologies, depending on the requirements. This is encouraging new ways of working from home, collaboration and information sharing in many industries. An accountant sitting at home can audit the books of a company based in another country, on a server situated in a third country that is remotely maintained by IT specialists in a fourth. These accounts could have been created by home-working bookkeepers, in other remote locations, based on information emailed to them from offices all over the world. Some of these things were possible before the widespread use of the Internet, but the cost of private leased lines would have made many of them infeasible in practice. An office worker away from their desk, perhaps on the other side of the world on a business trip or a holiday, can access their emails, access their data using cloud computing, or open a remote desktop session into their office PC using a secure virtual private network (VPN) connection on the Internet. This can give the worker complete access to all of their normal files and data, including email and other applications, while away from the office. It has been referred to among system administrators as the Virtual Private Nightmare, because it extends the secure perimeter of a corporate network into remote locations and its employees' homes.

Social Networking and Entertainment

Many people use the World Wide Web to access news, weather and sports reports, to plan and book vacations and to pursue their personal interests. People use chat, messaging and email to make and stay in touch with friends worldwide, sometimes in the same way as some previously had pen pals. Social networking websites such as Facebook, Twitter, and Myspace have created new ways to socialize and interact. Users of these sites are able to add a wide variety of information to pages, to pursue common interests, and to connect with others. It is also possible to find existing acquaintances, to allow communication among existing groups of people. Sites like LinkedIn foster commercial and business connections. YouTube and Flickr specialize in users' videos and photographs. While social networking sites were initially for individuals only, today they are widely used by businesses and other organizations to promote their brands, to market to their customers and to encourage posts to "go viral". "Black hat" social media techniques are also employed by some organizations, such as spam accounts and astroturfing.

A risk for both individuals and organizations writing posts (especially public posts) on social networking websites, is that especially foolish or controversial posts occasionally lead to an unexpected and possibly large-scale backlash on social media from other Internet users. This is also a risk in relation to controversial *offline* behavior, if it is widely made known. The nature of this backlash can range widely from counter-arguments and public mockery, through insults and hate speech, to, in extreme cases, rape and death threats. The online disinhibition effect describes the tendency of many individuals to behave more stridently or offensively online than they would in person.

A significant number of feminist women have been the target of various forms of harassment in response to posts they have made on social media, and Twitter in particular has been criticised in the past for not doing enough to aid victims of online abuse.

For organizations, such a backlash can cause overall brand damage, especially if reported by the media. However, this is not always the case, as any brand damage in the eyes of people with an opposing opinion to that presented by the organization could sometimes be outweighed by strengthening the brand in the eyes of others. Furthermore, if an organization or individual gives in to demands that others perceive as wrong-headed, that can then provoke a counter-backlash.

Some websites, such as Reddit, have rules forbidding the posting of personal information of individuals (also known as doxxing), due to concerns about such postings leading to mobs of large numbers of Internet users directing harassment at the specific individuals thereby identified. In particular, the Reddit rule forbidding the posting of personal information is widely understood to imply that all identifying photos and names must be censored in Facebook screenshots posted to Reddit. However, the interpretation of this rule in relation to public Twitter posts is less clear, and in any case, like-minded people online have many other ways they can use to direct each other's attention to public social media posts they disagree with.

Children also face dangers online such as cyberbullying and approaches by sexual predators, who sometimes pose as children themselves. Children may also encounter material which they may find upsetting, or material which their parents consider to be not age-appropriate. Due to naivety, they may also post personal information about themselves online, which could put them or their families at risk unless warned not to do so. Many parents choose to enable Internet filtering, and/or supervise their children's online activities, in an attempt to protect their children from inappropriate material on the Internet. The most popular social networking websites, such as Facebook and Twitter, commonly forbid users under the age of 13. However, these policies are typically trivial to circumvent by registering an account with a false birth date, and a significant number of children aged under 13 join such sites anyway. Social networking sites for younger children, which claim to provide better levels of protection for children, also exist.

The Internet has been a major outlet for leisure activity since its inception, with entertaining social experiments such as MUDs and MOOs being conducted on university servers, and humor-related Usenet groups receiving much traffic. Today, many Internet forums have sections devoted to games and funny videos. Over 6 million people use blogs or message boards as a means of communication and for the sharing of ideas. The Internet pornography and online gambling industries have taken advantage of the World Wide Web, and often provide a significant source of advertising revenue for other websites. Although many governments have attempted to restrict both industries' use of the Internet, in general, this has failed to stop their widespread popularity.

Another area of leisure activity on the Internet is multiplayer gaming. This form of recreation creates communities, where people of all ages and origins enjoy the fast-paced world of multiplayer games. These range from MMORPG to first-person shooters, from role-playing video games to online gambling. While online gaming has been around since the 1970s, modern modes of online gaming began with subscription services such as GameSpy and MPlayer. Non-subscribers were limited to certain types of game play or certain games. Many people use the Internet to access and download music, movies and other works for their enjoyment and relaxation. Free and fee-based services exist for all of these activities, using centralized servers and distributed peer-to-peer technologies. Some of these sources exercise more care with respect to the original artists' copyrights than others.

Internet usage has been correlated to users' loneliness. Lonely people tend to use the Internet as an outlet for their feelings and to share their stories with others, such as in the "I am lonely will anyone speak to me" thread.

Cybersectarianism is a new organizational form which involves: "highly dispersed small groups of practitioners that may remain largely anonymous within the larger social context and operate in relative secrecy, while still linked remotely to a larger network of believers who share a set of practices and texts, and often a common devotion to a particular leader. Overseas supporters provide funding and support; domestic practitioners distribute tracts, participate in acts of resistance, and share information on the internal situation with outsiders. Collectively, members and practitioners of such sects construct viable virtual communities of faith, exchanging personal testimonies and engaging in the collective study via email, on-line chat rooms, and web-based message boards." In particular, the British government has raised concerns about the prospect of young British Muslims being indoctrinated into Islamic extremism by material on the Internet, being persuaded to join terrorist groups such as the so-called "Islamic State", and then potentially committing acts of terrorism on returning to Britain after fighting in Syria or Iraq.

Cyberslacking can become a drain on corporate resources; the average UK employee spent 57 minutes a day surfing the Web while at work, according to a 2003 study by Peninsula Business Services. Internet addiction disorder is excessive computer use that interferes with daily life. Psychologist, Nicolas Carr believe that Internet use has other effects on individuals, for instance improving skills of scan-reading and interfering with the deep thinking that leads to true creativity.

Electronic Business

Electronic business (*e-business*) encompasses business processes spanning the entire value chain: purchasing, supply chain management, marketing, sales, customer service, and business relationship. E-commerce seeks to add revenue streams using the Internet to build and enhance relationships with clients and partners. According to International Data Corporation, the size of worldwide e-commerce, when global business-to-business and -consumer transactions are combined, equate to $16 trillion for 2013. A report by Oxford Economics adds those two together to estimate the total size of the digital economy at $20.4 trillion, equivalent to roughly 13.8% of global sales.

While much has been written of the economic advantages of Internet-enabled commerce, there is also evidence that some aspects of the Internet such as maps and location-aware services may serve to reinforce economic inequality and the digital divide. Electronic commerce may be responsible for consolidation and the decline of mom-and-pop, brick and mortar businesses resulting in increases in income inequality.

Author Andrew Keen, a long-time critic of the social transformations caused by the Internet, has recently focused on the economic effects of consolidation from Internet businesses. Keen cites a 2013 Institute for Local Self-Reliance report saying brick-and-mortar retailers employ 47 people for every $10 million in sales while Amazon employs only 14. Similarly, the 700-employee room rental start-up Airbnb was valued at $10 billion in 2014, about half as much as Hilton Hotels, which employs 152,000 people. And car-sharing Internet startup Uber employs 1,000 full-time employees and is valued at $18.2 billion, about the same valuation as Avis and Hertz combined, which together employ almost 60,000 people.

Telecommuting

Telecommuting is the performance within a traditional worker and employer relationship when it is facilitated by tools such as groupware, virtual private networks, conference calling, videoconferencing, and voice over IP (VOIP) so that work may be performed from any location, most conveniently the worker's home. It can be efficient and useful for companies as it allows workers to communicate over long distances, saving significant amounts of travel time and cost. As broadband Internet connections become commonplace, more workers have adequate bandwidth at home to use these tools to link their home to their corporate intranet and internal communication networks.

Crowdsourcing

The Internet provides a particularly good venue for crowdsourcing, because individuals tend to be more open in web-based projects where they are not being physically judged or scrutinized and thus can feel more comfortable sharing.

Collaborative Publishing

Wikis have also been used in the academic community for sharing and dissemination of information across institutional and international boundaries. In those settings, they have been found useful for collaboration on grant writing, strategic planning, departmental documentation, and committee work. The United States Patent and Trademark Office uses a wiki to allow the public to collaborate on finding prior art relevant to examination of pending patent applications. Queens, New York has used a wiki to allow citizens to collaborate on the design and planning of a local park. The English Wikipedia has the largest user base among wikis on the World Wide Web and ranks in the top 10 among all Web sites in terms of traffic.

Politics and Political Revolutions

Banner in Bangkok during the 2014 Thai coup d'état, informing the Thai public that 'like' or 'share' activities on social media could result in imprisonment (observed June 30, 2014).

The Internet has achieved new relevance as a political tool. The presidential campaign of Howard Dean in 2004 in the United States was notable for its success in soliciting donation via the Internet. Many political groups use the Internet to achieve a new method of organizing for carrying out their mission, having given rise to Internet activism, most notably practiced by rebels in the Arab Spring. *The New York Times* suggested that social media websites, such as Facebook and Twitter, helped people organize the political revolutions in Egypt, by helping activists organize protests, communicate grievances, and disseminate information.

The potential of the Internet as a civic tool of communicative power was explored by Simon R. B. Berdal in his 2004 thesis:

As the globally evolving Internet provides ever new access points to virtual discourse forums, it also promotes new civic relations and associations within which communicative power may flow and accumulate. Thus, traditionally ... national-embedded peripheries get entangled into greater, international peripheries, with stronger combined powers... The Internet, as a consequence, changes the topology of the "centre-periphery" model, by stimulating conventional peripheries to interlink into "super-periphery" structures, which enclose and "besiege" several centres at once.

Berdal, therefore, extends the Habermasian notion of the *public sphere* to the Internet, and underlines the inherent global and civic nature that interwoven Internet technologies provide. To limit the growing civic potential of the Internet, Berdal also notes how "self-protective measures" are put in place by those threatened by it:

If we consider China's attempts to filter "unsuitable material" from the Internet, most of us would agree that this resembles a self-protective measure by the system against the growing civic potentials of the Internet. Nevertheless, both types represent limitations to "peripheral capacities". Thus, the Chinese government tries to prevent communicative power to build up and unleash (as the 1989 Tiananmen Square uprising suggests, the government may find it wise to install "upstream measures"). Even though limited, the Internet is proving to be an empowering tool also to the Chinese periphery: Analysts believe that Internet petitions have influenced policy implementation in favour of the public's online-articulated will ...

Incidents of politically motivated Internet censorship have now been recorded in many countries, including western democracies.

Philanthropy

The spread of low-cost Internet access in developing countries has opened up new possibilities for peer-to-peer charities, which allow individuals to contribute small amounts to charitable projects for other individuals. Websites, such as DonorsChoose and GlobalGiving, allow small-scale donors to direct funds to individual projects of their choice. A popular twist on Internet-based philanthropy is the use of peer-to-peer lending for charitable purposes. Kiva pioneered this concept in 2005, offering the first web-based service to publish individual loan profiles for funding. Kiva raises funds for local intermediary microfinance organizations which post stories and updates on behalf of the borrowers. Lenders can contribute as little as $25 to loans of their choice, and receive their money back as borrowers repay. Kiva falls short of being a pure peer-to-peer charity, in that loans are disbursed before being funded by lenders and borrowers do not communicate with lenders themselves.

However, the recent spread of low-cost Internet access in developing countries has made genuine international person-to-person philanthropy increasingly feasible. In 2009, the US-based nonprofit Zidisha tapped into this trend to offer the first person-to-person microfinance platform to link lenders and borrowers across international borders without intermediaries. Members can fund loans for as little as a dollar, which the borrowers then use to develop business activities that improve their families' incomes while repaying loans to the members with interest. Borrowers access the Internet via public cybercafes, donated laptops in village schools, and even smart phones, then create their own profile pages through which they share photos and information about themselves and their businesses. As they repay their loans, borrowers continue to share updates and dialogue with lenders via their profile pages. This direct web-based connection allows members themselves to take on many of the communication and recording tasks traditionally performed by local organizations, bypassing geographic barriers and dramatically reducing the cost of microfinance services to the entrepreneurs.

Security

Internet resources, hardware, and software components are the target of criminal or malicious attempts to gain unauthorized control to cause interruptions, commit fraud, engage in blackmail or access private information. Such attempts include computer viruses which copy with the help of humans, computer worms which copy themselves automatically, denial of service attacks, ransomware, botnets, and spyware that reports on the activity and typing of users. Usually, these activities constitute cybercrime. Defense theorists have also speculated about the possibilities of cyber warfare using similar methods on a large scale.

Surveillance

The vast majority of computer surveillance involves the monitoring of data and traffic on the Internet. In the United States for example, under the Communications Assistance For Law Enforcement Act, all phone calls and broadband Internet traffic (emails, web traffic, instant messaging, etc.) are required to be available for unimpeded real-time monitoring by Federal law enforcement agencies. Packet capture is the monitoring of data traffic on a computer network. Computers communicate over the Internet by breaking up messages (emails, images, videos, web pages, files, etc.) into small chunks called "packets", which are routed through a network of computers, until they reach their destination, where they are assembled back into a complete "message" again. Packet Capture Appliance intercepts these packets as they are traveling through the network, in order to examine their contents using other programs. A packet capture is an information *gathering* tool, but not an *analysis* tool. That is it gathers "messages" but it does not analyze them and figure out what they mean. Other programs are needed to perform traffic analysis and sift through intercepted data looking for important/useful information. Under the Communications Assistance For Law Enforcement Act all U.S. telecommunications providers are required to install packet sniffing technology to allow Federal law enforcement and intelligence agencies to intercept all of their customers' broadband Internet and voice over Internet protocol (VoIP) traffic.

The large amount of data gathered from packet capturing requires surveillance software that filters and reports relevant information, such as the use of certain words or phrases, the access of certain types of web sites, or communicating via email or chat with certain parties. Agencies, such as the Information Awareness Office, NSA, GCHQ and the FBI, spend billions of dollars per year to develop, purchase, implement, and operate systems for interception and analysis of data. Similar systems are operated by Iranian secret police to identify and suppress dissidents. The required hardware and software was allegedly installed by German Siemens AG and Finnish Nokia.

Censorship

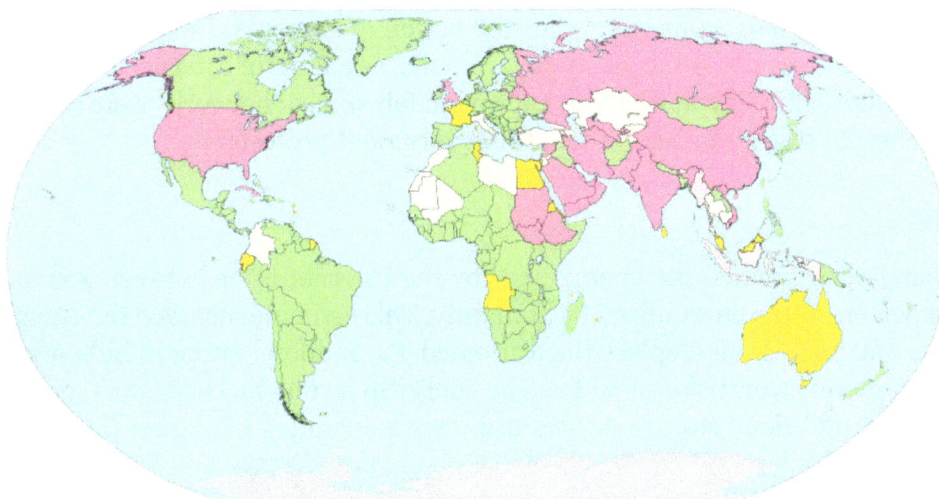

Internet censorship and surveillance by country

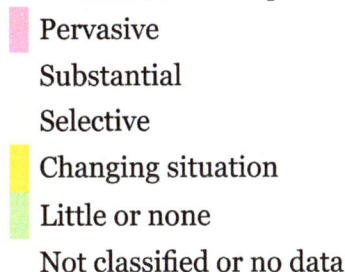

▮ Pervasive

Substantial

Selective

▮ Changing situation

▮ Little or none

Not classified or no data

Some governments, such as those of Burma, Iran, North Korea, the Mainland China, Saudi Arabia and the United Arab Emirates restrict access to content on the Internet within their territories, especially to political and religious content, with domain name and keyword filters.

In Norway, Denmark, Finland, and Sweden, major Internet service providers have voluntarily agreed to restrict access to sites listed by authorities. While this list of forbidden resources is supposed to contain only known child pornography sites, the content of the list is secret. Many countries, including the United States, have enacted laws against the possession or distribution of certain material, such as child pornography, via the Internet, but do not mandate filter software. Many free or commercially available software programs, called content-control software are available to users to block offensive websites on individual computers or networks, in order to limit access by children to pornographic material or depiction of violence.

Performance

As the Internet is a heterogeneous network, the physical characteristics, including for example the data transfer rates of connections, vary widely. It exhibits emergent phenomena that depend on its large-scale organization.

Outages

An Internet blackout or outage can be caused by local signalling interruptions. Disruptions of submarine communications cables may cause blackouts or slowdowns to large areas, such as in the 2008 submarine cable disruption. Less-developed countries are more vulnerable due to a small number of high-capacity links. Land cables are also vulnerable, as in 2011 when a woman digging for scrap metal severed most connectivity for the nation of Armenia. Internet blackouts affecting almost entire countries can be achieved by governments as a form of Internet censorship, as in the blockage of the Internet in Egypt, whereby approximately 93% of networks were without access in 2011 in an attempt to stop mobilization for anti-government protests.

Energy Use

In 2011, researchers estimated the energy used by the Internet to be between 170 and 307 GW, less than two percent of the energy used by humanity. This estimate included the energy needed to build, operate, and periodically replace the estimated 750 million laptops, a billion smart phones and 100 million servers worldwide as well as the energy that routers, cell towers, optical switches, Wi-Fi transmitters and cloud storage devices use when transmitting Internet traffic.

Wireless Network

A wireless network is any type of computer network that uses wireless data connections for connecting network nodes

Wireless networking is a method by which homes, telecommunications networks and enterprise (business) installations avoid the costly process of introducing cables into a building, or as a connection between various equipment locations. Wireless telecommunications networks are generally implemented and administered using radio communication. This implementation takes place at the physical level (layer) of the OSI model network structure.

Wireless icon

Examples of wireless networks include cell phone networks, Wireless local networks, wireless sensor networks, satellite communication networks, and terrestrial microwave networks.

History

Wireless Links

- *Terrestrial microwave* – Terrestrial microwave communication uses Earth-based transmitters and receivers resembling satellite dishes. Terrestrial microwaves are in the low gigahertz range, which limits all communications to line-of-sight. Relay stations are spaced approximately 48 km (30 mi) apart.

- *Communications satellites* – Satellites communicate via microwave radio waves, which are not deflected by the Earth's atmosphere. The satellites are stationed in space, typically in geosynchronous orbit 35,400 km (22,000 mi) above the equator. These Earth-orbiting systems are capable of receiving and relaying voice, data, and TV signals.

- *Cellular and PCS systems* use several radio communications technologies. The systems divide the region covered into multiple geographic areas. Each area has a low-power transmitter or radio relay antenna device to relay calls from one area to the next area.

- *Radio and spread spectrum technologies* – Wireless local area networks use a high-frequency radio technology similar to digital cellular and a low-frequency radio technology. Wireless LANs use spread spectrum technology to enable communication between multiple devices in a limited area. IEEE 802.11 defines a common flavor of open-standards wireless radio-wave technology known as Wifi.

- *Free-space optical communication* uses visible or invisible light for communications. In most cases, line-of-sight propagation is used, which limits the physical positioning of communicating devices.

Types of Wireless Networks

Wireless PAN

Wireless personal area networks (WPANs) interconnect devices within a relatively small area, that is generally within a person's reach. For example, both Bluetooth radio and invisible infrared light provides a WPAN for interconnecting a headset to a laptop. ZigBee also supports WPAN applications. Wi-Fi PANs are becoming commonplace (2010) as equipment designers start to integrate Wi-Fi into a variety of consumer electronic devices. Intel "My WiFi" and Windows 7 "virtual Wi-Fi" capabilities have made Wi-Fi PANs simpler and easier to set up and configure.

Wireless LAN

A wireless local area network (WLAN) links two or more devices over a short distance using a wireless distribution method, usually providing a connection through an access point for internet access. The use of spread-spectrum or OFDM technologies may allow users to move around within a local coverage area, and still remain connected to the network.

Products using the IEEE 802.11 WLAN standards are marketed under the Wi-Fi brand name. Fixed wireless technology implements point-to-point links between computers or networks at two distant locations, often using dedicated microwave or modulated laser light beams over line of sight paths. It is often used in cities to connect networks in two or more buildings without installing a wired link.

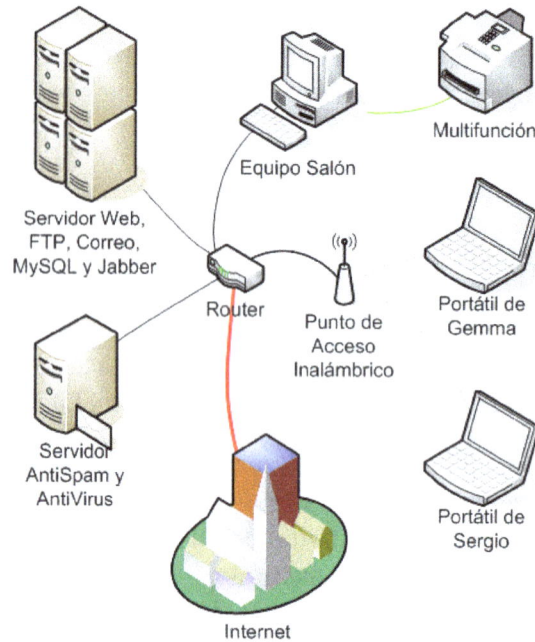

Wireless LANs are often used for connecting to local resources and to the Internet

Wireless Mesh Network

A wireless mesh network is a wireless network made up of radio nodes organized in a mesh topology. Each node forwards messages on behalf of the other nodes. Mesh networks can "self-heal", automatically re-routing around a node that has lost power.

Wireless MAN

Wireless metropolitan area networks are a type of wireless network that connects several wireless LANs.

- WiMAX is a type of Wireless MAN and is described by the IEEE 802.16 standard.

Wireless WAN

Wireless wide area networks are wireless networks that typically cover large areas, such as between neighbouring towns and cities, or city and suburb. These networks can be used to connect branch offices of business or as a public Internet access system. The wireless connections between access points are usually point to point microwave links using parabolic dishes on the 2.4 GHz band, rather than omnidirectional antennas used with smaller networks. A typical system contains base station gateways, access points and wireless bridging relays. Other configurations are mesh systems where each access point acts as a relay also. When combined with renewable energy systems such as photovoltaic solar panels or wind systems they can be stand alone systems.

Global Area Network

A global area network (GAN) is a network used for supporting mobile across an arbitrary number of wireless LANs, satellite coverage areas, etc. The key challenge in mobile communications is handing off user communications from one local coverage area to the next. In IEEE Project 802, this involves a succession of terrestrial wireless LANs.

Space Network

Space networks are networks used for communication between spacecraft, usually in the vicinity of the Earth. The example of this is NASA's Space Network.,

Different Uses

Some examples of usage include cellular phones which are part of everyday wireless networks, allowing easy personal communications. Another example, Intercontinental network systems, use radio satellites to communicate across the world. Emergency services such as the police utilize wireless networks to communicate effectively as well. Individuals and businesses use wireless networks to send and share data rapidly, whether it be in a small office building or across the world.

Properties

General

In a general sense, wireless networks offer a vast variety of uses by both business and home users.

"Now, the industry accepts a handful of different wireless technologies. Each wireless technology is defined by a standard that describes unique functions at both the Physical and the Data Link layers of the OSI model. These standards differ in their specified signaling methods, geographic ranges, and frequency usages, among other things. Such differences can make certain technologies better suited to home networks and others better suited to network larger organizations."

Performance

Each standard varies in geographical range, thus making one standard more ideal than the next depending on what it is one is trying to accomplish with a wireless network. The performance of wireless networks satisfies a variety of applications such as voice and video. The use of this technology also gives room for expansions, such as from 2G to 3G and, most recently, 4G technology, which stands for the fourth generation of cell phone mobile communications standards. As wireless networking has become commonplace, sophistication increases through configuration of network hardware and software, and greater capacity to send and receive larger amounts of data, faster, is achieved.

Space

Space is another characteristic of wireless networking. Wireless networks offer many advantages when it comes to difficult-to-wire areas trying to communicate such as across a street or river, a warehouse on the other side of the premises or buildings that are physically separated but operate as one. Wireless networks allow for users to designate a certain space which the network will be able to communicate with other devices through that network. Space is also created in homes as a result of eliminating clutters of wiring. This technology allows for an alternative to installing physical network mediums such as TPs, coaxes, or fiber-optics, which can also be expensive.

Home

For homeowners, wireless technology is an effective option compared to Ethernet for sharing printers, scanners, and high-speed Internet connections. WLANs help save the cost of installation of cable mediums, save time from physical installation, and also creates mobility for devices connected to the network. Wireless networks are simple and require as few as one single wireless access point connected directly to the Internet via a router.

Wireless Network Elements

The telecommunications network at the physical layer also consists of many interconnected wireline network elements (NEs). These NEs can be stand-alone systems or products that are either supplied by a single manufacturer or are assembled by the service provider (user) or system integrator with parts from several different manufacturers.

Wireless NEs are the products and devices used by a wireless carrier to provide support for the backhaul network as well as a mobile switching center (MSC).

Reliable wireless service depends on the network elements at the physical layer to be protected against all operational environments and applications (.

What are especially important are the NEs that are located on the cell tower to the base station (BS) cabinet. The attachment hardware and the positioning of the antenna and associated closures and cables are required to have adequate strength, robustness, corrosion resistance, and resistance against wind, storms, icing, and other weather conditions. Requirements for individual components, such as hardware, cables, connectors, and closures, shall take into consideration the structure to which they are attached.

Difficulties

Interferences

Compared to wired systems, wireless networks are frequently subject to electromagnetic interference. This can be caused by other networks or other types of equipment that generate radio waves that are within, or close, to the radio bands used for communication. Interference can degrade the signal or cause the system to fail.

Absorption and Reflection

Some materials cause absorption of electromagnetic waves, preventing it from reaching the receiver, in other cases, particularly with metallic or conductive materials reflection occurs. This can cause dead zones where no reception is available. Aluminium foiled thermal isolation in modern homes can easily reduce indoor mobile signals by 10 dB frequently leading to complaints about the bad reception of long-distance rural cell signals.

Multipath Fading

In multipath fading two or more different routes taken by the signal, due to reflections, can cause the signal to cancel out at certain locations, and to be stronger in other places (upfade).

Hidden Node Problem

The hidden node problem occurs in some types of network when a node is visible from a wireless access point (AP), but not from other nodes communicating with that AP. This leads to difficulties in media access control.

Shared Resource Problem

The wireless spectrum is a limited resource and shared by all nodes in the range of its transmitters. Bandwidth allocation becomes complex with multiple participating users. Often users are not aware that advertised numbers (e.g., for IEEE 802.11 equipment or LTE networks) are not their capacity, but shared with all other users and thus the individual user rate is far lower. With increasing demand, the capacity crunch is more and more likely to happen. User-in-the-loop (UIL) may be an alternative solution to ever upgrading to newer technologies for over-provisioning.

Capacity

Channel

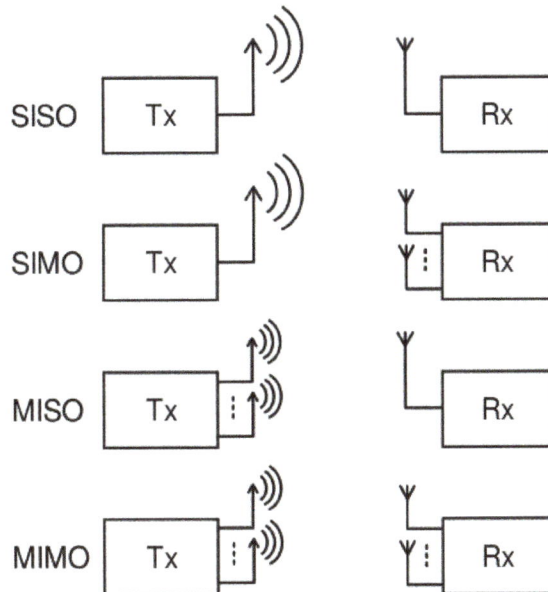

Understanding of SISO, SIMO, MISO and MIMO. Using multiple antennas and transmitting in different frequency channels can reduce fading, and can greatly increase the system capacity.

Shannon's theorem can describe the maximum data rate of any single wireless link, which relates to the bandwidth in hertz and to the noise on the channel.

One can greatly increase channel capacity by using MIMO techniques, where multiple aerials or multiple frequencies can exploit multiple paths to the receiver to achieve much higher throughput – by a factor of the product of the frequency and aerial diversity at each end.

Under Linux, the Central Regulatory Domain Agent (CRDA) controls the setting of channels.

Network

The total network bandwidth depends on how dispersive the medium is (more dispersive medium generally has better total bandwidth because it minimises interference), how many frequencies are available, how noisy those frequencies are, how many aerials are used and whether a directional antenna is in use, whether nodes employ power control and so on.

Cellular wireless networks generally have good capacity, due to their use of directional aerials, and their ability to reuse radio channels in non-adjacent cells. Additionally, cells can be made very small using low power transmitters this is used in cities to give network capacity that scales linearly with population density.

Safety

Wireless access points are also often close to humans, but the drop off in power over distance is fast, following the inverse-square law. The position of the United Kingdom's Health Protection Agency (HPA) is that "...radio frequency (RF) exposures from WiFi are likely to be lower than those from mobile phones." It also saw "...no reason why schools and others should not use WiFi equipment." In October 2007, the HPA launched a new "systematic" study into the effects of WiFi networks on behalf of the UK government, in order to calm fears that had appeared in the media in a recent period up to that time". Dr Michael Clark, of the HPA, says published research on mobile phones and masts does not add up to an indictment of WiFi.

Public Switched Telephone Network

The public switched telephone network (PSTN) is the aggregate of the world's circuit-switched telephone networks that are operated by national, regional, or local telephony operators, providing infrastructure and services for public telecommunication. The PSTN consists of telephone lines, fiber optic cables, microwave transmission links, cellular networks, communications satellites, and undersea telephone cables, all interconnected by switching centers, thus allowing most telephones to communicate with each other. Originally a network of fixed-line analog telephone systems, the PSTN is now almost entirely digital in its core network and includes mobile and other networks, as well as fixed telephones.

The technical operation of the PSTN adheres to the standards created by the ITU-T. These standards allow different networks in different countries to interconnect seamlessly. The E.163 and E.164 standards provide a single global address space for telephone numbers. The combination of the interconnected networks and the single numbering plan allow telephones around the world to dial each other.

History (America)

The first telephones had no network but were in private use, wired together in pairs. Users who wanted to talk to different people had as many telephones as necessary for the purpose. A user who wished to speak whistled loudly into the transmitter until the other party heard.

However, a bell was added soon for signaling, so an attendant no longer need wait for the whistle, and then a switch hook. Later telephones took advantage of the exchange principle already employed in telegraph networks. Each telephone was wired to a local telephone exchange, and the exchanges were wired together with trunks. Networks were connected in a hierarchical manner until they spanned cities, countries, continents and oceans. This was the beginning of the PSTN though the term was not used for many decades.

Automation introduced pulse dialing between the phone and the exchange, and then among exchanges, followed by more sophisticated address signaling including multi-frequency, culminating in the SS7 network that connected most exchanges by the end of the 20th century.

The growth of the PSTN meant that teletraffic engineering techniques needed to be deployed to deliver quality of service (QoS) guarantees for the users. The work of A. K. Erlang established the mathematical foundations of methods required to determine the capacity requirements and configuration of equipment and the number of personnel required to deliver a specific level of service.

In the 1970s the telecommunications industry began implementing packet switched network data services using the X.25 protocol transported over much of the end-to-end equipment as was already in use in the PSTN.

In the 1980s the industry began planning for digital services assuming they would follow much the same pattern as voice services, and conceived a vision of end-to-end circuit switched services, known as the Broadband Integrated Services Digital Network (B-ISDN). The B-ISDN vision has been overtaken by the disruptive technology of the Internet.

At the turn of the 21st century, the oldest parts of the telephone network still use analog technology for the last mile loop to the end user. However, digital technologies such as DSL, ISDN, FTTx, and cable modems have become more common in this portion of the network.

Several large private telephone networks are not linked to the PSTN, usually for military purposes. There are also private networks run by large companies which are linked to the PSTN only through limited gateways, such as a large private branch exchange (PBX).

Operators

The task of building the networks and selling services to customers fell to the network operators. The first company to be incorporated to provide PSTN services was the Bell Telephone Company in the United States.

In some countries, however, the job of providing telephone networks fell to government as the investment required was very large and the provision of telephone service was increasingly becoming an essential public utility. For example, the General Post Office in the United Kingdom brought together a number of private companies to form a single nationalized company. In recent decades however, these state monopolies were broken up or sold off through privatization.

Regulation

In most countries, the central has a regulator dedicated to monitoring the provision of PSTN services in that country. Their tasks may be for example to ensure that end customers are not overcharged for services where monopolies may exist. They may also regulate the prices charged between the operators to carry each other's traffic.

Technology

Network topology

The PSTN network architecture had to evolve over the years to support increasing numbers of subscribers, calls, connections to other countries, direct dialing and so on. The model developed by the United States and Canada was adopted by other nations, with adaptations for local markets.

The original concept was that the telephone exchanges are arranged into hierarchies, so that if a call cannot be handled in a local cluster, it is passed to one higher up for onward routing. This reduced the number of connecting trunks required between operators over long distances and also kept local traffic separate.

However, in modern networks the cost of transmission and equipment is lower and, although hierarchies still exist, they are much flatter, with perhaps only two layers.

Digital Channels

As described above, most automated telephone exchanges now use digital switching rather than mechanical or analog switching. The trunks connecting the exchanges are also digital, called circuits or channels. However analog two-wire circuits are still used to connect the last mile from the exchange to the telephone in the home (also called the local loop). To carry a typical phone call from a calling party to a called party, the analog audio signal is digitized at an 8 kHz sample rate with 8-bit resolution using a special type of nonlinear pulse code modulation known as G.711. The call is then transmitted from one end to another via telephone exchanges. The call is switched using a call set up protocol (usually ISUP) between the telephone exchanges under an overall routing strategy.

The call is carried over the PSTN using a 64 kbit/s channel, originally designed by Bell Labs. The name given to this channel is Digital Signal 0 (DS0). The DS0 circuit is the basic granularity of circuit switching in a telephone exchange. A DS0 is also known as a timeslot because DS0s are aggregated in time-division multiplexing (TDM) equipment to form higher capacity communication links.

A Digital Signal 1 (DS1) circuit carries 24 DS0s on a North American or Japanese T-carrier (T1) line, or 32 DS0s (30 for calls plus two for framing and signaling) on an E-carrier (E1) line used in most other countries. In modern networks, the multiplexing function is moved as close to the end user as possible, usually into cabinets at the roadside in residential areas, or into large business premises.

These aggregated circuits are conveyed from the initial multiplexer to the exchange over a set of equipment collectively known as the access network. The access network and inter-exchange transport use synchronous optical transmission, for example, SONET and Synchronous Digital Hierarchy (SDH) technologies, although some parts still use the older PDH technology.

Within the access network, there are a number of reference points defined. Most of these are of interest mainly to ISDN but one – the V reference point – is of more general interest. This is the reference point between a primary multiplexer and an exchange. The protocols at this reference point were standardized in ETSI areas as the V5 interface.

Impact on IP Standards

Voice quality over PSTN networks was used as the benchmark for the development of the Telecommunications Industry Association's TIA-TSB-116 standard on voice-quality recommendations for IP telephony, to determine acceptable levels of audio delay and echo.

Packet Switching

Packet switching is a digital networking communications method that groups all transmitted data into suitably sized blocks, called *packets*, which are transmitted via a medium that may be shared by multiple simultaneous communication sessions. Packet switching increases network efficiency, robustness and enables technological convergence of many applications operating on the same network.

Packets are composed of a header and payload. Information in the header is used by networking hardware to direct the packet to its destination where the payload is extracted and used by application software.

Starting in the late 1950s, American computer scientist Paul Baran developed the concept *Distributed Adaptive Message Block Switching* with the goal to provide a fault-tolerant, efficient routing method for telecommunication messages as part of a research program at the RAND Corporation, funded by the US Department of Defense. This concept contrasted and contradicted then-established principles of pre-allocation of network bandwidth, largely fortified by the development of telecommunications in the Bell System. The new concept found little resonance among network implementers until the independent work of British computer scientist Donald Davies at the National Physical Laboratory (United Kingdom) in the late 1960s. Davies is credited with coining the modern name *packet switching* and inspiring numerous packet switching networks in Europe in the decade following, including the incorporation of the concept in the early ARPANET in the United States.

Concept

A simple definition of packet switching is:

The routing and transferring of data by means of addressed packets so that a channel is occupied during the transmission of the packet only, and upon completion of the transmission the channel is made available for the transfer of other traffic

Packet switching features delivery of variable bit rate data streams, realized as sequences of packets, over a computer network which allocates transmission resources as needed using statistical multiplexing or dynamic bandwidth allocation techniques. As they traverse network nodes, such as switches and routers, packets are received, buffered, queued, and transmitted (stored and forwarded), resulting in variable latency and throughput depending on the link capacity and the traffic load on the network.

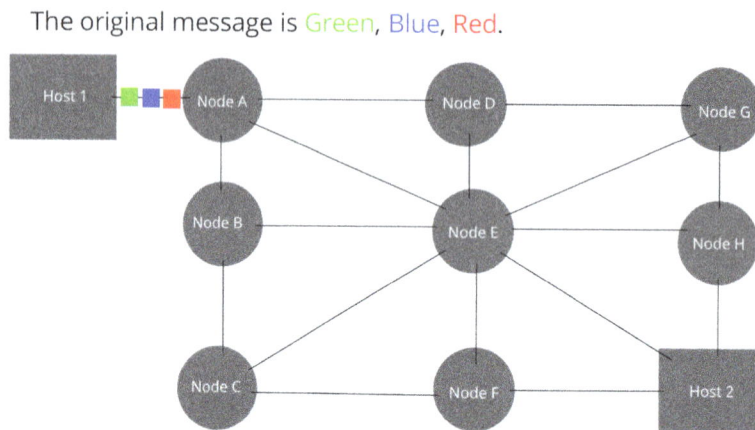

An animation demonstrating data packet switching across a network

Packet switching contrasts with another principal networking paradigm, circuit switching, a method which pre-allocates dedicated network bandwidth specifically for each communication session, each having a constant bit rate and latency between nodes. In cases of billable services, such as cellular communication services, circuit switching is characterized by a fee per unit of connection time, even when no data is transferred, while packet switching may be characterized by a fee per unit of information transmitted, such as characters, packets, or messages.

Packet mode communication may be implemented with or without intermediate forwarding nodes (packet switches or routers). Packets are normally forwarded by intermediate network nodes asynchronously using first-in, first-out buffering, but may be forwarded according to some scheduling discipline for fair queuing, traffic shaping, or for differentiated or guaranteed quality of service, such as weighted fair queuing or leaky bucket. In case of a shared physical medium (such as radio or 10BASE5), the packets may be delivered according to a multiple access scheme.

History

In the late 1950s, the US Air Force established a wide area network for the Semi-Automatic Ground Environment (SAGE) radar defense system. They sought a system that might survive a nuclear attack to enable a response, thus diminishing the attractiveness of the first strike advantage by enemies.

Leonard Kleinrock conducted early research in queueing theory which proved important in packet switching, and published a book in the related field of digital message switching (without the packets) in 1961; he also later played a leading role in building and management of the world's first packet-switched network, the ARPANET.

The concept of switching small blocks of data was first explored independently by Paul Baran at the RAND Corporation in the US and Donald Davies at the National Physical Laboratory (NPL) in the UK in the early to mid-1960s.

Baran developed the concept of *distributed adaptive message block switching* during his research at the RAND Corporation for the US Air Force into communications networks, that could survive nuclear wars, first presented to the Air Force in the summer of 1961 as briefing B-265, later published as RAND report P-2626 in 1962, and finally in report RM 3420 in 1964. Report P-2626 described a general architecture for a large-scale, distributed, survivable communications network. The work focuses on three key ideas: use of a decentralized network with multiple paths between any two points, dividing user messages into *message blocks*, later called packets, and delivery of these messages by store and forward switching.

Baran's work was known to Robert Taylor and J.C.R. Licklider at the Information Processing Technology Office, who advocated wide area networks, and it influenced Lawrence Roberts to adopt the technology in the development of the ARPANET.

Starting in 1965, Donald Davies at the National Physical Laboratory, UK, independently developed the same message routing methodology as developed by Baran. He called it *packet switching*, a more accessible name than Baran's, and proposed to build a nationwide network in the UK. He gave a talk on the proposal in 1966, after which a person from the Ministry of Defence (MoD) told him about Baran's work. A member of Davies' team (Roger Scantlebury) met Lawrence Roberts at the 1967 ACM Symposium on Operating System Principles and suggested it for use in the ARPANET.

Davies had chosen some of the same parameters for his original network design as did Baran, such as a packet size of 1024 bits. In 1966, Davies proposed that a network should be built at the laboratory to serve the needs of NPL and prove the feasibility of packet switching. The NPL Data Communications Network entered service in 1970.

The first computer network and packet switching network deployed for computer resource sharing was the Octopus Network at the Lawrence Livermore National Laboratory that began connecting four Control Data 6600 computers to several shared storage devices (including an IBM 2321 Data Cell in 1968 and an IBM Photostore in 1970) and to several hundred Teletype Model 33 ASR terminals for time sharing use starting in 1968.

In 1973, Vint Cerf and Bob Kahn wrote the specifications for Transmission Control Protocol (TCP), an internetworking protocol for sharing resources using packet-switching among the nodes.

Connectionless and Connection-oriented Modes

Packet switching may be classified into connectionless packet switching, also known as datagram switching, and connection-oriented packet switching, also known as virtual circuit switching.

Examples of connectionless protocols are Ethernet, Internet Protocol (IP), and the User Datagram Protocol (UDP). Connection-oriented protocols include X.25, Frame Relay, Multiprotocol Label Switching (MPLS), and the Transmission Control Protocol (TCP).

In connectionless mode each packet includes complete addressing information. The packets are routed individually, sometimes resulting in different paths and out-of-order delivery. Each packet is labeled with a destination address, source address, and port numbers. It may also be labeled with the sequence number of the packet. This precludes the need for a dedicated path to help the packet find its way to its destination, but means that much more information is needed in the packet header, which is therefore larger, and this information needs to be looked up in power-hungry content-addressable memory. Each packet is dispatched and may go via different routes; potentially, the system has to do as much work for every packet as the connection-oriented system has to do in connection set-up, but with less information as to the application's requirements. At the destination, the original message/data is reassembled in the correct order, based on the packet sequence number. Thus a virtual connection, also known as a virtual circuit or byte stream is provided to the end-user by a transport layer protocol, although intermediate network nodes only provides a connectionless network layer service.

Connection-oriented transmission requires a setup phase in each involved node before any packet is transferred to establish the parameters of communication. The packets include a connection identifier rather than address information and are negotiated between endpoints so that they are delivered in order and with error checking. Address information is only transferred to each node during the connection set-up phase, when the route to the destination is discovered and an entry is added to the switching table in each network node through which the connection passes. The signaling protocols used allow the application to specify its requirements and discover link parameters. Acceptable values for service parameters may be negotiated. Routing a packet requires the node to look up the connection id in a table. The packet header can be small, as it only needs to contain this code and any information, such as length, timestamp, or sequence number, which is different for different packets.

Packet Switching in Networks

Packet switching is used to optimize the use of the channel capacity available in digital telecommunication networks such as computer networks, to minimize the transmission latency (the time it takes for data to pass across the network), and to increase robustness of communication.

The best-known use of packet switching is the Internet and most local area networks. The Internet is implemented by the Internet Protocol Suite using a variety of Link Layer technologies. For example, Ethernet and Frame Relay are common. Newer mobile phone technologies (e.g., GPRS, i-mode) also use packet switching.

X.25 is a notable use of packet switching in that, despite being based on packet switching methods, it provided virtual circuits to the user. These virtual circuits carry variable-length packets. In 1978, X.25 provided the first international and commercial packet switching network, the International Packet Switched Service (IPSS). Asynchronous Transfer Mode (ATM) also is a virtual circuit technology, which uses fixed-length cell relay connection oriented packet switching.

Datagram packet switching is also called connectionless networking because no connections are established. Technologies such as Multiprotocol Label Switching (MPLS) and the Resource Reservation Protocol (RSVP) create virtual circuits on top of datagram networks. Virtual circuits are especially useful in building robust failover mechanisms and allocating bandwidth for delay-sensitive applications.

MPLS and its predecessors, as well as ATM, have been called "fast packet" technologies. MPLS, indeed, has been called "ATM without cells". Modern routers, however, do not require these technologies to be able to forward variable-length packets at multigigabit speeds across the network.

X.25 vs. Frame Relay

Both X.25 and Frame Relay provide connection-oriented operations. X.25 provides it via the network layer of the OSI Model, whereas Frame Relay provides it via level two, the data link layer. Another major difference between X.25 and Frame Relay is that X.25 requires a handshake between the communicating parties before any user packets are transmitted. Frame Relay does not define any such handshakes. X.25 does not define any operations inside the packet network. It only operates at the user-network-interface (UNI). Thus, the network provider is free to use any procedure it wishes inside the network. X.25 does specify some limited re-transmission procedures at the UNI, and its link layer protocol (LAPB) provides conventional HDLC-type link management procedures. Frame Relay is a modified version of ISDN's layer two protocol, LAPD and LAPB. As such, its integrity operations pertain only between nodes on a link, not end-to-end. Any retransmissions must be carried out by higher layer protocols. The X.25 UNI protocol is part of the X.25 protocol suite, which consists of the lower three layers of the OSI Model. It was widely used at the UNI for packet switching networks during the 1980s and early 1990s, to provide a standardized interface into and out of packet networks. Some implementations used X.25 within the network as well, but its connection-oriented features made this setup cumbersome and inefficient. Frame relay operates principally at layer two of the OSI Model. However, its address field (the Data Link Connection ID, or DLCI) can be used at the OSI network layer, with a minimum set of procedures. Thus, it rids itself of many X.25 layer 3 encumbrances, but still has the DLCI as an ID beyond a node-to-node layer two link protocol. The simplicity of Frame Relay makes it faster and more efficient than X.25. Because Frame relay is a data link layer protocol, like X.25 it does not define internal network routing operations. For X.25 its packet IDs---the virtual circuit and virtual channel numbers have to be correlated to network addresses. The same is true for Frame Relays DLCI. How this is done is up to the network provider. Frame Relay, by virtue of having no network layer procedures is connection-oriented at layer two, by using the HDLC/LAPD/LAPB Set Asynchronous Balanced Mode (SABM). X.25 connections are typically established for each communication session, but it does have a feature allowing a limited amount of traffic to be passed across the UNI without the connection-oriented handshake. For a while, Frame Relay was used to interconnect LANs across wide area networks. However, X.25 and well as Frame Relay have been supplanted by the Internet Protocol (IP) at the network layer, and the Asynchronous Transfer Mode (ATM) and or versions of Multi-Protocol Label Switching (MPLS) at layer two. A typical configuration is to run IP over ATM or a version of MPLS. <Uyless Black, X.25 and Related Protocols, IEEE Computer Society, 1991> <Uyless Black, Frame Relay Networks, McGraw-Hill, 1998> <Uyless Black, MPLS and Label Switching Networks, Prentice Hall, 2001> < Uyless Black, ATM, Volume I, Prentice Hall, 1995>

Packet-switched Networks

The history of packet-switched networks can be divided into three overlapping eras: early networks before the introduction of X.25 and the OSI model, the X.25 era when many postal, telephone, and telegraph companies introduced networks with X.25 interfaces, and the Internet era.

Early Networks

ARPANET and SITA HLN became operational in 1969. Before the introduction of X.25 in 1973, about twenty different network technologies had been developed. Two fundamental differences involved the division of functions and tasks between the hosts at the edge of the network and the network core. In the datagram system, the hosts have the responsibility to ensure orderly delivery of packets. The User Datagram Protocol (UDP) is an example of a datagram protocol. In the virtual call system, the network guarantees sequenced delivery of data to the host. This results in a simpler host interface with less functionality than in the datagram model. The X.25 protocol suite uses this network type.

Appletalk

AppleTalk was a proprietary suite of networking protocols developed by Apple Inc. in 1985 for Apple Macintosh computers. It was the primary protocol used by Apple devices through the 1980s and 90s. AppleTalk included features that allowed local area networks to be established *ad hoc* without the requirement for a centralized router or server. The AppleTalk system automatically assigned addresses, updated the distributed namespace, and configured any required inter-network routing. It was a plug-n-play system.

AppleTalk versions were also released for the IBM PC and compatibles, and the Apple IIGS. AppleTalk support was available in most networked printers, especially laser printers, some file servers and routers. AppleTalk support was terminated in 2009, replaced by TCP/IP protocols.

ARPANET

The ARPANET was a progenitor network of the Internet and the first network to run the TCP/IP suite using packet switching technologies.

BNRNET

BNRNET was a network which Bell Northern Research developed for internal use. It initially had only one host but was designed to support many hosts. BNR later made major contributions to the CCITT X.25 project.

CYCLADES

The CYCLADES packet switching network was a French research network designed and directed by Louis Pouzin. First demonstrated in 1973, it was developed to explore alternatives to the early ARPANET design and to support network research generally. It was the first network to make the hosts responsible for reliable delivery of data, rather than the network itself, using unreliable datagrams and associated end-to-end protocol mechanisms. Concepts of this network influenced later ARPANET architecture.

DECnet

DECnet is a suite of network protocols created by Digital Equipment Corporation, originally released in 1975 in order to connect two PDP-11 minicomputers. It evolved into one of the first peer-to-peer network architectures, thus transforming DEC into a networking powerhouse in the 1980s. Initially built with three layers, it later (1982) evolved into a seven-layer OSI-compliant networking protocol. The DECnet protocols were designed entirely by Digital Equipment Corporation. However, DECnet Phase II (and later) were open standards with published specifications, and several implementations were developed outside DEC, including one for Linux.

DDX-1

This was an experimental network from Nippon PTT. It mixed circuit switching and packet switching. It was succeeded by DDX-2.

EIN née COST II

European Informatics Network was a project to link several national networks. It became operational in 1976.

EPSS

The Experimental Packet Switching System (EPSS) was an experiment of the UK Post Office. Ferranti supplied the hardware and software. The handling of link control messages (acknowledgements and flow control) was different from that of most other networks.

GEIS

As General Electric Information Services (GEIS), General Electric was a major international provider of information services. The company originally designed a telephone network to serve as its internal (albeit continent-wide) voice telephone network.

In 1965, at the instigation of Warner Sinback, a data network based on this voice-phone network was designed to connect GE's four computer sales and service centers (Schenectady, New York, Chicago, and Phoenix) to facilitate a computer time-sharing service, apparently the world's first commercial online service. (In addition to selling GE computers, the centers were computer service bureaus, offering batch processing services. They lost money from the beginning, and Sinback, a high-level marketing manager, was given the job of turning the business around. He decided that a time-sharing system, based on Kemney's work at Dartmouth—which used a computer on loan from GE—could be profitable. Warner was right.)

After going international some years later, GEIS created a network data center near Cleveland, Ohio. Very little has been published about the internal details of their network. (Though it has been stated by some that Tymshare copied the GEIS system to create their network, Tymnet.) The design was hierarchical with redundant communication links.

IPSANET

IPSANET was a semi-private network constructed by I. P. Sharp Associates to serve their time-sharing customers. It became operational in May 1976.

IPX/SPX

The Internetwork Packet Exchange (IPX) and Sequenced Packet Exchange (SPX) are Novell networking protocols derived from Xerox Network Systems' IDP and SPP protocols, respectively. They were used primarily on networks using the Novell NetWare operating systems.

Merit Network

Merit Network, Inc., an independent non-profit 501(c)(3) corporation governed by Michigan's public universities, was formed in 1966 as the Michigan Educational Research Information Triad to explore computer networking between three of Michigan's public universities as a means to help the state's educational and economic development. With initial support from the State of Michigan and the National Science Foundation (NSF), the packet-switched network was first demonstrated in December 1971 when an interactive host to host connection was made between the IBM mainframe computer systems at the University of Michigan in Ann Arbor and Wayne State University in Detroit. In October 1972 connections to the CDC mainframe at Michigan State University in East Lansing completed the triad. Over the next several years in addition to host to host interactive connections the network was enhanced to support terminal to host connections, host to host batch connections (remote job submission, remote printing, batch file transfer), interactive file transfer, gateways to the Tymnet and Telenet public data networks, X.25 host attachments, gateways to X.25 data networks, Ethernet attached hosts, and eventually TCP/IP and additional public universities in Michigan join the network. All of this set the stage for Merit's role in the NSFNET project starting in the mid-1980s.

NPL

Donald Davies of the National Physical Laboratory, UK made many important contributions to the theory of packet switching. NPL built a single node network to connect sundry hosts at NPL.

OCTOPUS

Octopus was a local network at Lawrence Livermore National Laboratory. It connected sundry hosts at the lab to interactive terminals and various computer peripherals including a bulk storage system.

Philips Research

Philips Research Laboratories in Redhill, Surrey developed a packet switching network for internal use. It was a datagram network with a single switching node.

PUP

PARC Universal Packet (PUP or Pup) was one of the two earliest internetwork protocol suites; it was created by researchers at Xerox PARC in the mid-1970s. The entire suite provided routing and packet delivery, as well as higher level functions such as a reliable byte stream, along with numerous applications. Further developments led to Xerox Network Systems (XNS).

RCP

RCP was an experimental network created by the French PTT. It was used to gain experience with packet switching technology before the specification of Transpac was frozen. RCP was a virtual-circuit network in contrast to CYCLADES which was based on datagrams. RCP emphasised terminal to host and terminal to terminal connection; CYCLADES was concerned with host-to-host communication. TRANSPAC was introduced as an X.25 network. RCP influenced the specification of X.25

RETD

Red Especial de Transmisión de Datos was a network developed by Compañía Telefónica Nacional de España. It became operational in 1972 and thus was the first public network.

SCANNET

"The experimental packet-switched Nordic telecommunication network SCANNET was implemented in Nordic technical libraries in 70's, and it included first Nordic electronic journal Extemplo. Libraries were also among first ones in universities to accommodate microcomputers for public use in early 80's."

SITA HLN

SITA is a consortium of airlines. Their High Level Network became operational in 1969 at about the same time as ARPANET. It carried interactive traffic and message-switching traffic. As with many non-academic networks very little has been published about it.

IBM Systems Network Architecture

IBM Systems Network Architecture (SNA) is IBM's proprietary networking architecture created in 1974. An IBM customer could acquire hardware and software from IBM and lease private lines from a common carrier to construct a private network.

Telenet

Telenet was the first FCC-licensed public data network in the United States. It was founded by former ARPA IPTO director Larry Roberts as a means of making ARPANET technology public. He had tried to interest AT&T in buying the technology, but the monopoly's reaction was that this was incompatible with their future. Bolt, Beranack and Newman (BBN) provided the financing. It initially used ARPANET technology but changed the host interface to X.25 and the terminal interface to X.29. Telenet designed these protocols and helped standardize them in the CCITT. Telenet was incorporated in 1973 and started operations in 1975. It went public in 1979 and was then sold to GTE.

Tymnet

Tymnet was an international data communications network headquartered in San Jose, CA that utilized virtual call packet switched technology and used X.25, SNA/SDLC, BSC and ASCII interfaces to connect host computers (servers)at thousands of large companies, educational institutions, and government agencies. Users typically connected via dial-up connections or dedicated async connections. The business consisted of a large public network that supported dial-up users and a private network business that allowed government agencies and large companies (mostly banks and airlines) to build their own dedicated networks. The private networks were often connected via gateways to the public network to reach locations not on the private network. Tymnet was also connected to dozens of other public networks in the U.S. and internationally via X.25/X.75 gateways. (Interesting note: Tymnet was not named after Mr. Tyme. Another employee suggested the name.)

XNS

Xerox Network Systems (XNS) was a protocol suite promulgated by Xerox, which provided routing and packet delivery, as well as higher level functions such as a reliable stream, and remote procedure calls. It was developed from PARC Universal Packet (PUP).

X.25 Era

There were two kinds of X.25 networks. Some such as DATAPAC and TRANSPAC were initially implemented with an X.25 external interface. Some older networks such as TELENET and TYMNET were modified to provide a X.25 host interface in addition to older host connection schemes. DATAPAC was developed by Bell Northern Research which was a joint venture of Bell Canada (a common carrier) and Northern Telecom (a telecommunications equipment supplier). Northern Telecom sold several DATAPAC clones to foreign PTTs including the Deutsche Bundespost. X.75 and X.121 allowed the interconnection of national X.25 networks. A user or host could call a host on a foreign network by including the DNIC of the remote network as part of the destination address.

AUSTPAC

AUSTPAC was an Australian public X.25 network operated by Telstra. Started by Telecom Australia in the early 1980s, AUSTPAC was Australia's first public packet-switched data network, supporting applications such as on-line betting, financial applications — the Australian Tax Office made use of AUSTPAC — and remote terminal access to academic institutions, who maintained their connections to AUSTPAC up until the mid-late 1990s in some cases. Access can be via a dial-up terminal to a PAD, or, by linking a permanent X.25 node to the network.

ConnNet

ConnNet was a packet-switched data network operated by the Southern New England Telephone Company serving the state of Connecticut.

Datanet 1

Datanet 1 was the public switched data network operated by the Dutch PTT Telecom (now known as KPN). Strictly speaking Datanet 1 only referred to the network and the connected users via leased lines (using the X.121 DNIC 2041), the name also referred to the public PAD service *Telepad* (using the DNIC 2049). And because the main Videotex service used the network and modified PAD devices as infrastructure the name Datanet 1 was used for these services as well. Although this use of the name was incorrect all these services were managed by the same people within one department of KPN contributed to the confusion.

Datapac

DATAPAC was the first operational X.25 network (1976). It covered major Canadian cities and was eventually extended to smaller centres.

Datex-P

Deutsche Bundespost operated this national network in Germany. The technology was acquired from Northern Telecom.

Eirpac

Eirpac is the Irish public switched data network supporting X.25 and X.28. It was launched in 1984, replacing Euronet. Eirpac is run by Eircom.

HIPA-NET

Hitachi designed a private network system for sale as a turnkey package to multi-national organizations. In addition to providing X.25 packet switching, message switching software was also included. Messages were buffered at the nodes adjacent to the sending and receiving terminals. Switched virtual calls were not supported, but through the use of "logical ports" an originating terminal could have a menu of pre-defined destination terminals.

Iberpac

Iberpac is the Spanish public packet-switched network, providing X.25 services. Iberpac is run by Telefonica.

JANET

JANET was the UK academic and research network, linking all universities, higher education establishments, publicly funded research laboratories. The X.25 network was based mainly on GEC 4000 series switches, and run X.25 links at up to 8 Mbit/s in its final phase before being converted to an IP based network. The JANET network grew out of the 1970s SRCnet (later called SERCnet) network.

PSS

Packet Switch Stream (PSS) was the UK Post Office (later to become British Telecom) national X.25 network with a DNIC of 2342. British Telecom renamed PSS under its GNS (Global Network Service) name, but the PSS name has remained better known. PSS also included public dial-up PAD access, and various InterStream gateways to other services such as Telex.

Transpac

Transpac was the national X.25 network in France. It was developed locally at about the same time as DataPac in Canada. The development was done by the French PTT and influenced by the experimental RCP network. It began operation in 1978.

VENUS-P

VENUS-P was an international X.25 network that operated from April 1982 through March 2006. At its subscription peak in 1999, VENUS-P connected 207 networks in 87 countries.

Venepaq

Venepaq is the national X.25 public network in Venezuela. It is run by Cantv and allow direct connection and dial up connections. Provides nationalwide access at very low cost. It provides national and international access. Venepaq allow connection from 19.2 kbit/s to 64 kbit/s in direct connections, and 1200, 2400 and 9600 bit/s in dial up connections.

Internet Era

When Internet connectivity was made available to anyone who could pay for an ISP subscription, the distinctions between national networks blurred. The user no longer saw network identifiers such as the DNIC. Some older technologies such as circuit switching have resurfaced with new names such as fast packet switching. Researchers have created some experimental networks to complement the existing Internet.

CSNET

The Computer Science Network (CSNET) was a computer network funded by the U.S. National Science Foundation (NSF) that began operation in 1981. Its purpose was to extend networking benefits, for computer science departments at academic and research institutions that could not be directly connected to ARPANET, due to funding or authorization limitations. It played a significant role in spreading awareness of, and access to, national networking and was a major milestone on the path to development of the global Internet.

Internet2

Internet2 is a not-for-profit United States computer networking consortium led by members from the research and education communities, industry, and government. The Internet2 community, in partnership with Qwest, built the first Internet2 Network, called Abilene, in 1998 and was a prime investor in the National LambdaRail (NLR) project. In 2006, Internet2 announced a partnership with Level 3 Communications to launch a brand new nationwide network, boosting its capacity from 10 Gbit/s to 100 Gbit/s. In October, 2007, Internet2 officially retired Abilene and now refers to its new, higher capacity network as the Internet2 Network.

NSFNET

The National Science Foundation Network (NSFNET) was a program of coordinated, evolving projects sponsored by the National Science Foundation (NSF) beginning in 1985 to promote advanced research and education networking in the United States. NSFNET was also the name given to several nationwide backbone networks operating at speeds of 56 kbit/s, 1.5 Mbit/s (T1), and 45 Mbit/s (T3) that were constructed to support NSF's networking initiatives from 1985-1995. Initially created to link researchers to the nation's NSF-funded supercomputing centers, through further public funding and private industry partnerships it developed into a major part of the Internet backbone.

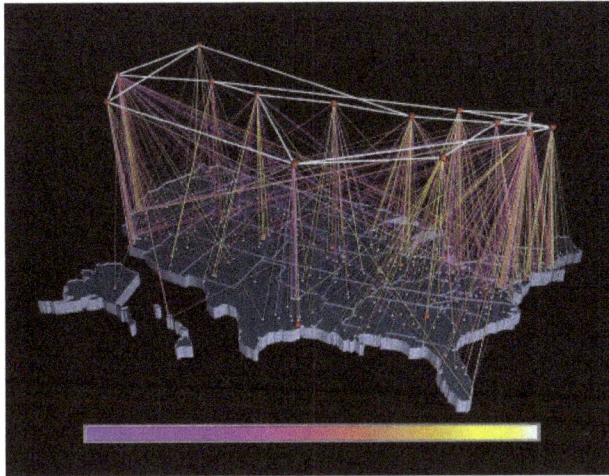

NSFNET Traffic 1991, NSFNET backbone nodes are shown at the top, regional networks below, traffic volume is depicted from purple (zero bytes) to white (100 billion bytes), visualization by NCSA using traffic data provided by the Merit Network.

NSFNET Regional Networks

In addition to the five NSF supercomputer centers, NSFNET provided connectivity to eleven regional networks and through these networks to many smaller regional and campus networks in the United States. The NSFNET regional networks were:

- BARRNet, the Bay Area Regional Research Network in Palo Alto, California;

- CERFNET, California Education and Research Federation Network in San Diego, California, serving California and Nevada;

- CICNet, the Committee on Institutional Cooperation Network via the Merit Network in Ann Arbor, Michigan and later as part of the T3 upgrade via Argonne National Laboratory outside of Chicago, serving the Big Ten Universities and the University of Chicago in Illinois, Indiana, Michigan, Minnesota, Ohio, and Wisconsin;

- Merit/MichNet in Ann Arbor, Michigan serving Michigan, formed in 1966, still in operation as of 2016;

- MIDnet in Lincoln, Nebraska serving Arkansas, Iowa, Kansas, Missouri, Nebraska, Oklahoma, and South Dakota;

- NEARNET, the New England Academic and Research Network in Cambridge, Massachusetts, added as part of the upgrade to T3, serving Connecticut, Maine, Massachusetts, New Hampshire, Rhode Island, and Vermont, established in late 1988, operated by BBN under contract to MIT, BBN assumed responsibility for NEARNET on 1 July 1993;

- NorthWestNet in Seattle, Washington, serving Alaska, Idaho, Montana, North Dakota, Oregon, and Washington, founded in 1987;

- NYSERNet, New York State Education and Research Network in Ithaca, New York;

- JVNCNet, the John von Neumann National Supercomputer Center Network in Princeton, New Jersey, serving Delaware and New Jersey;

- SESQUINET, the Sesquicentennial Network in Houston, Texas, founded during the 150th anniversary of the State of Texas;

- SURAnet, the Southeastern Universities Research Association network in College Park, Maryland and later as part of the T3 upgrade in Atlanta, Georgia serving Alabama, Florida, Georgia, Kentucky, Louisiana, Maryland, Mississippi, North Carolina, South Carolina, Tennessee, Virginia, and West Virginia, sold to BBN in 1994; and

- Westnet in Salt Lake City, Utah and Boulder, Colorado, serving Arizona, Colorado, New Mexico, Utah, and Wyoming.

National LambdaRail

The National LambdaRail was launched in September 2003. It is a 12,000-mile high-speed national computer network owned and operated by the U.S. research and education community that runs over fiber-optic lines. It was the first transcontinental 10 Gigabit Ethernet network. It operates with high aggregate capacity of up to 1.6 Tbit/s and a high 40 Gbit/s bitrate, with plans for 100 Gbit/s.

TransPAC, TransPAC2, and TransPAC3

TransPAC2 and TransPAC3, continuations of the TransPAC project, a high-speed international Internet service connecting research and education networks in the Asia-Pacific region to those in the US. TransPAC is part of the NSF's International Research Network Connections (IRNC) program.

Very High-speed Backbone Network Service (vBNS)

The Very high-speed Backbone Network Service (vBNS) came on line in April 1995 as part of a National Science Foundation (NSF) sponsored project to provide high-speed interconnection between NSF-sponsored supercomputing centers and select access points in the United States. The network was engineered and operated by MCI Telecommunications under a cooperative agreement with the NSF. By 1998, the vBNS had grown to connect more than 100 universities and research and engineering institutions via 12 national points of presence with DS-3 (45 Mbit/s), OC-3c (155 Mbit/s), and OC-12c (622 Mbit/s) links on an all OC-12c backbone, a substantial engineering feat for that time. The vBNS installed one of the first ever production OC-48c (2.5 Gbit/s) IP links in February 1999 and went on to upgrade the entire backbone to OC-48c.

In June 1999 MCI WorldCom introduced vBNS+ which allowed attachments to the vBNS network by organizations that were not approved by or receiving support from NSF. After the expiration of the NSF agreement, the vBNS largely transitioned to providing service to the government. Most universities and research centers migrated to the Internet2 educational backbone. In January 2006, when MCI and Verizon merged, vBNS+ became a service of Verizon Business.

Radio Network

There are two types of radio networks currently in use around the world: the one-to-many broadcast network commonly used for public information and mass media entertainment; and the two-way radio type used more commonly for public safety and public services such as police, fire, taxicabs, and delivery services. Cell Phones are able to send and receive simultaneously by using two different frequencies at the same time. Many of the same components and much of the same basic technology applies to all three.

The Two-way type of radio network shares many of the same technologies and components as the Broadcast type radio network but is generally set up with fixed broadcast points (transmitters) with co-located receivers and mobile receivers/transmitters or Transceivers. In this way both the fixed and mobile radio units can communicate with each other over broad geographic regions ranging in size from small single cities to entire states/provinces or countries. There are many ways in which multiple fixed transmit/receive sites can be interconnected to achieve the range of coverage required by the jurisdiction or authority implementing the system: conventional wireless links in numerous frequency bands, fibre-optic links, or micro-wave links. In all of these cases the signals are typically backhauled to a central switch of some type where the radio message is processed and resent (repeated) to all transmitter sites where it is required to be heard.

In contemporary two-way radio systems a concept called trunking is commonly used to achieve better efficiency of radio spectrum use and provide very wide ranging coverage with no switching of channels required by the mobile radio user as it roams throughout the system coverage. Trunking of two-way radio is identical to the concept used for cellular phone systems where each fixed and mobile radio is specifically identified to the system Controller and its operation is switched by the controller.

Broadcasting Networks

The Broadcast type of radio network is a network system which distributes programming to multiple stations simultaneously, or slightly delayed, for the purpose of extending total coverage beyond the limits of a single broadcast signal. The resulting expanded audience for radio programming or information essentially applies the benefits of mass-production to the broadcasting enterprise. A radio network has two sales departments, one to package and sell programs to radio stations, and one to sell the audience of those programs to advertisers.

Most radio networks also produce much of their programming. Originally, radio networks owned some or all of the stations that broadcast the network's radio format programming. Presently however, there are many networks that do not own any stations and only produce and/or distribute programming. Similarly station ownership does not always indicate network affiliation. A company might own stations in several different markets and purchase programming from a variety of networks.

Radio networks rose rapidly with the growth of regular broadcasting of radio to home listeners in the 1920s. This growth took various paths in different places. In Britain the BBC was developed with public funding, in the form of a broadcast receiver license, and a broadcasting monopoly in its early decades. In contrast, in the United States various competing commercial broadcasting networks arose funded by advertising revenue. In that instance, the same corporation that owned or operated the network often manufactured and marketed the listener's radio.

Major technical challenges to be overcome when distributing programs over long distances are maintaining signal quality and managing the number of switching/relay points in the signal chain. Early on, programs were sent to remote stations (either owned or affiliated) by various methods, including leased telephone lines, pre-recorded gramophone records and audio tape. The world's first all-radio, non-wireline network was claimed to be the Rural Radio Network, a group of six upstate New York FM stations that began operation in June 1948. Terrestrial microwave relay, a technology later introduced to link stations, has been largely supplanted by coaxial cable, fiber, and satellite, which usually offer superior cost-benefit ratios.

Television Network

A television network is a telecommunications network for distribution of television program content, whereby a central operation provides programming to many television stations or pay television providers. Until the mid-1980s, television programming in most countries of the world was dominated by a small number of broadcast networks. Many early television networks (such as the BBC, NBC or CBC) evolved from earlier radio networks.

Overview

In countries where most networks broadcast identical, centrally originated content to all of their stations and where most individual television transmitters therefore operate only as large "repeater stations", the terms "television network", "television channel" (a numeric identifier or radio frequency) and "television station" have become mostly interchangeable in everyday language, with professionals in television-related occupations continuing to make a differentiation between them. Within the industry, a tiering is sometimes created among groups of networks based on whether their programming is simultaneously originated from a central point, and whether the network master control has the technical and administrative capability to take over the programming of their affiliates in real-time when it deems this necessary – the most common example being during national breaking news events.

In North America in particular, many television networks available via cable and satellite television are branded as "channels" because they are somewhat different from traditional networks in the sense defined above, as they are singular operations – they have no affiliates or component stations, but instead are distributed to the public via cable or direct-broadcast satellite providers. Such networks are commonly referred to by terms such as "specialty channels" in Canada or "cable networks" in the U.S.

A network may or may not produce all of its own programming. If not, production companies (such as Warner Bros. and Sony Pictures Television) can distribute their content to the various networks, and it is common that a certain production firm may have programs that air on two or more rival networks. Similarly, some networks may import television programs from other countries, or use archived programming to help complement their schedules.

Some stations have the capability to interrupt the network through the local insertion of television commercials, station identifications and emergency alerts. Others completely break away from the network for their own programming, a method known as regional variation. This is common where small networks are members of larger networks. The majority of commercial television stations are self-owned, even though a variety of these instances are the property of an owned-and-operated television network. The commercial television stations can also be linked with a noncommercial educational broadcasting agency. It is also important to note that some countries have launched national television networks, so that individual television stations can act as common repeaters of nationwide programs.

On the other hand, television networks also undergo the impending experience of major changes related to cultural varieties. The emergence of cable television has made available in major media markets, programs such as those aimed at American bi-cultural Latinos. Such a diverse captive audience presents an occasion for the networks and affiliates to advertise the best programming that needs to be aired.

This is explained by author Tim P. Vos in his abstract *A Cultural Explanation of Early Broadcast*, where he determines targeted group/non-targeted group representations as well as the cultural specificity employed in the television network entity. Vos notes that policymakers did not expressly intend to create a broadcast order dominated by commercial networks. In fact, legislative attempts were made to limit the network's preferred position.

As to individual stations, modern network operations centers usually use broadcast automation to handle most tasks. These systems are not only used for programming and for video server playout, but use exact atomic time from Global Positioning Systems or other sources to maintain perfect synchronization with upstream and downstream systems, so that programming appears seamless to viewers.

Global

A major international television network is the British Broadcasting Corporation (BBC), which is perhaps most well known for its news agency BBC News. Owned by the Crown, the BBC operates primarily in the United Kingdom. It is funded by the television licence paid by British residents that watch terrestrial television and as a result, no commercial advertising appears on its networks. Outside of the UK, advertising is broadcast because the licence fee only applies to the BBC's British operations. 23,000 people worldwide are employed by the BBC and its subsidiary, BBC Worldwide.

United States

Television in the United States had long been dominated by the Big Three television networks, ABC, CBS and NBC; however Fox, which launched in October 1986, has gained prominence and is now considered part of the "Big Four." The Big Three provide a significant amount of programs to each of their affiliates, including newscasts, prime time, daytime and sports programming, but still reserve periods during each day where their affiliate can air local programming, such as local news or syndicated programs. Since the creation of Fox, the number of American television networks has increased, though the amount of programming they provide is often much less: for example, The CW only provides ten hours of primetime programming each week (along with six hours on Saturdays and five hours a week during the daytime), leaving its affiliates to fill time periods where network programs are not broadcast with a large amount of syndicated programming. Other networks are dedicated to specialized programming, such as religious content or programs presented in languages other than English, particularly Spanish.

The largest television network in the United States, however, is the Public Broadcasting Service (PBS), a non-profit, publicly owned, non-commercial educational service. In comparison to the commercial television networks, there is no central unified arm of broadcast programming, meaning that each PBS member station has a significant amount of freedom to schedule television shows as they consent to. Some public television outlets, such as PBS, carry separate digital subchannel networks through their member stations (for example, Georgia Public Broadcasting; in fact, some programs airing on PBS were branded on other channels as coming from GPB Kids and PBS World).

This works as each network sends its signal to many local affiliated television stations across the country. These local stations then carry the "network feed," which can be viewed by millions of households across the country. In such cases, the signal is sent to as many as 200+ stations or as little as just a dozen or fewer stations, depending on the size of the network.

With the adoption of digital television, television networks have also been created specifically for distribution on the digital subchannels of television stations (including networks focusing on classic television series and films operated by companies like Weigel Broadcasting (owners of Movies! and Me-TV) and Tribune Broadcasting (owners of This TV and Antenna TV), along with networks focusing on music, sports and other niche programming).

Cable and satellite providers pay the networks a certain rate per subscriber (the highest charge being for ESPN, in which cable and satellite providers pay a rate of more than $5.00 per subscriber to ESPN). The providers also handle the sale of advertising inserted at the local level during national programming, in which case the broadcaster and the cable/satellite provider may share revenue. Networks that maintain a home shopping or infomercial format may instead pay the station or cable/satellite provider, in a brokered carriage deal. This is especially common with low-power television stations, and in recent years, even more so for stations that used this revenue stream to finance their conversion to digital broadcasts, which in turn provides them with several additional channels to transmit different programming sources.

History

Television broadcasting in the United States was heavily influenced by radio. Early individual experimental radio stations in the United States began limited operations in the 1910s. In November 1920, Westinghouse signed on "the world's first commercially licensed radio station", KDKA in Pittsburgh, Pennsylvania. Other companies built early radio stations in Detroit, Boston, New York

City and other areas. Radio stations received permission to transmit through broadcast licenses obtained through the Federal Radio Commission (FRC), a government entity that was created in 1926 to regulate the radio industry. With few exceptions, radio stations east of the Mississippi River received official call signs beginning with the letter "W"; those west of the Mississippi were assigned calls beginning with a "K". The amount of programs that these early stations aired was often limited, in part due to the expense of program creation. The idea of a network system which would distribute programming to many stations simultaneously, saving each station the expense of creating all of their own programs and expanding the total coverage beyond the limits of a single broadcast signal, was devised.

NBC set up the first permanent coast-to-coast radio network in the United States by 1928, using dedicated telephone line technology. The network physically linked individual radio stations, nearly all of which were independently owned and operated, in a vast chain, NBC's audio signal thus transmitted from station to station to listeners across the United States. Other companies, including CBS and the Mutual Broadcasting System, soon followed suit, each network signed hundreds of individual stations on as affiliates: stations which agreed to broadcast programs from one of the networks.

As radio prospered throughout the 1920s and 1930s, experimental television stations, which broadcast both an audio and a video signal, began sporadic broadcasts. Licenses for these experimental stations were often granted to experienced radio broadcasters, and thus advances in television technology closely followed breakthroughs in radio technology. As interest in television grew, and as early television stations began regular broadcasts, the idea of networking television signals (sending one station's video and audio signal to outlying stations) was born. However, the signal from an electronic television system, containing much more information than a radio signal (6 MHz), required a broadband transmission medium. Transmission by a nationwide series of radio relay towers would be possible but extremely expensive.

Researchers at AT&T subsidiary Bell Telephone Laboratories patented coaxial cable in 1929, primarily as a telephone improvement device. Its high capacity (transmitting 240 telephone calls simultaneously) also made it ideal for long-distance television transmission, where it could handle a frequency band of 1 MHz. German television first demonstrated such an application in 1936 by relaying televised telephone calls from Berlin to Leipzig, 180 km (110 mi) away, by cable.

AT&T laid the first L-carrier coaxial cable between New York City and Philadelphia, with automatic signal booster stations every 10 miles (16 km), and in 1937 it experimented with transmitting televised motion pictures over the line. Bell Labs gave demonstrations of the New York–Philadelphia television link in 1940 and 1941. AT&T used the coaxial link to transmit the Republican National Convention in June 1940 from Philadelphia to New York City, where it was televised to a few hundred receivers over the NBC station W2XBS (which evolved into WNBC) as well as seen in Schenectady, New York via W2XB (which evolved into WRGB) via off-air relay from the New York station.

NBC had earlier demonstrated an inter-city television broadcast on February 1, 1940, from its station in New York City to another in Schenectady, New York by General Electric relay antennas, and began transmitting some programs on an irregular basis to Philadelphia and Schenectady in 1941. Wartime priorities suspended the manufacture of television and radio equipment for civilian use from April 1, 1942 to October 1, 1945, temporarily shutting down expansion of television networking. However, in 1944 a short film, "Patrolling the Ether", was broadcast simultaneously over three stations as an experiment.

The DuMont Television Network in 1949. DuMont's network of stations stretched from Boston to St. Louis. These stations were linked together via AT&T's coaxial cable feed, allowing the network to broadcast live television programming to all the stations at the same time. Stations not yet connected received kinescope recordings via physical delivery.

AT&T made its first postwar addition in February 1946, with the completion of a 225-mile (362 km) cable between New York City and Washington, D.C., although a blurry demonstration broadcast showed that it would not be in regular use for several months. The DuMont Television Network, which had begun experimental broadcasts before the war, launched what *Newsweek* called "the country's first permanent commercial television network" on August 15, 1946, connecting New York City with Washington. Not to be outdone, NBC launched what it called "the world's first regularly operating television network" on June 27, 1947, serving New York City, Philadelphia, Schenectady and Washington. Baltimore and Boston were added to the NBC television network in late 1947. DuMont and NBC would be joined by CBS and ABC in 1948.

In the 1940s, the term "chain broadcasting" was used when discussing network broadcasts, as the television stations were linked together in long chains along the East Coast. But as the television networks expanded westward, the interconnected television stations formed major networks of connected affiliate stations. In January 1949, with the sign-on of DuMont's WDTV in Pittsburgh, the Midwest and East Coast networks were finally connected by coaxial cable (with WDTV airing the best shows from all four networks). By 1951, the four networks stretched from coast to coast, carried on the new microwave radio relay network of AT&T Long Lines. Only a few local television stations remained independent of the networks.

Each of the four major television networks originally only broadcast a few hours of programs a week to their affiliate stations, mostly between 8:00 and 11:00 p.m. Eastern Time, when most viewers were watching television. Most of the programs broadcast by the television stations were still locally produced. As the networks increased the number of programs that they aired, however, officials at the Federal Communications Commission (FCC) grew concerned that local television might disappear altogether. Eventually, the federal regulator enacted the Prime Time Access Rule, which restricted the amount of time that the networks could air programs; officials hoped that the rules would foster the development of quality local programs, but in practice, most local stations did not want to bear the burden of producing many of their own programs, and instead chose to purchase programs from independent producers. Sales of television programs to individual local stations are done through a method called "broadcast syndication," and today nearly every television station in the United States obtains syndicated programs in addition to network-produced fare.

Late in the 20th century, cross-country microwave radio relays were replaced by fixed-service satellites. Some terrestrial radio relays remained in service for regional connections.

After the failure and shutdown of DuMont in 1956, several attempts at new networks were made between the 1950s and the 1970s, with little success. The Fox Broadcasting Company, founded by the Rupert Murdoch-owned News Corporation (now owned by 21st Century Fox), was launched on October 9, 1986 after the company purchased the television assets of Metromedia; it would eventually ascend to the status of the fourth major network by 1994. Two other networks launched within a week of one another in January 1995: The WB Television Network, a joint venture between Time Warner and the Tribune Company, and the United Paramount Network (UPN), formed through a programming alliance between Chris-Craft Industries and Paramount Television (whose parent, Viacom, which later acquire half and later all of the network over the course of its existence). In September 2006, The CW was launched as a "merger" of The WB and UPN (in actuality, a consolidation of each respective network's higher-rated programs onto one schedule); MyNetworkTV, a network formed from affiliates of UPN and The WB that did not affiliate with The CW, launched at the same time.

Regulation

FCC regulations in the United States restricted the number of television stations that could be owned by any one network, company or individual. This led to a system where most local television stations were independently owned, but received programming from the network through a franchising contract, except in a few major cities that had owned-and-operated stations (O&O) of a network and independent stations. In the early days of television, when there were often only one or two stations broadcasting in a given market, the stations were usually affiliated with multiple networks and were able to choose which programs would air. Eventually, as more stations were licensed, it became common for each station to be exclusively affiliated with only one network and carry all of the "prime-time" programs that the network offered. Local stations occasionally break from regularly scheduled network programming however, especially when a breaking news or severe weather situation occurs in the viewing area. Moreover, when stations return to network programming from commercial breaks, station identifications are displayed in the first few seconds before switching to the network's logo.

Another FCC regulation, the Prime Time Access Rule, restricted the number of hours of network programming that could be broadcast on the local affiliate stations. This was done to encourage the development of local programming, and to give local residents access to broadcast time. More often, the result included a substantial amount of syndicated programming, usually consisting of older movies, independently produced and syndicated shows, and reruns of network programs. Occasionally, these shows were presented by a local host, especially programs that showed cartoons and comedy shorts intended for children.

Canada

A number of different definitions of "network" are used by government agencies, industry, and the general public. Under the Broadcasting Act, a network is defined as "any operation where control over all or any part of the programs or program schedules of one or more broadcasting undertakings is delegated to another undertaking or person," and must be licensed by the Canadian Radio-television and Telecommunications Commission (CRTC).

Only three national over-the-air television networks are currently licensed by the CRTC: government-owned CBC Television (English) and Ici Radio-Canada Télé (French), French-language private network TVA, and a network focused on Canada's indigenous peoples. A third French-language service, V, is licensed as a provincial network within Quebec, but is not licensed or locally distributed (outside of carriage on the digital tiers of pay television providers) on a national basis.

Currently, licensed national or provincial networks must be carried by all cable providers (in the country or province, respectively) with a service area above a certain population threshold, as well as all satellite providers. However, they are no longer necessarily expected to achieve over-the-air coverage in all areas (APTN, for example, only has terrestrial coverage in parts of northern Canada).

In addition to these licensed networks, the two main private English-language over-the-air services, CTV and Global, are also generally considered to be "networks" by virtue of their national coverage, although they are not officially licensed as such. CTV was previously a licensed network, but relinquished this licence in 2001 after acquiring most of its affiliates, making operating a network licence essentially redundant (per the above definition).

Smaller groups of stations with common branding are often categorized by industry watchers as television systems, although the public and the broadcasters themselves will often refer to them as "networks" regardless. Some of these systems, such as CTV Two and the now-defunct E!, essentially operate as mini-networks, but have reduced geographical coverage. Others, such as Omni Television or the Crossroads Television System, have similar branding and a common programming focus, but schedules may vary significantly from one station to the next. City originally began operating as a television system in 2002 when CKVU-TV in Vancouver started to carry programs originating from and adopted the then "Citytv" branding used by CITY-TV in Toronto, but gradually became a network by virtue of national coverage through expansions into other markets west of Atlantic Canada between 2005 and 2013.

Most local television stations in Canada are now owned and operated directly by their network, with only a small number of stations still operating as affiliates.

Europe, Asia, Africa and South America

Most television services outside North America are national networks established by a combination of publicly funded broadcasters and commercial broadcasters. Most nations established television networks in a similar way: the first television service in each country was operated by a public broadcaster, often funded by a television licensing fee, and most of them later established a second or even third station providing a greater variety of content. Commercial television services also became available when private companies applied for television broadcasting licenses. Often, each new network would be identified with their channel number, so that individual stations would often be numbered "One," "Two," "Three," and so forth.

United Kingdom

The first television network in the United Kingdom was operated by the BBC. On 2 November 1936 the BBC opened the world's first regular high-definition television service, from a 405 lines transmitter at Alexandra Palace. The BBC remained dominant until eventually on 22 September 1955, commercial broadcasting was established in order to create a second television network. Rather than creating a single network with local stations owned and operated by a single company (as is the case with the BBC), each local area had a separate television station that was independently owned and operated, although most of these stations shared a number of programmes, particularly during peak evening viewing hours. These stations formed the ITV network.

When the advent of UHF broadcasting allowed a greater number of television stations to broadcast, the BBC launched a second network, BBC2 (with the original service being renamed BBC1). A fourth national commercial service was launched, Channel 4, although Wales instead introduced a Welsh-language service, S4C. These were later followed by the launch of a fifth network, Channel 5. Since the introduction of digital television, the BBC, ITV, Channel 4 and Channel 5 each introduced a number of digital-only networks. BSkyB operates a large number of networks including Sky1, Sky Living and Sky Atlantic; as does UKTV, which operates networks like Dave, Gold, Watch and Yesterday.

Netherlands

Until 1989, Netherlands Public Broadcasting was the only television network in the Netherlands, with three stations, Nederland 1, Nederland 2 and Nederland 3. Rather than having a single production arm, there are a number of public broadcasting organizations that create programming for each of the three stations, each working relatively independently. Commercial broadcasting in the Netherlands is currently operated by two networks, RTL Nederland and SBS Broadcasting, which together broadcast seven commercial stations.

Russia

Soviet Era

The one television network in the Soviet Union launched on 7 July 1938 when Petersburg – Channel 5 of Leningrad Television became a unionwide network. The two television network in the Soviet Union launched on 22 March 1951 when Channel One of USSR Central Television became a unionwide network. Until 1989, there were six television networks, all owned by the USSR Gosteleradio. This changed during former president Mikhail Gorbachev's Perestroika when the first independent television network, 2×2, was launched.

1990s

Following the breakup of the Soviet Union, USSR Gosteleradio ceased to exist as well as its six networks. Only Channel One had a smooth transition and survived as a network, becoming Ostankino Channel One. The other five networks were operated by Ground Zero. This free airwave space allowed many private television networks like NTV and TV-6 to launch in the mid-1990s.

2000s

The 2000s were marked by the increased state intervention in Russian television. On April 14, 2001 NTV experienced management changes following the expulsion of former oligarch and NTV founder Vladimir Gusinsky. As a result, most of the prominent reporters featured on NTV left the network. Later on January 22, 2002, the second largest private television network TV-6, where the former NTV staff took refuge, was shut down allegedly because of its editorial policy. Five months later on June 1, TVS was launched, mostly employing NTV/TV-6 staff, only to cease operations the following year. Since then, the four largest television networks (Channel One, Russia 1, NTV and Russia 2) have been state-owned.

Still, the 2000s saw a rise of several independent television networks such as REN (its coverage increased vastly allowing it to become a federal network), Petersburg – Channel Five (overall the same), the relaunched 2×2. The Russian television market is mainly shared today by five major companies: Channel One, Russia 1, NTV, TNT and CTC.

Brazil

The major television network in Brazil is Rede Globo, which was founded in 1965. It grew to become the largest and most successful media conglomerate in the country, having a dominating presence in various forms of media including television, radio, print (newspapers and magazines) and the Internet.

Australia

Australia has two national public networks, ABC Television and SBS. The ABC operates eight stations as part of its main network ABC1, one for each state and territory, as well as three digital-only networks, ABC2, ABC3 and ABC News 24. SBS currently operates two stations, SBS One and SBS Two.

The first commercial networks in Australia involved commercial stations that shared programming in Sydney, Melbourne, Brisbane, Adelaide and later Perth, with each network forming networks based on their allocated channel numbers: TCN-9 in Sydney, GTV-9 in Melbourne, QTQ-9 in Brisbane, NWS-9 in Adelaide and STW-9 in Perth together formed the Nine Network; while their equivalents on VHF channels 7 and 10 respectively formed the Seven Network and Network Ten. Until 1989, areas outside of these main cities had access to only a single commercial station, and these rural stations often formed small networks such as Prime Television. Beginning in 1989, however, television markets in rural areas began to aggregate, allowing these rural networks to broadcast over a larger area, often an entire state, and become full-time affiliates to one specific metropolitan network.

As well as these Free-to-air channels, there are other's on Australia's Pay television network Foxtel.

New Zealand

New Zealand has one public network, Television New Zealand (TVNZ), which consists of two main networks: TVNZ 1 is the network's flagship network which carries news, current affairs and sports programming as well as the majority of the locally produced shows broadcast by TVNZ and imported shows. TVNZ's second network, TV2, airs mostly imported shows with some locally produced programs such as *Shortland Street*. TVNZ also operates a network exclusive to pay television services, TVNZ Heartland, available on providers such as Sky. TVNZ previously operated a non-commercial public service network, TVNZ 7, which ceased operations in June 2012 and was replaced by the timeshift channel TV One Plus 1. The network operated by Television New Zealand has progressed from operating as four distinct local stations within the four main centers in the 1960s, to having the majority of the content produced from TVNZ's Auckland studios at present.

New Zealand also has several privately owned television networks with the largest being operated by MediaWorks. MediaWorks' flagship network is TV3, which competes directly with both TVNZ broadcast networks. MediaWorks also operates a second network, FOUR, which airs mostly imported programmes with children's shows airing in the daytime and shows targeted at teenagers and adult between 15 and 39 years of age during prime time. MediaWorks also operates a timeshift network, TV3 + 1, and a 24-hour music network, C4.

All relevision networks in New Zealand air the same programming across the entire country with the only regional deviations being for local advertising; a regional news service existed in the 1980s, carrying a regional news programme from TVNZ's studios in New Zealand's four largest cities, Auckland, Wellington, Christchurch and Dunedin.

In the 1960s, the service operated at the time by the New Zealand Broadcasting Corporation was four separate television stations – AKTV2 in Auckland, WNTV1 in Wellington, CHTV3 in Christchurch and DNTV2 in Dunedin – which each ran their own newscast and produced some in-house programmes, with other shows being shared between the stations. Programmes and news footage were distributed via mail, with a programme airing in one region being mailed to another region for broadcast the following week. A network was finally established in 1969, with the same programmes being relayed to all regions simultaneously. From the 1970s to the 1990s, locally produced programmes that aired on TV One and TV2 were produced out of one of the four main studios, with TVNZ's network hub based in Wellington. Today, most locally produced programmes that are aired by both TVNZ and other networks are not actually produced in-house, instead they are often produced by a third party company (for example, the TV2 programme *Shortland Street* is produced by South Pacific Pictures). The networks produce their own news and current affairs programs, with most of the content filmed in Auckland.

New Zealand also operates several regional television stations, which are only available in individual markets. The regional stations will typically air a local news programme, produce some shows in-house and cover local sports events; the majority of programming on the regional stations will be imported from various sources.

Philippines

In the Philippines, in practice, the terms "network," "station" and "channel" are used interchangeably as programming lineups are mostly centrally planned from the networks' main offices, and since provincial/regional stations usually just relay the broadcast from their parent network's flagship station (usually based in the Mega Manila area). As such, networks made up of VHF stations are sometimes informally referred to by their over-the-air channel number in the Mega Manila area (for example, Channel 2 or *Dos* for ABS-CBN, Channel 5 or *Singko* for TV5, and Channel 7 or *Siyete* for GMA Network), while some incorporate their channel numbers in the network's name (for example, TV5, Studio 23 and Net 25, which respectively broadcast on VHF channel 5, and UHF channels 23 and 25).

Unlike the United States, where networks receive programmes produced by various production companies, the two largest networks in the Philippines produce all of their prime time programmes except for Asianovelas. Other networks adopt block-time programming, which utilizes programming arrangements similar to the relationship between a U.S. network and station.

References

- Simmonds, A; Sandilands, P; van Ekert, L (2004). "An Ontology for Network Security Attack". Lecture Notes in Computer Science. Lecture Notes in Computer Science. 3285: 317–323. doi:10.1007/978-3-540-30176-9_41. ISBN 978-3-540-23659-7.

- L.A Lievrouw - Handbook of New Media: Student Edition (p.253) (edited by L.A Lievrouw, S.M. Livingstone), published by SAGE 2006 (abridged, reprint, revised), 475 pages, ISBN 1412918731 [Retrieved 2015-08-15]

- R. Oppliger. Internet and Intranet Security (p.12). Artech House, 1 Jan 2001, 403 pages, Artech House computer security series, ISBN 1580531660. Retrieved 2015-08-15.

- (ed. by H. Bidgoli). The Internet Encyclopedia, G Â– O. published by John Wiley & Sons 11 May 2004, 840 pages, ISBN 0471689963, Volume 2 of The Internet Encyclopedia. Retrieved 2015-08-15.

- Isaacson, Walter (2014). The Innovators: How a Group of Hackers, Geniuses, and Geeks Created the Digital Revolution. Simon & Schuster. p. 237. ISBN 9781476708690.

- M. Ziewitz & I. Brown (2013). Research Handbook on Governance of the Internet. Edward Elgar Publishing. p. 7. ISBN 1849805040. Retrieved 2015-08-16.

- Fritz E. Froehlich; Allen Kent (1990). "ARPANET, the Defense Data Network, and Internet". The Froehlich/Kent Encyclopedia of Telecommunications. 1. CRC Press. pp. 341–375. ISBN 978-0-8247-2900-4.

- Abbate, Janet (11 June 1999). Inventing the Internet. Cambridge, MA: MIT Press. ASIN B003VPWY6E. ISBN 0-262-01172-7.

- Gerald W. Brock (September 25, 2003). The Second Information Revolution. Harvard University Press. p. 151. ISBN 0-674-01178-3.

- Douglas E. Comer (2000). Internetworking with TCP/IP – Principles, Protocols and Architecture (4th ed.). Prentice Hall. ISBN 0-13-018380-6. 2.4.9 – Ethernet Hardware Addresses, p. 29, explains the filtering.

- Heinz-Gerd Hegering; Alfred Lapple (1993). Ethernet: Building a Communications Infrastructure. Addison-Wesley. ISBN 0-201-62405-2.

- Urs von Burg (2001). The Triumph of Ethernet: technological communities and the battle for the LAN standard. Stanford University Press. p. 175. ISBN 0-8047-4094-1.

- Allan, David; Bragg, Nigel (2012). 802.1aq Shortest Path Bridging Design and Evolution : The Architects' Perspective. New York: Wiley. ISBN 978-1-118-14866-2.

- Kim, Byung-Keun (2005). Internationalising the Internet the Co-evolution of Influence and Technology. Edward Elgar. pp. 51–55. ISBN 1845426754.

- Mueller, Milton L. (2010). Networks and States: The Global Politics of Internet Governance. MIT Press. p. 61. ISBN 978-0-262-01459-5.

- Mossberger, Karen. "Digital Citizenship – The Internet, Society and Participation" By Karen Mossberger, Caroline J. Tolbert, and Ramona S. McNeal." 23 Nov. 2011. ISBN 978-0-8194-5606-9

- The Shallows: What the Internet Is Doing to Our Brains, Nicholas Carr, W. W. Norton, 7 June 2010, 276 pp., ISBN 0-393-07222-3, ISBN 978-0-393-07222-8

- MM Wanderley; D Birnbaum; J Malloch (2006). New Interfaces For Musical Expression. IRCAM – Centre Pompidou. p. 180. ISBN 2-84426-314-3.

- Access Controlled: The Shaping of Power, Rights, and Rule in Cyberspace, Ronald J. Deibert, John G. Palfrey, Rafal Rohozinski, and Jonathan Zittrain (eds), MIT Press, April 2010, ISBN 0-262-51435-4, ISBN 978-0-262-51435-4

- Guowang Miao, Jens Zander, Ki Won Sung, and Ben Slimane, Fundamentals of Mobile Data Networks, Cambridge University Press, ISBN 1107143217, 2016.

- Inside AppleTalk, Second Edition, Gursharan Sidhu, Richard Andrews and Alan Oppenheiner, Addison-Wesley, 1989, ISBN 0-201-55021-0

Permissions

All chapters in this book are published with permission under the Creative Commons Attribution Share Alike License or equivalent. Every chapter published in this book has been scrutinized by our experts. Their significance has been extensively debated. The topics covered herein carry significant information for a comprehensive understanding. They may even be implemented as practical applications or may be referred to as a beginning point for further studies.

We would like to thank the editorial team for lending their expertise to make the book truly unique. They have played a crucial role in the development of this book. Without their invaluable contributions this book wouldn't have been possible. They have made vital efforts to compile up to date information on the varied aspects of this subject to make this book a valuable addition to the collection of many professionals and students.

This book was conceptualized with the vision of imparting up-to-date and integrated information in this field. To ensure the same, a matchless editorial board was set up. Every individual on the board went through rigorous rounds of assessment to prove their worth. After which they invested a large part of their time researching and compiling the most relevant data for our readers.

The editorial board has been involved in producing this book since its inception. They have spent rigorous hours researching and exploring the diverse topics which have resulted in the successful publishing of this book. They have passed on their knowledge of decades through this book. To expedite this challenging task, the publisher supported the team at every step. A small team of assistant editors was also appointed to further simplify the editing procedure and attain best results for the readers.

Apart from the editorial board, the designing team has also invested a significant amount of their time in understanding the subject and creating the most relevant covers. They scrutinized every image to scout for the most suitable representation of the subject and create an appropriate cover for the book.

The publishing team has been an ardent support to the editorial, designing and production team. Their endless efforts to recruit the best for this project, has resulted in the accomplishment of this book. They are a veteran in the field of academics and their pool of knowledge is as vast as their experience in printing. Their expertise and guidance has proved useful at every step. Their uncompromising quality standards have made this book an exceptional effort. Their encouragement from time to time has been an inspiration for everyone.

The publisher and the editorial board hope that this book will prove to be a valuable piece of knowledge for students, practitioners and scholars across the globe.

Index

www.ingramcontent.com/pod-product-compliance
Lightning Source LLC
Chambersburg PA
CBHW061320190326
41458CB00011B/3846